Atomic and Molecular Beams

Production and Collimation

Atomic and Molecular Beams
Production and Collimation

C B Lucas
Visiting Fellow
Department of Physical Sciences
The Open University

CRC Press
Taylor & Francis Group
Boca Raton London New York

CRC Press is an imprint of the
Taylor & Francis Group, an **informa** business

The front cover photograph is of a lamp originally belonging to Sir John Ambrose Fleming FRS, professor of Engineering at University College London. The carbon filament is attached with copper electrodes within the evacuated envelope. In 1883 Professor Fleming explained that copper was deposited on to the inside of the bulb except for a shadow of the filament produced opposite the connection at the higher resistance. He therefore realised that atoms travel in straight lines. Hence the kinetic theory of gases was verified experimentally and his observation has led to many applications of atomic and molecular beams in fundamental research internationally. *The photograph is reproduced with permission of the UCL Engineering Collection, 2012.*

CRC Press
Taylor & Francis Group
6000 Broken Sound Parkway NW, Suite 300
Boca Raton, FL 33487-2742

First issued in paperback 2017

© 2013 by Taylor & Francis Group, LLC
CRC Press is an imprint of Taylor & Francis Group, an Informa business

No claim to original U.S. Government works

ISBN-13: 978-1-4665-6103-8 (hbk)
ISBN-13: 978-1-138-19887-6 (pbk)

Library of Congress Cataloging-in-Publication Data

Lucas, C. B. (Cyril Bernard)
  Atomic and molecular beams : production and collimation / Cyril Bernard Lucas.
    pages cm
  "A CRC title."
  Includes bibliographical references and index.
  ISBN 978-1-4665-6103-8 (hardcover : alk. paper)
  1. Electron beams. 2. Kinetic theory of gases. I. Title.

QC793.5.E622L83 2014
539.7'3--dc23   2013045719

**Visit the Taylor & Francis Web site at
http://www.taylorandfrancis.com**

**and the CRC Press Web site at
http://www.crcpress.com**

Contents

List of Figures

List of Tables

Preface

Although the topic of atomic and molecular beam applications has received substantial and ongoing review, as Section 1.2 indicates, in the more recent work there has been little attention either to detailed experimental techniques or to the comparison of theoretical developments. This book is aimed at the research worker in the physical or biological sciences who is already familiar with vacuum systems but who needs to know the approaches that are most likely to lead to the successful production of atomic and molecular beams of high quality. An important feature in the following chapters is the grouping together of related work into topics.

Following an introductory chapter, the basic expressions of the kinetic theory of gases and those needed for the calculation of properties of beams formed by orifices are covered in Chapter 2. Chapter 3 aims to cover experimental techniques and materials for the construction and use of ovens and furnaces and also the production of circular apertures and slits. Their application for producing beams is divided into two chapters. Chapter 4 covers the alkali metals and their salts whereas Chapter 5 considers all other elements and compounds. However, the dissociation of molecules to form atomic beams is discussed in Chapter 6. This is followed by a short chapter on techniques for recycling gases and the use of atomic beams for levitation. The theoretical treatment of atomic and molecular beams formed by tubes is more involved than that of orifices and this is considered in Chapter 8. Chapter 9 gives expressions that can readily be employed to calculate the properties of beams and suggestions for designing them. The many techniques that have been employed to produce collimating arrays in a variety of materials are detailed in Chapter 10. Chapter 11 compares the measurements of beam properties with what is expected from the expressions given in Chapter 9. Some relevant topics that do not naturally fit into other chapters are discussed in Chapter 12. Finally in Chapter 13 some concluding observations are collected together.

Despite the availability of search and alerting tools, obtaining a comprehensive coverage of all relevant publications has proved to be difficult. Searching using the terms 'atomic beams' or 'molecular beams' produces very many results. However, many papers that discuss research that employs atomic beams either give no details of their production or possibly state no more than that the beam was produced by a thermally heated oven. Many papers that are included have been found by other means, including checking the relevant citations of other authors. Some selection has of course been necessary, and the aim has been to include those papers with the most useful detail for readers. So for those who feel that their contributions have been overlooked, we can only apologise and suggest they inform us of their work directly.

Acknowledgments

I am grateful for the award of an Alexander von Humboldt Research Fellowship at the Physics Institute of the University of Tübingen, where I received my first practical experience of atomic beams. I acknowledge the invaluable assistance of university libraries both there and at the Universities of Münster and York, colleges of the University of London, and the Open University as well as the British Library. The early stages of this work benefited from valuable discussions with Dr. Christine Davies of the mathematics department of Royal Holloway, which were facilitated by a grant from the University of London Research Fund. I gladly acknowledge that she made her notes available, which formed the basis of the original theoretical approach. I am grateful to Professors André Baldy, E.R. Collins, R.H. Jones, John G. King and D. Washington for the provision of original photographs and diagrams and also to Mr. Nicholas Booth for his assistance with Fleming's lamps. Photographs have been processed using 'Gimp', with diagrams drawn with 'Inkscape'. The encouragement of Professors D.W.O. Heddle and the late M.R.C. McDowell is gratefully acknowledged. Heartfelt thanks are particularly due to Professor N.J. Mason OBE of the Open University for arranging visiting research fellowships for me.

E-mail: C.B.Lucas@physics.org

Author

C.B. Lucas studied physics at University College London and obtained his PhD by measuring the polarisation of electron impact radiation in helium. He was a Harwell Research Fellow at the UKAEA Culham Laboratory, where he studied electron collisions with helium ions. At the University of Tübingen he was an Alexander von Humboldt Research Fellow working with beams of atomic hydrogen. He designed gaseous atomic beam experiments at the University of York using a focussing capillary array. At the University of Münster he used focussed atomic beams to measure the angular distribution and polarisation of elastically scattered electrons from the inert gases, which he compared with his computations. He cooperated with the Physics Institute at the University of Belgrade and also developed an atomic beam experiment for the teaching laboratory to measure the velocity distribution of a beam. At Royal Holloway College of the University of London he joined a group studying the polarisation of electrons produced by laser ionisation of a beam of sodium atoms.

The remainder of Lucas' full-time career was in administrative posts, but he kept in contact with the academic world as a part-time tutor in the undergraduate and postgraduate programmes of the Open University and particularly in continuing with the research needed to produce this book, which has continued in his retirement.

1 Introduction

1.1 ATOMIC AND MOLECULAR BEAM BASICS

When a gas or vapour at a higher pressure passes through an impedance to the flow into a region of lower pressure, then in general a beam is formed. Whether the beam is composed mainly of atoms or of molecules, the beam-forming process is the same, and without loss of generality; we will mainly use the term 'atomic beam' to mean either an atomic or a molecular beam. We justify this preference since atomic beams are much more common than molecular beams. However, we note that many authors use 'molecular beam' to cover atomic beams as well and 'particle' is also used. It is also more usual to refer to gas flow through a bounded space such as a tube as 'molecular flow,' and 'molecular beam epitaxy' (MBE) is a well-established expression.

This work is concerned with the means of production of a thermal neutral beam from any material, whether initially in solid, liquid or gaseous form. There is treatment of both the theory and practice of beams formed by impedances in the form of circular orifices and to a lesser extent rectangular slits. This is extended to the formation of well-collimated beams with as little material as possible wasted by not being utilised for the purposes for which the beam was designed. Such beams require more detailed theoretical treatment and involve more technical considerations in order that they achieve their potential.

Atomic beams have made important contributions to many fields of science, and any attempt to detail these is likely to miss important fields of application. Compared with atoms in a gas, those in a beam are effectively collision free, so that atoms can be studied without disturbance from other atoms, either within the beam or with ambient gas. Beams can be used to study the interaction of atoms with surfaces, with electric and magnetic fields, and with other charged or neutral particles and photons. These latter collision studies are usually known as crossed beam techniques. The directional nature of a beam compared with the random motion of ambient gas means that the Doppler broadening caused by the thermal atomic motion is much reduced. The broadening has been mentioned independently by Heddle and Keesing (1968) and Kuyatt (1968) who all refer to Bethe (1937), who derived the expression for neutron scattering. An application of a quite different nature is the use of a beam for the levitation of microspheres so that they can be uniformly metallised. This aspect is discussed in Section 7.3. Further details of applications are available in almost all the works cited in Section 1.2. The titles of papers are given in the reference section, which should also give an indication of the measurements being made when only technical information is quoted here.

Our approach is to review beam production under experimental conditions such that the kinetic theory of gases is applicable. These are therefore thermal beams, since the distribution of velocities of the atoms in the beam is Maxwellian. We exclude the production of nonthermal beams by the very different techniques of supersonic flow, which are also known as nozzle or gas-dynamic sources, jets, or now usually as free jet sources. They are divided into those of the Fenn type and those of the Campargue type (Campargue 1984). However, some of the techniques used to produce a free-jet source may also be adaptable for thermal beams and so we include details that might be useful for their production. It is of course of interest to compare the beams produced by the best of both methods since this is important when the main requirement is to produce the maximum beam intensity in the region where it is required coupled with the minimum of background gas in the same region. Free jet sources have specialised purposes, such as the production of clusters of atoms and producing atoms with a narrow distribution of velocities that cannot be produced by thermal beams without the use of velocity selectors. Atomic beams can also be produced by the technique known as sputtering, and we exclude these because they are not thermal. Duncan (2012) has made a comprehensive review of cluster sources. Sometimes beams formed from gases are required that have a lower velocity than that corresponding to laboratory temperature by the use of a cooled source, and we consider any special techniques for producing a beam at low temperatures.

Atomic beams can also be produced when atoms are not in the ground state, for example, metastable atoms, but we do not specifically consider these methods, since the general means of production are initially the same as for a beam with atoms in their ground state. No separate discussion is given of producing beams from radioactive atoms, since no difference in beam-forming technique is involved. However, the recycling of expensive gases because they are rare isotopes is reviewed in Section 7.2.

Methods of producing thin films by evaporation and MBE require uniform distribution of atoms over a large area in general, rather than the production of a small pencil of atoms and the techniques for each are quite different. The apparatus for MBE is frequently available commercially. The source may be a heated cylinder closed at one end with a length that is little more than the diameter, known as a Knudsen cell. Clearly comparing the measured beam properties with theoretical treatments is complicated by the fact that the apparent cell dimensions change as the material evaporates and also the molten material is likely to wet the cell material and so be evaporated from the whole inside surface of the cell. An additional problem is that the cell length of only a few diameters makes comparison with the theory of beam formation more difficult. However, we include in Chapter 5 any evaporation details that refer to thin films and MBE, particularly if no other information is available. So that they are collected together, we refer here, rather than in Section 1.2, to reviews of beam production for MBE by Herman (1982), Joyce (1985) and Bauer and Springholz (1992). Books with experimental details include those by Pamplin (1980), Parker (1980), Kasper and Bean (1988), Herman and Sitter (1989), Panish and Temkin (1993), Farrow (1995) and Foord et al. (1997). More specific reviews are considered in the relevant sections of Chapter 5.

We mention in the opening paragraph that a beam is produced when gas or vapour flows through an impedance into a region of ambient lower pressure gas. According to the Oxford English Dictionary Online, T. Graham introduced the term 'effusion' of gases in 1850 'by which I express their passage into a vacuum by a small aperture in a thin plate.' Hence the term strictly applies only to thin apertures such as orifices and slits, but it is now more generally applied to the flow of gases through any sort of impedance, such as a channel or a tube. Without loss of generality we will refer for brevity to the source of atoms as gas, even when the material forming the beam is solid or liquid at laboratory temperature, and so is strictly a vapour. However, we will make the distinction when it is necessary to do so. A vapour is a gas at a temperature below its critical temperature, so that it can be condensed to the liquid or solid state by pressure alone. We will also usually refer to the lower-pressure ambient region as vacuum. It is important to remember that part of the art of forming a good beam is for that vacuum to be as high as possible. This is both so that collisions between the beam and the background gas are minimised and that the effect to be observed is caused as much as possible by the beam and not the ambient gas. Early papers, such as Kratzenstein (1935) mentioned in Section 4.5, usually refer to a 'cloud' being formed in front of the orifice or slit, and this was almost certainly due to collisions taking place between the atoms in the beam and the ambient gas when the pumping speed in the path of the beam was inadequate.

One of the chief problems with the production of a gaseous beam is therefore to provide sufficient pumping speed to handle those atoms from the source that do not form part of the beam. If a beam is formed from a vapour, the surfaces surrounding it may offer a much higher effective pumping speed by condensation than that of vacuum pumps. However, the choice of surface may be important. As mentioned in Section 4.6.2, Bacal et al. (1982) used a pyrolytic graphite lining of their interaction chamber when producing a caesium beam. Pyrolytic graphite is produced commercially by chemical vapour deposition. Alleau et al. (1967) reviewed work that showed that potassium and rubidium should also be gettered by carbon. Bhaskar and Kahla (1990) studied the gettering of caesium by polycrystalline synthetic graphite and also by colloidal graphite. They found that 80% to 90% of the impinging atoms stuck to the former if this was maintained at temperatures up to 400 K. Getters are normally used to reduce the residual atmospheric gases in sealed vacuum systems such as electronic valves and so information on the gettering of nonatmospheric vapours has not proved to be readily available. The producers of gas beams may envy the ease with which their colleagues with vapour beams can maintain the vacuum desired, but the negative side is the contamination of the impedance and the effort required to clean the apparatus.

It is important to consider the possible toxic and environmental hazards of the use of beams. Laboratory safety officers should be consulted concerning all possible hazards. These might include safe handling of compressed gases and vacuum pump exhausts, cleaning the apparatus and the disposal of solid and liquid waste materials. Possible fire risks in the event of accidental conditions such as loss of vacuum causing exposure of highly reactive molten materials or boiling flammable substances to the oxygen in the atmosphere will also need to be considered. Section 4.2 refers to some possible explosive reactions. Some high-temperature atomic beam sources utilise high-energy electron bombardment and shielding of X-rays may be necessary in these cases. If high-power radiofrequency and microwave sources are employed, dangerous radiation levels may also be present. Some beams require the use of high temperatures, which like lasers, may be an optical hazard. Directly heated ovens usually employ very heavy electric currents, which could lead to dangerous conditions if electrical short circuits occur. In experiments that require cooling water in large quantities, the effects of its sudden failure must be considered.

The history of atomic beams dates back to observations in the nineteenth century. In a short note Fleming (1883) describes his observations of Edison lamps, which had a horseshoe-shaped carbon filament. The two extremities of the carbon loop were clamped into small copper clamps on the ends of the platinum wire lead throughs. The ends of the loop were electroplated with copper where they were connected to each clamp in order to make good electrical contacts. If the point of greatest resistance occurred at the clamp, copper was evaporated and was deposited on the inside of the glass envelope. Fleming observed a narrow line on the glass envelope in the plane of the loop on the opposite side to the point of evaporation, which effectively was a shadow of the loop. He concluded that: 'Molecules are shot off in straight lines'. Preece (1885) reports on the discolouration of the evacuated glass envelope of carbon filament lamps made for him in the USA by Edison. In particular, when the filament was mounted on copper electrodes, there was a clear line of demarcation between the carbon and copper deposit, indicating that travel was in straight lines. Fleming (1885a,b) showed that by doubling the operating voltage of any lamp for a short period, the same shadow could be produced. Fleming (1890a,b) refers to Preece's work and indicated that the mean free path of the atoms (Section 2.2) was up to 100 mm at the vacuum in the lamp, which was believed to be about 100 mPa, so the phenomenon he described confirmed 'in a very beautiful manner the deductions of the kinetic theory of gases'. The front cover of this book shows one of Fleming's lamps that is preserved in the Electrical Engineering Collection at University College London.

The first use of an apparatus to demonstrate the properties of an atomic beam is credited to Dunoyer (1911a,b). He evaporated sodium and potassium in an evacuated glass vessel and proved the rectilinear propagation of atoms. Although Dunoyer did not use his beam to perform any experiments, he realised the potential of the atomic beam for doing so. Maire (1961) gives a diagram of Dunoyer's apparatus. Experiments with atomic beams date from Stern's (1920a,b) preliminary measurements of the speed of a beam of silver atoms. Subsequent work on speed measurements is reviewed by Miller and Kusch (1955). It is not considered further in this book since there are insufficient oven details to merit inclusion. According to Gerlach and Cilliers (1924), experiments to determine the magnetic moments of atoms date back to Gerlach and Stern's experiments published in 1921. The earliest techniques for producing atomic beams for these experiments are described, when appropriate, in Chapter 5 under the relevant element. Hence we discontinue the historical approach following this brief introduction, since it is more useful to those wishing to design a beam to group experiments basically in terms of elements and compounds rather than chronological order.

In the first atomic beam experiments, the impedance was simply a circular orifice, a fraction of a millimetre in diameter in a sheet that was thinner than this. It follows from the kinetic theory of gases that the beam formed by such an orifice ideally has a cosine distribution as shown in Figure 1.1. This follows because atoms that pass through the orifice can do so when travelling in all directions, so those

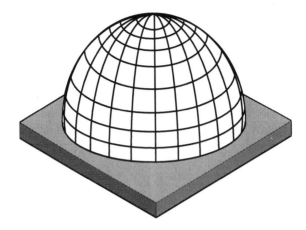

FIGURE 1.1   The beam formed by an orifice.

travelling perpendicular to its plane have a greater probability of passing through it than those at near grazing incidence. This probability is proportional to the cosine of the angle between the path of the atom and the normal to the orifice. Section 2.3 gives the relevant equations needed to design a beam formed by an orifice.

Soon after the first atomic beams were produced, the orifice was replaced by a slit in the many experiments dating from 1921 and mentioned above. This was partly a means of obtaining higher densities of atoms in the beam, since a slit can be considered to be a row of orifices. Another reason is the same as an optical spectrometer having a slit source, namely, that an effect is to be studied in a direction mutually perpendicular to the axis of the beam and the slit, so a line source is preferable to a point source.

The cosine distribution of beam intensity from the source is clearly much wider than is desirable in almost all, if not all, applications catered to here. The next development was Mayer's (1929) experiments following those on the angular distribution of beams of various gases passing through apertures in which he studied the beam formed by single capillary tubes that were 350 μm diameter and up to 100 mm long. He found that these produced a beam with a much narrower angular distribution than the cosine distribution he found with apertures. We are using the term 'collimated beam' to distinguish any narrow beam from that produced by a thin aperture, even though the beam is not strictly parallel.

If the gas pressure is so low that atoms do not collide with each other within a tube, it follows that some atoms will pass straight through it without colliding with the walls. Some atoms will collide with the tube walls quite close to the tube entrance, and these have a chance of then leaving the tube with a consequently slightly wider distribution. However, they might collide with the walls slightly nearer the exit to join those that have reached the walls directly from the entrance, and these may leave the tube through a wider cone of angles. Those that collide with the tube walls in its exit plane may leave it at any forward angle. Hence a near cusp-shaped atom beam distribution is formed by a tube, which is a great improvement in shape over that produced by an orifice. It should also be noted that atoms may also return to the source after making one or more collisions with the walls of the tube. Some paths are shown in Figure 1.2, which introduces the terms V_p for the source of atoms and V_0 where they exit a tube. When atoms collide with other atoms in the tube, clearly more complicated paths are possible, but the end result, namely, that an atom either returns back to the source or leaves the tube exit, remains the same.

The shape of a beam formed by a single tube is shown in Figure 1.3, which shows an obvious improvement compared with the orifice shown in Figure 1.1 even though it is drawn for a tube with Γ, the ratio of length to diameter, that is fairly small. This is necessary in order to show more than a letter 'L' of revolution. We introduce the terminology to describe beams in Section 2.5.

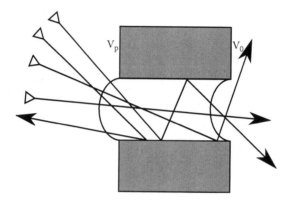

FIGURE 1.2   Some paths of atoms in a tube.

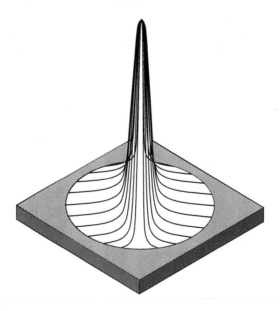

FIGURE 1.3   The beam formed by a tube.

Although it does not immediately seem important for atomic beam calculations, the concept of transmission probability, which is developed further in Section 8.4, is introduced here. It is also frequently referred to as the Clausing factor, being the probability W that an atom entering a tube eventually leaves it without returning to the source. Clearly for an ideal orifice under collisionless conditions $W = 1$. As Γ increases, W will decrease since an atom makes many more collisions with the tube walls, which increase its chance of returning to the source. However, as those with leaks in vacuum systems will be well aware, the transmission probability never reaches zero.

Unfortunately the condition that no atom collisions with other atoms were permissible within the tube means that the input gas pressure has to be sufficiently low that their mean free path between collisions, which is inversely proportional to the pressure, is greater than the tube length, rather than its diameter, as in the case of an orifice. Hence it follows that, in these circumstances, the axial intensity of the beam will be much lower than that formed by an orifice. As we discuss in Chapter 8, the theory of flow through a tube has been extended by several authors to include collisions, and numerical evaluation of these theories confirms measurements that show that, as the gas pressure is increased from a value such that the mean free path is equal to the tube length, the axial intensity initially increases linearly with pressure but the beam halfwidth hardly increases. Hence atom collisions within the tube enhance the beam quality rather than degrade it. Unfortunately, as we discuss in Section 1.2, several authors of reviews have independently stated incorrectly without proof or justification that a single tube does not produce a superior beam to an orifice under any input pressure conditions. These errors appear to date back to Zacharias and Haun (1954).

The next development in atomic beam production was the realisation that arrays of tubes or tube-like structures, which in general are usually referred to as canals or more usually channels, would produce most of the advantages of the single tube but with a much improved beam intensity. The original arrays of channels were made of metal and became known as crinkly sheet by their developers (Zacharias and Haun 1954) but also as crinkly foil arrays. Such multichannel arrays have been made of many metals, which are available in the form of foil. We discuss the many different techniques that have been used to make capillary arrays in Chapter 10. Arrays have successfully been made in metals by various methods that improve upon the crinkly foil arrays. Various other materials including plastics and glass have been successfully employed.

The arrays described above consist essentially of near-parallel and near-circular channels. By producing a fan-shaped array of channels that converge in the direction of the atoms, it was shown by Stanley (1966) that it is possible to form a focussed beam in one dimension with an obvious improvement in the intensity since a line focus is produced. Two-dimensional focussing arrays were produced shortly afterward with the axes of all channels directed to one point. They are described by Aubert et al. (1971a,b). These arrays yield the highest beam intensity if the convergence of the beam can be tolerated. They are discussed further in Sections 10.7 and 11.4.

Following the introduction of the expressions needed for atomic beam calculations in Section 2.2, the equations for orifice flow are covered in Section 2.3. Section 2.4

discusses the calculation of atomic and molecular diameters, which are needed in calculations.

Whatever form of beam collimation is used, if a beam is to be produced from a vapour, the construction of the oven or furnace may require considerable technical skills. A boiling pure material is often extremely reactive, so the materials with which it may come into contact have to be selected with care. Ovens are usually heated either with a resistive heating wire, or directly, by passing a current through the oven itself, or by electron bombardment. Direct electron, photon or laser beam heating has also been employed as well as induction heating. It is usually necessary to maintain the beam-forming impedance at a higher temperature than that of the evaporation to prevent its clogging, both during operation and in the heating and cooling phases. This temperature gradient may cause problems since the surface tensions of metals in the liquid state are usually quite high and liquid metals 'wet' many other materials; that is to say, the contact angles are 90° or less. Surface tension decreases with increasing temperature; hence the surface tension decreases toward the hotter impedance. Since surface tension is simply described as a kind of skin that holds the liquid together, it follows that the forces acting on the liquid decrease with increasing temperature. Hence metal can creep toward the hotter impedance sufficiently to block it partially, so causing instabilities in the beam. Special techniques can be used to reduce these thermocapillary effects. They are commonly observed when evaporating metals from a filament or boat, which will be hottest away from the electric current leads. The last drop of liquid will evaporate from the hottest point.

General aspects of oven and furnace design are discussed in Chapter 3. An alternative approach to the conventional oven for producing beams from high-melting-point materials is direct electron beam heating of a pool of liquid material in a water-cooled container. This technique has been employed to produce beams of several metals and is described in many sections of Chapter 5. The method is probably the closest approach to the universal oven for producing beams of materials that are solid or liquid at laboratory temperature. It solves both problems of a reaction of the material with the oven and impurities formed by its evaporation. However, the method only produces poorly collimated beams.

A particularly important application of atomic beams is in the study of radicals, which are dissociated species that are normally molecular. They are also known as free radicals. Since atoms formed from molecules readily recombine on collision with most surfaces, atom beams form one of the best ways of studying them. There are more papers about the dissociation of molecular hydrogen to form beams of atomic hydrogen than for any other element. This is because of its fundamental importance as the simplest element, its abundance in the universe and more recently its use in surface preparation and studies. Atomic hydrogen, oxygen and atomic halogens can be produced by thermal dissociation in a furnace. Any molecule can be at least partially dissociated in an electrical discharge at any frequency from direct current to microwave. Producing beams of atoms from molecules is discussed in Chapter 6. It is fortunately usually possible to produce a well-collimated beam of dissociated atoms.

In general, producing a beam of molecules tends not to be associated with any particular experimental difficulty. However, some boiling metallic compounds may cause problems because they are extremely corrosive, and this is mentioned under the appropriate atom, for example, in Section 4.4. In Section 5.12 we include experimental techniques particularly where knowledge of these might be useful because of the unusual nature of the molecules, for example, fullerene (C_{60}) and DNA bases.

Whether a beam is required that is atomic or molecular, it is usually important that the molecular composition of the beam is known. For example, atoms may associate to form dimers, trimers or higher molecular species and molecules thought to be diatomic might either dissociate into atoms or associate into larger molecules. There are no general rules that can be applied, other than that the beam composition is probably temperature dependent, so each will be discussed under each substance in Chapters 4 and 5.

The fact that the beam emerging from the impedance may have a different composition from the material placed in the oven can also be used to advantage as a means of obtaining a beam from an atom that is unstable in air without the use of special materials handling. For example, to obtain a beam of caesium, the stable compounds caesium chromate and silicon powder may be placed in the oven, which then emits a beam of pure caesium. This approach is particularly important for the alkali metals and so the methods are grouped together in Section 4.2. It is less common for other elements and so the methods are discussed in the appropriate sections of Chapter 5.

A special type of oven that has been employed for the alkali atoms enables material that does not pass through the first beam-forming impedance to be returned to the oven. Several designs for recirculating ovens are discussed in Section 4.6.

Whereas the theory of atomic beam formation by an ideal orifice is straightforward, well established and confirmed by experiment, beam formation by a tube, especially when atom collisions occur within the tube, is much more uncertain. This theory is discussed in stages. First the rate of impingement of atoms on the walls of a tube is reviewed in Sections 8.2 and 8.3. This is the first stage in the development of the calculation of the properties of the beam including its angular distribution. This is followed by the treatment of transmission probability in Section 8.4, which also enables the throughput or total flow to be calculated. The equations of flow through a tube are presented in Section 8.5. In Section 8.5.2 collision-free flow is considered, and interatomic collisions are included in Section 8.5.4. Many measurements of beam properties for many different materials have been made, and it is important to compare these with the theoretical treatments to assess their reliability. This discussion is in Chapter 11, where we also make a comparison with quoted measurements on free jet sources in view of the interest in comparing such widely different approaches to forming a beam.

We have purposely described the beam-forming device as an impedance, since clearly if its value is decreased, an increased beam intensity will be obtained, provided that the vacuum can still be maintained low enough for collisions to remain unimportant. It therefore follows that describing a beam merely in terms of its intensity is inadequate. Chapter 9 presents equations for the calculation of beam properties that are easy to apply. This is followed by a discussion of the way in which the quality of a beam can be described. We also consider the way in which the beam-forming qualities of arrays can be compared. The equations describing the flow of gas through the impedance are such that it is not obvious how to make the best choice for the particular purpose when there is a choice of input pressure, tube length, tube diameter, and, if an array of tubes is used, its open area. Hence we discuss what defines the quality of an array and then use it in Chapter 10 to compare the various impedances that have been employed in practice.

Our approach in the following chapters that describe experimental techniques is also to include quite simple designs. These may be useful in preliminary experiments and also for use in designing experiments for undergraduate teaching laboratories.

Finally in this section we need to emphasise the importance of improving the definition of a beam by placing at least one aperture between it and the region of experiment. If an orifice or slit source is used, the amount of material not going in the desired forward direction is large, and even with a collimated beam, some material is emitted at angles of up to 90° to the normal. Hence an aperture is placed fairly close to the beam-forming impedance to cut off the unwanted part of the beam, as shown in Figure 1.4. The volume between the impedance and the aperture requires pumping, particularly if a gaseous beam is being formed. Because the aperture itself is also pumping this intermediate chamber from the direction of the experimental chamber, the process is known as differential pumping. It is discussed in detail by Fluendy and Lawley (1973). This is usually neglected in books on vacuum technology despite being particularly important for reducing the background gas pressure where a gaseous beam is used. The ambient gas in this intermediate region forms the source of an atomic beam with the aperture as the impedance, so a low-intensity near-cosine distribution is produced in the experimental chamber from the gas in the intermediate chamber. However, if the beam entering this intermediate chamber is narrow rather than a cosine distribution, it is much easier to obtain a high-quality final beam. This intermediate vacuum chamber needs careful design so that high pumping speed is obtained without the impedance being situated too far from this aperture. This is because atomic beams from each channel are always divergent and so the beam density decreases with distance from the source unless a focussing array is employed.

Further stages of differential pumping can be used in order to obtain a beam in the experimental chamber with less ambient gas, but these require careful design, since the resulting improved vacuum is offset by the loss of atom density in the beam, because it has had to be transported a greater distance from the beam-forming impedance.

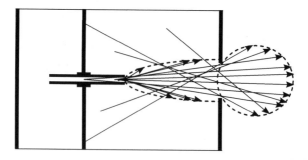

FIGURE 1.4   An atomic beam with differential pumping.

1.2   A REVIEW OF REVIEWS

Reviews of thermal atomic beam techniques have appeared as books, as chapters in books, and as review papers. We review these briefly in order of publication and normally alphabetically according to the first author when more than one is published in any year. We concentrate on those aspects covered in this work, but it can be assumed, unless stated otherwise, that reviews discuss both experimental methods beyond the actual formation of the beam and results obtained by these methods. Hence the reviews should provide a source of details not only of techniques applicable after the beam has been produced but also of applications. We also include those reviews that consider nonthermal and radioactive beams. The early reviews are of course mainly of historical interest, but they assist in following the development of improvements in producing beams and calculating their properties. We will note any misconceptions about the formation of beams by channels and multichannel arrays. We have mentioned some reviews of MBE in the previous section, so they have not been repeated.

Fraser's (1931) discussion of beam-forming impedances is concerned with the problem of mean free paths being less than the diameter of an orifice, when a quasi-hydrodynamic flow occurs leading to the production of a turbulent gas jet rather than an atomic beam. At that time, there was still controversy over the problem of whether a true beam was being formed or a cloud was produced at the exit of the orifice, and Fraser discusses this. No mention was made of the possibility of using channels rather than thin orifices, but it was realised that slits were generally preferable to circular orifices since the mean free path still had to be greater than the slit width for free-atomic flow rather than its much greater height. It was recognised that increasing the diameter of an orifice led to no gain in beam intensity, since the pressure in the source had to be reduced in proportion, in order to increase the mean free path proportionally. However, with an increase in the width of a slit, with no change in height, there is a gain since only one dimension is changed.

Rodebush (1931) gives a brief introduction to the theory of atomic beam formation, including the effective pressure when it impinges on a target and the design of multiple slit systems. He does not consider oven design, but does consider the results of experiments with beams. Taylor (1931) also looks briefly at the design concepts of an atomic beam apparatus. His review of applications includes determination of vapour pressures, the degree of dissociation of molecules, surface studies, and atomic velocities and mean free paths.

The short book by Fraser (1937) is entitled 'Molecular Beams' whereas his first book (1931) was 'Molecular Rays,' which indicates that 'beams' were then becoming established as the terminology in preference to 'rays'. Fraser (1937) only has space to devote a few pages to the production of beams, so he does not discuss advances in the theory and practice of their formation.

Bessey and Simpson (1942) consider both the production and detection of atomic beams. They mention how early measurements were concerned with verifying the rectilinear propagation of particles, their velocity distribution and the concept of the mean free path as well as the wave nature of uncharged heavy particles. Only beams formed by slits are considered, for which no theory is given.

In a review paper, Estermann (1946) discusses the theory of effusion through both circular apertures and slits. The design of some ovens for the production of beams from solids or liquids is considered. This includes a design using baffles to prevent slits being clogged by creep. As well as oven designs incorporating heating wires, Estermann also gives designs for electron bombardment ovens. This review introduces the employment of single canals of circular cross sections rather than slits and mentions the cloud formation in front of the beam-forming impedance when the input pressure to it is too high for the mean free path to be greater than the smallest dimension. In the following article, Kellogg and Millman (1946) discuss work up to 1941 on the determination of nuclear magnetic moments and spins as well as some other fundamental quantities such as hyperfine structure. They describe in detail a typical atomic beam apparatus where the beam formed by slits travels nearly one metre to the detector. A beam was usually obtained from a molecule containing the atom of interest, and this appeared not to have presented any technical difficulty. Kuhn (1946) discussed early beam applications.

Massey and Burhop (1952) mention orifice and single capillary sources and correctly state that the mean free path must be greater than the diameter of the capillary for free atomic flow, which they also refer to as effusive for the capillary case, by comparison with a turbulent jet.

The revision by Smith (1955) of Fraser's (1937) book, mentioned above, is of a similar length to the original and so adds nothing to details of beam formation.

As well as discussing the formation of beams by slits, King and Zacharias (1956) discuss the use of arrays of capillaries and the production of crinkly foil collimators, including their use with ovens. Means of dissociating molecules to form atomic beams are discussed briefly.

Ramsey (1956, 1990a) wrote a more substantial volume on molecular beams than the books already mentioned. In a chapter on gas kinetics, he discusses the theory of beams formed by orifices and slits and also by long channels. The theory at the time the book was first written was only applicable to the case where there are no interatomic collisions in the channel, which is assumed to be when the mean free path is greater than the channel length. Ramsey also considers the shape of the beam when the beam-forming slit is followed by another slit that is used to cut off the wings of the cosine distribution. Following numerous chapters on application techniques, Ramsey discusses molecular-beam design principles with a detailed example for molecular hydrogen. The final chapter, on molecular-beam techniques, discusses those applicable when the beam is formed as well as their production in ovens.

The production of beams from radioactive materials is discussed by Nierenberg (1957). He gives a list in order of atomic number of the elements that had been investigated using atomic beams with references to over 300 papers. Schlier (1957) discusses the production and detection of atomic beams and their focussing in multipole fields. Unfortunately Schlier erroneously states without justification that the mean free path of the atoms must be longer than the length of a channel in a multicapillary array.

Estermann (1959a) edited a volume dedicated to Otto Stern on the occasion of his 70th birthday. The opening chapter, by Estermann (1959b) himself, provides a brief historical account of experiments in Stern's laboratory in Hamburg and subsequent developments in the USA. Although most other papers are concerned with the physics gained from the use of beams, Hurlbut (1959) describes his experiments of molecular scattering at solid surfaces. This is of importance when discussing whether specular reflection influences the flow through capillaries. In the same volume, Marcus and McFee (1959) present their measurements of velocity distributions in potassium beams. Frisch (1959) follows an introduction to the early history of the subject with details of apparatus employing slits to form the beam, including what he calls a 'channel slit,' which produces a beam with a narrower halfwidth than a slit in a very thin foil. The review forms a very readable account of some of the early applications of beam techniques. In the same year, some details of oven construction that are not included in the original papers are provided by Kusch and Hughes (1959). Unfortunately they also state that for canals, the mean free path of the atoms must be longer than the canal length, again without giving a theoretical or experimental justification for their assertion. Production of beams of atomic hydrogen as well as of atoms in metastable and optically excited states is also considered. Techniques for the use of velocity selectors to measure the velocity distribution of beams are also discussed with results presented for several atoms and molecules. In the latter case, it is shown that the presence of dimers and trimers in the beam produce a velocity distribution that differs from that to be expected if only monomers are present.

Maire (1961) outlines atomic beam theory and then reviews over 200 papers covering the speeds of atoms, surface effects and the results of measurements of beams in electric and magnetic fields. Pauly (1961) reviews molecular scattering and the determination of intermolecular potentials. Experimental techniques are considered only briefly, but these do include the production and selection of beams above thermal energies.

Both the techniques for the production and detection of beams, including the use of 'canal slits,' are reviewed by Trischka (1962). Unfortunately he states that the mean free path must be of the same order as the length of the canal, which, as already mentioned, is incorrect.

Fite and Datz (1963) do not cover experimental techniques in their review, which is concerned with chemical applications. They discuss reactive and nonreactive collisions of neutral particles with other neutral species as well as collisions of ions with neutral particles and gas interactions with surfaces.

Knuth (1964) mentions thermal atomic beams briefly and discusses over 20 supersonic beam facilities. He reviews many publications that discuss the properties of supersonic beams, including beam collimation and speed distribution. There is also a section on vacuum techniques and anticipated applications. Leonas (1964) discusses work on beam sources including multichannel arrays and free jet sources. He quotes a formula for the intensity of free jet sources compared with an effusion source and concludes that the intensity of the former is typically two orders of magnitude better than the latter, without, however, comparing the throughputs and angular distributions. Leonas also considers charge exchange and sputtering sources. A 'resource letter' for students was produced by Zorn (1964), which gives brief details of relevant publications, which are classified according to their subject area and suitability for various levels of study.

The techniques of atomic beam production both by thermal means and by free jet sources have been discussed by Anderson et al. (1965). They discuss critically thermal beams produced by circular orifices, slits, capillaries and capillary arrays. They do not, however, give full details of construction techniques of the latter. Unfortunately, their conclusions concerning the optimum length of a capillary or capillary arrays in terms of an effective length are misleading, since operating at an input pressure such that the mean free path is less than the tube length has subsequently proved to be an advantage. In the same volume, the study of intermolecular potentials with molecular beams at thermal energies is reviewed by Pauly and Toennies (1965). Their article includes a short discussion of beam formation by orifices and multichannel arrays produced in the laboratory, although they felt that there was 'little room for further development' of them. Since then there has been considerable progress, both in the development of the prediction of beam properties and the commercial production of far superior multichannel arrays, including focussing arrays.

Anderson et al. (1966) present a detailed review of the production and application of free jet sources. They make a theoretical comparison of the supersonic beam produced by expanding high-pressure gas into a vacuum with that produced by an effusive source and conclude that 'in principle' the total beam intensity is 470 times greater at a Mach number of 10 and this ratio increases with Mach number. As we discuss later in this section by reference to Kudryavtsev et al. (1993) and also in Section 11.7, not only should comparison be made with the best beam produced by a capillary array, rather than the effusive beam produced by an orifice or slit, but also beam intensity is an insufficient basis to be used for comparison. Leonas (1966) discusses the techniques of studying chemical reactions using molecular beams once they have been produced and also the results that have been achieved.

As part of a chapter on sources of atomic particles, Lew (1967) discusses the production of atomic beams. Effusion through circular apertures and slits is discussed, as well as the use of long canals to form beams and the extension to multicapillary arrays made from several materials. Ovens for the production of beams from solid materials and furnaces for the dissociation of atomic hydrogen are considered. Lew's article is the most detailed which appeared on oven and collimator design up to that time, since it is devoted entirely to beam production. In the same volume, Lipworth (1967) discusses the techniques used to produce beams of radioactive isotopes.

Pauly and Toennies (1968) review the interaction of neutral particles at thermal energies from which interaction potentials can be determined. They discuss both thin-walled orifices and multichannel arrays but incorrectly conclude that the latter are rarely used in scattering experiments because they do not provide greater axial beams. They also consider the use of multichannel glass capillary arrays but conclude without reference that the theoretical gain in relative intensity and directivity over metal foil arrays is only about a factor two. Pauly and Toennies also discuss free jet sources. The use of crossed molecular beams to study chemical reactions is reviewed by Toennies (1968). He mentions the use of multichannel arrays and notes that the mean free path of atoms may be less than the channel length.

Mueller's (1970) review is particularly concerned with molecules. Following a brief introduction to beam formation and detection, he considers the behaviour of molecules in homogeneous and inhomogeneous electric and magnetic fields. Examples of applications are given. Schlier (1970) edited the proceedings of an

Enrico Fermi international school on molecular beams and reaction kinetics. To quote from his introduction:

> 'Some people may miss lectures on the technical aspects of molecular beam experiments. After some consideration, the conviction that lectures on constructional details are liable to be boring has succeeded in keeping them off the schedule.'

We hope that readers of this work will strongly disagree!

The production of molecular beams labelled with short-lived radioactive nuclides using capillary arrays is discussed by Grover et al. (1971). In crossed beam experiments, for example, in the study of reactive scattering, use of species with a half-life comparable to the beam path length considerably increases the detection efficiency. Massey (1971) briefly discusses the essentials of atomic beam formation, including from multichannel arrays, but only for the case where the mean free path is comparable to the length of the channels.

Fluendy and Lawley (1973) also discuss the formation of beams by canals but unfortunately state that for these, the mean free path must be comparable to the tube length. However, they do recognise that the ratio of useful atoms in a beam to the total number emitted is an important factor to consider in designing a beam, rather than just the intensity. As mentioned in Section 1.1, Fluendy and Lawley also discuss differential pumping. Free jet sources are also considered.

Anderson (1974) gives a brief introduction to the history of thermal atomic beam production and then discusses how this led to the development of nozzle sources, which he discusses in detail. Childs (1974) reviews magnetic resonance studies on beams of metastable atoms and includes brief details of their production and detection to provide data on atomic structure. English and Zorn (1974) briefly discuss flow through orifices, slits and channels, including the Maxwellian velocity distribution in a beam. They consider both measurements of the angular and velocity distributions of beams. They noted that sometimes deviations from the expected cosine and velocity distributions have been measured. They suggest that the advantage of multichannel arrays is that high beam intensity can be obtained while still maintaining the mean free path greater than the length of an individual tube in the array by keeping the length short, but this is not the best approach, as we discuss in Chapter 9. English and Zorn only discuss ovens briefly but also review free jet sources. Toennies (1974) discusses crossed molecular beam methods used to study elastic, inelastic and reactive scattering and gives a brief discussion of both thermal beam production and free jet sources.

Vályi (1977) devotes his book to details of sources of atoms and ions, but unfortunately, only part of a chapter is devoted to thermal atomic beam production.

The techniques for producing beams of refractory atoms and also metastable atoms for use with the magnetic resonance method have been reviewed by Penselin (1978), who also discusses the results achieved on both stable and radioactive isotopes. Steckelmacher et al. (1978) have reviewed the theory of atomic beam formation by orifices, single tubes and capillary arrays useful for ion–atom collision studies but only under collisionless conditions.

The use of supersonic nozzles and lasers to study reactive scattering is briefly discussed by Leonas (1979).

Clough and Geddes (1981) discuss the importance of using molecular beams in studying chemical reactions and briefly describe experimental techniques used to study various collision processes.

In a mainly theoretical discussion, Engel and Rieder (1982) briefly mention effusive sources and some early work with capillary arrays before discussing free jet sources. They state that the higher intensity of a nozzle source is obtained at the expense of a much-increased throughput.

Sonntag and Wuilleumier (1983) review some designs of ovens for photoemission studies from atoms and molecules.

Steckelmacher (1986) reviews the progress in the theory of atomic beams since Knudsen's early work, including Monte-Carlo calculations. He also reviews the problem of flow at higher pressures such that the mean free path is less than the diameter of an orifice or tube.

In his 1986 Nobel Lecture, Herschbach (1987) began his review of reactive scattering with some personal anecdotal accounts of his and others' early thermal atomic beam experiments and also discusses beam production by supersonic nozzles. The means of producing both thermal and supersonic beams is also briefly reviewed by Pendlebury and Smith (1987). In addition to beams of atoms and molecules, they also discuss producing beams of clusters of atoms and fast atom beams produced by the neutralisation of ions.

A two-volume work edited by Scoles (1988, 1992) mainly discusses methods once the beam has been produced, but Miller (1988) discusses free jet sources, Gentry (1988) considers their extension to pulsed beams and Pauly (1988a) reviews the theory and practice of thermal atomic beam production, including the production of radicals and excited atoms, but concentrating on experiments using beams in Germany and neighbouring countries. Pauly (1988b) also reviews the production of fast beams. Cluster molecular beams are reviewed by Kappes and Leutwyler (1988).

Ramsey (1990 b) discusses the development of his atomic beam experiments that have led to many precision measurements of atomic and nuclear properties and to standards, particularly of time.

Kudryavtsev et al. (1993) have made a wide-ranging review of more recent techniques for the production of beams. Arrays only receive a very brief mention. However, they extend the discussion of ovens to include evaporation by direct electron bombardment of the material and laser heating. They also include both pulsed and steady state free jet sources and other gas dynamic sources. Various means of producing beams of atoms from molecules, particularly atomic oxygen, are considered. Unfortunately their conclusion that free jet beams are superior to thermal beams is inconclusive. First they compare intensities and as we have already mentioned in the previous section, we show in Chapter 9 that the quality of a beam cannot be represented merely by its intensity. However, as the authors point out, gas dynamic sources require much higher pumping speeds than thermal beams. This indicates that the throughput needs to be considered in each case. Secondly they compare with beams formed by orifices. Since beams formed by capillary arrays are far superior to these, their conclusion that gas dynamic sources 'provide beams with intensities far above those for effusion sources' is misleading. The question of whether a jet-powered vehicle accelerates on a runway faster than one with an internal combustion engine can neither be answered by comparing a modern fighter aircraft with an early car nor the first jet aircraft with the latest grand prix racing car.

Ross and Sonntag (1995) present details of many ovens used for the production of atomic beams of metals that are either liquid or solid at laboratory temperatures. In addition to the use of various heaters, electron beam and induction heating are also considered. The authors also discuss the measurement of oven temperatures and beam properties. Their references are mainly to the use of beams in photoelectron spectroscopy. Practical experimental details are particularly abundant, and references, many of which are to unpublished work, are often supported by the authors' own experiences.

Ramsey (1996) contributes a short article that does little to update previous reviews.

In his book on electron spectroscopy, Schmidt (1997) includes a section on the formation of a gas beam by capillaries.

Casavecchia (2000) reviews the application of beams to reactive scattering studies. Following a historical introduction, Pauly (2000a) gives the theoretical background to the kinetic theory of gases as applied to thermal atomic beams. He also treats the theory of hydrodynamic flow and means of beam detection. The treatment of atomic beam production is less detailed and practical than considered here. Pauly (2000b) considers other types of beam production outside the scope of this work.

Campargue (2001) has edited a large volume on atomic and molecular beams. The preface allocates one asterisk to the then 11 Nobel Prize winners in physics and two asterisks to the four in chemistry who gained their prizes for their researches in atomic and molecular beams, but this should not be taken as an indication of a bias in the scope of the book. The nearly 70 topics discussed on the development of atomic and molecular beams unfortunately do not include one on the developments in producing and collimating atomic beams.

Brunger and Buckman (2002) give a short review of recent experimental techniques for forming atomic beams, mainly by the use of capillary arrays, for use in crossed beam experiments with electrons. Theoretical work on beam formation is mentioned briefly.

A review of the various means of production of pulsed atomic and molecular beams over a wide energy range has been given by Makarov (2003).

2 Kinetic Theory of Gases and Atomic Beam Terminology

2.1 INTRODUCTION

The theoretical work on the flow of gases at low pressures through orifices and tubes has been discussed in many books on the kinetic theory of gases. Review articles by Steckelmacher (1966, 1974, 1986) and Venema (1973) have also covered this field. In the next section, we give without proof some general results from the kinetic theory of gases that are needed in the design of atomic beams. The theory of the derivations may also be found in textbooks on heat and thermodynamics that cover the kinetic theory of gases and also in books on vacuum physics.

In all cases we give expressions in their most useful form for calculating atomic beam parameters. The equations for the formation of an atomic beam by an orifice are presented in Section 2.3. We leave the discussion of flow through channels to Chapter 8, since this involves a much more detailed treatment. Also the formation of a beam by slits is not treated here, since they are now rarely used. The lack of symmetry means that the expressions are more complicated, but they have been discussed by King and Zacharias (1956) and Ramsey (1956, 1990a). If a rectangular slit is employed instead of a circular aperture, it is assumed that the mean free path (Section 2.2) still has to be maintained only greater than the slit width, rather than its height. In Section 2.4 we discuss the determination of atomic diameters, since these are needed in atomic beam calculations. Finally we introduce terminology used in describing atomic beams.

2.2 KINETIC THEORY EXPRESSIONS

We first quote some general expressions we require from the kinetic theory of gases. This defines the mean free path of an atom between collisions λ (m) which is usually given by

$$\lambda = kT/\left(\sqrt{2}\pi p\sigma^2\right),$$

where k is Boltzmann's constant, T the absolute temperature (K), p the pressure (Pa) and σ the atomic diameter (m). For accurate conversion of Celsius temperatures (°C) to degrees K, 273.15 is added to T in °C. To five significant figures, $k = 1.3807 \times 10^{-23}$ J K^{-1}. As discussed in Section 2.4, atomic diameters can be derived from the gaseous viscosity. It is important to note that λ is inversely proportional to the pressure p. Substituting known values for the constants and expressing σ in pm (10^{-12} m) we have the following general equation:

$$\lambda = 3.11 \times 10^{3} T/(p\sigma^{2}) \text{ mm.} \qquad (2.1)$$

Other units of pressure still in use are the torr, where 1 torr = 133.322 Pa, and the bar, where 1 B = 100 kPa, or more usually in vacuum work, 1 mB = 100 Pa. In very approximate work, 1 mB is often used in place of 1 torr. The advantage of using Pascal for the units of pressure is that equations such as the above then require no pressure conversion. Atmospheric pressure, which is also known as the normal or standard pressure, is 101.3 kPa or approximately 10^{5} Pa. We express pressures in Pascal in accordance with BS ISO 80000-1 (2009).

For beams at laboratory temperature (295 K) the equation becomes

$$\lambda = 9.17 \times 10^{5}/(p\sigma^{2}) \text{ mm.}$$

As an example for molecular nitrogen at $T = 295$ K, for which the molecular diameter $\sigma = 373$ pm:

$$\lambda = 6.59/p \text{ mm.}$$

This means that if the condition $\lambda > d$ is to apply for nitrogen in a capillary array with pore internal diameters $d = 10$ μm, the input pressure $p < 660$ Pa. If the beam is required to travel 500 mm in the ambient gas with negligible collisions, then $\lambda > 500$ mm and $p < 13$ mPa, which is not a particularly stringent vacuum requirement.

The distribution of the free path is of negative exponential form, so that free paths shorter than the mean are much more probable than those that are longer, but fortunately only the mean is normally used without apparently causing any problems.

An important quantity that is obtained from all atomic-beam calculations is the beam intensity $I(\theta)$ at an angle θ to the normal to the impedance, expressed in atoms s^{-1} sr^{-1}. The axial intensity is thus $I(0)$. The number density ρ (m^{-3}) at a distance R (m) is derived from this with the following expression:

$$\rho = I(0)/(R^{2}\bar{c}),$$

where $\bar{c}$ is the mean velocity (m s^{-1}). When the number density is expressed in mm^{-3} and the distance in mm, with the mean velocity remaining in m s^{-1},

$$\rho = 10^{-3} I(0)/(R^{2}\bar{c}) \text{ mm}^{-3}. \qquad (2.2)$$

In the future, unless there is any ambiguity we will follow common practice and refer to ρ for brevity just as 'density' or 'atom density'. It follows that $I\,(0)$ is a more fundamental quantity than ρ, which involves knowledge of the distance from the impedance to the place of measurement. To obtain the density we require an expression for the mean velocity of the atoms in the beam. This is discussed further in Section 12.3. The Maxwellian velocity distribution in the oven or in ambient gas, leads to a mean velocity given by

$$\bar{c} = \sqrt{8kT/(\pi m)},$$

where m (kg) is the mass of the atom. If known values of constants are substituted and M is the atomic or molecular weight of the substance and using the unified atomic mass constant, which is 1.6605×10^{-27} kg to five significant figures, then

$$\bar{c} = 145.5\sqrt{T/M} \ \text{m s}^{-1}, \tag{2.3}$$

and at laboratory temperature

$$\bar{c} = 2.50 \times 10^3 / \sqrt{M} \ \text{m s}^{-1}.$$

As an example, for nitrogen ($M = 28.01$) at laboratory temperature

$$\bar{c} = 472 \ \text{m s}^{-1}.$$

Another quantity we need to derive from the axial intensity is the rate of impingement per mm^2 on a surface. Since $I\,(0)$ is the axial rate of impingement s^{-1} sr^{-1}, at a distance R (mm) from the source, the rate of impingement is $I\,(0)/4\,R^2$ mm^{-2} s^{-1}.

The velocity of impingement is given by

$$\bar{c}_{imp} = \sqrt{9\pi kT/(8m)} \tag{2.4}$$

which is also the mean beam velocity likely to be measured in time of flight measurements.

Substituting for known constants as before

$$\bar{c}_{imp} = 171\sqrt{T/M} \ \text{m s}^{-1},$$

and at laboratory temperature

$$\bar{c}_{imp} = 2.94 \times 10^3 / \sqrt{M} \ \text{m s}^{-1},$$

so for nitrogen at laboratory temperature

$$\bar{c}_{imp} = 556 \text{ m s}^{-1}.$$

The total gas flowing through the impedance into the vacuum is termed the throughput N. Total flow is also used. This is important for three purposes.

For any state of matter, the rate of usage of material Q is given by

$$Q = N \, m \, M \text{ kg s}^{-1}.$$

So on substituting constants,

$$Q = 5.98 \times 10^{-24} N M \text{ kg h}^{-1}, \tag{2.5}$$

so if $N = 10^{18}$ atoms s^{-1} then, for example, silver would be used at about 6.5 mg h^{-1} which means that a piece the size of a pea is the typical amount loaded into an oven. Mungall et al. (1981) estimate that 1 g of caesium loaded into their atomic clock oven should last 25 years.

To obtain the rate of usage of gas, we need the generally useful relation for the number of atoms n mm^{-3} when the gas is at a pressure p Pa and temperature T K. This is obtained from Boltzmann's constant k, in the simple relationship $p = n k T$. Substituting for k and converting from m^3, we have

$$n = 7.24 \times 10^{13} \, p/T \text{ mm}^{-3}, \tag{2.6}$$

and when $T = 295$ K

$$n = 2.46 \times 10^{11} \, p \text{ mm}^{-3}.$$

Hence the flow rate G of gas, at 295 K in atmospheric l h^{-1} for any throughput N is obtained from the same equation, namely,

$$G = 1.45 \times 10^{-19} \, N \text{ atm l h}^{-1}, \tag{2.7}$$

so taking again the example of $N = 10^{18}$ atoms s^{-1}, less than 0.15 at l of gas would be used per hour, so that the smallest cylinder of gas would provide a beam for many hours.

If the gas volume around the beam is pumped with a speed S l s^{-1}, after allowing for any cold traps and the loss of pumping speed in connecting tubing, S is the quotient of the gas flow and the ambient pressure p. Substituting constants

$$p = 4.07 \times 10^{-18} \, N/S \text{ Pa}, \tag{2.8}$$

assuming that the gas is pumped at a laboratory temperature of 295 K. With the same throughput and a pumping speed in the beam region of 1000 l s^{-1}, the above equation shows that ambient pressure would be about 4 mPa.

More generally the conductance C is used mainly by those using the terminology of vacuum physics, for any gas flow (usually expressed in $1\ s^{-1}$) divided by the pressure. Hence S is a special case of the conductance. The time constant τ in the volume around the beam is required if the beam is chopped in order to separate a signal due to the beam from that of the background gas, in which case the background gas pressure should be insensitive to whether the beam is on or off. The quotient of volume (1) and speed S ($1\ s^{-1}$) gives τ (s) and this should be appreciably longer than the time for which the beam is on or off. Since a high pumping speed is desirable in order to reduce the background gas pressure, a fast chopping rate is indicated.

2.3   EQUATIONS OF FLOW THROUGH AN ORIFICE

For an orifice of area A, provided the mean free path λ is larger than or comparable to its diameter, the number of atoms, $dI\,(\theta)$, flowing through it per second into a solid angle $d\omega$ in a direction making an angle θ with the normal to the area is given by

$$dI(\theta) = \frac{d\omega}{4\pi} nA\bar{c} \cos\theta \qquad (2.9)$$

where n is the density of the gas and $\bar{c}$ is the mean speed of the atoms given in Section 2.2. This equation is the familiar cosine law for effusion through an orifice, from which it follows that the beam halfwidth $H = 2 \times \cos^{-1} 0.5 = 120°$. The terminology is discussed in more detail in Section 2.5.

Writing $d\omega = \sin\theta\ d\theta\ d\phi$ and integrating over the range $0 \le \theta \le \pi/2$, $0 \le \phi \le 2\pi$, we obtain the throughput, namely, the total number of atoms flowing through the orifice per second:

$$N = \int_{\omega} I(\theta)\,d\omega = nA\bar{c}/4. \qquad (2.10)$$

This is probably the best remembered equation of the whole of the kinetic theory of gases, but it is of little use until n is converted to p using Equation 2.6 and $\bar{c}$ is expressed using Equation 2.3. From Equations 2.9 and 2.10 the axial intensity when $\theta = 0$ is given by

$$I(0)d\omega = \frac{Nd\omega}{\pi}\ sr^{-1}\ s^{-1}. \qquad (2.11)$$

The axial density can then be obtained from Equation 2.2.

The conductance C was introduced in the previous section and for an orifice, C_O is given by

$$C_O = 10^{-3} A\bar{c}/4\ 1\ s^{-1}, \qquad (2.12)$$

when A is expressed in mm^2 and $\bar{c}$ in m s^{-1}. It is not needed in calculating beam properties other than to reduce some reported measurements, including those of flow through channels, to the terms we use. It is also of use in differential pumping calculations since it also represents the pumping speed of an orifice to vacuum or of a surface on which atoms are completely accommodated, so it can be compared with the pumping speed of vacuum pumps.

We have retained A in the equations so far since it is needed for calculating the throughput through a noncircular orifice such as a slit. For a circular aperture of diameter d mm making the substitutions in Equation 2.10 and provided that p is sufficiently low that λ is greater than the aperture diameter

$$N = 2.07 \times 10^{18} d^2 p / \sqrt{MT} \text{ s}^{-1}, \tag{2.13}$$

Hence

$$I(0) = 6.59 \times 10^{17} d^2 p / \sqrt{MT} \text{ sr}^{-1} \text{ s}^{-1}, \tag{2.14}$$

and when $T = 295$ K,

$$N = 1.20 \times 10^{17} d^2 p / \sqrt{M} \text{ s}^{-1}.$$

Hence

$$I(0) = 3.84 \times 10^{16} d^2 p / \sqrt{M} \text{ sr}^{-1} \text{ s}^{-1}.$$

As an example with argon

$$N = 1.91 \times 10^{16} d^2 p \text{ s}^{-1},$$

and the axial intensity is

$$I(0) = 6.07 \times 10^{15} d^2 p \text{ sr}^{-1} \text{ s}^{-1}.$$

Finally for a typical beam-forming aperture 100 μm in diameter, the laboratory temperature throughput for argon is given by

$$N = 1.91 \times 10^{14} p \text{ s}^{-1}.$$

Hence

$$I(0) = 6.07 \times 10^{13} p \text{ sr}^{-1} \text{ s}^{-1}.$$

At higher pressures, when λ is less than the diameter of the aperture, the equations are no longer valid. The gas can be imagined to form a blob in front of the aperture from which atoms emerge in all directions so that the angular distribution tends to be even broader, being largely independent of θ rather than proportional to $\cos\theta$. In practice, for the highest intensity p will be such that $\lambda = d$. So we can substitute $\lambda = d$ in Equation 2.1 to eliminate p from the above equations. So when σ *is* in pm

$$N = 6.43 \times 10^{21} (d/\sigma^2)\sqrt{T/M} \ \text{s}^{-1}.$$

Hence

$$I(0) = 2.05 \times 10^{21} (d/\sigma^2)\sqrt{T/M} \ \text{sr}^{-1} \ \text{s}^{-1}.$$

So at a temperature of 295 K

$$N = 1.10 \times 10^{23} d \Big/ \left(\sigma^2 \sqrt{T/M}\right) \ \text{s}^{-1}.$$

Hence

$$I(0) = 3.52 \times 10^{22} d \Big/ \left(\sigma^2 \sqrt{M}\right) \ \text{sr}^{-1} \ \text{s}^{-1}.$$

For argon flowing through a 100-μm diameter aperture

$$N = 1.34 \times 10^{16} \ \text{s}^{-1},$$

and $I(0) = 4.27 \times 10^{15} \ \text{sr}^{-1} \ \text{s}^{-1}$.

If we now use Equation 2.2 to obtain the density at a distance of 100 mm from the aperture

$$\rho = 9.17 \times 10^5 \ \text{mm}^{-3}.$$

We can compare this with the equivalent pressure of such a density in ambient gas, namely, 3.74×10^{-6} Pa. This rather small value is the reason that capillary arrays are used to obtain more intense atomic beams.

Another term that is occasionally used is the 'peaking factor', κ, which is discussed further in Section 11.5. It was introduced by Becker (1961a) and used subsequently by Jones et al. (1969) and several other authors. Becker defines it as the factor by which the axial intensity of a collimating tube is greater than that of an orifice for the same throughput. It therefore represents the improvement of the beam formed by a tube over an ideal orifice. The peaking factor has been defined by Beijerinck and Verster (1975), namely,

$$\kappa = \pi I(0)/N, \tag{2.15}$$

so that from Equation 2.11 $\kappa = 1$ for an ideal orifice.

2.4   DETERMINATION OF ATOMIC DIAMETER

The basic kinetic theory of gases enables atomic diameters to be obtained in terms of the transport coefficients of the gas. This is discussed by Chapman and Cowling (1970) and by Hirschfelder et al. (1964).

In particular, the viscosity η of a gas is given by

$$\eta = 5 \ (\pi m k T)^{1/2}/(16 \pi \sigma^2), \qquad (2.16)$$

where, as before, m is the mass of the atom, k is Boltzmann's constant, T the absolute temperature and σ is the gas kinetic or viscous atomic diameter. Hence the viscosities of gases and vapours increase with temperature. The above formula assumes that the gas is sufficiently rarefied so that only binary collisions occur and that the motion of atoms during a collision can be described by classical mechanics. Atoms are assumed to be smooth, rigid, elastic and spherical. The interatomic forces are assumed to act only between the fixed centres of the atoms so that the interatomic potential is spherically symmetric.

When known values are substituted for the constants, and atomic diameters expressed in pm, we have the expression for the atomic diameter:

$$\sigma = 163.39 \ (MT)^{1/4}/\eta^{1/2} \ \text{pm},$$

with T in Kelvin, where M is the atomic weight and η is in μPa s (μNsm^{-2}).

Some gaseous viscosity data are still tabulated in the cgs units of micropoise, and to use these data directly

$$\sigma = 516.68 \ (MT)^{1/4}/\eta^{1/2} \ \text{pm}$$

with T and M as above and η in micropoise.

As an example for a typical gaseous molecule, for nitrogen at 295 K, $M = 28.01$, $\eta = 17.4 \ \mu$Pa s and so $\sigma = 373$ pm to three significant figures.

Detailed discussion of the potentials of colliding atoms is given in the above two books and by Reid et al. (1987), who critically discuss the models. However, for obtaining values of σ, the above formula is usually used. More detailed formulations give very similar results in the temperature and pressure range of interest and involve more detailed evaluation, depending on the parameters of the theoretical model chosen. For example, in addition to the viscosity, some experimentally determined quantities at the critical point may be required. There are probably more extensive and more reliable measurements or estimates of viscosity for atoms and molecules than for any other quantity from which the gas kinetic diameters can be determined, and these are usually available over wide temperature ranges. Since for flow through tubes, other assumptions such as nonspecular reflection of atoms from surfaces are known to be invalid (Section 12.2) and for most design purposes an approximate value of the flow parameters through collimators is adequate, the use of the above formula is usually justified. However, if really accurate comparisons between theory and experiment are needed, a more accurate value of σ may be required. Experiments that lead more directly to the determination of gas kinetic diameters at temperatures used in forming beams are therefore clearly desirable.

Note that the gas kinetic dimension must not be confused with either the metallic, covalent, ionic or van der Waals radius. These are discussed in detail by Alcock (1990). Briefly the metallic radius is one half the distance between atoms in a metal. The covalent radius is the effective radius of an atom when it is covalently bonded, whereas the ionic radius refers to the charged atom in a crystal. The van der Waals radius is closer to what is needed, but it is still concerned with bonding and is closely related to the shape of the atomic potential. For example, Alcock tabulates an ionic radius of 90 pm, covalent radius of 137 pm, metallic radius of 159 pm, and van der Waals radius of 205 pm for antimony. Using the above equation and Yaws' (1997) data, the viscous atomic radius is calculated to be 199 pm at the lowest temperature for which viscosity data is available, namely, 1898 K, which is closest to the quoted van der Waals radius, as is to be expected.

To avoid confusion with other quantities, we shall always use diameters to refer to gas kinetic quantities. We note that 'cross section' is preferred by those studying collision phenomena but follow many authors including McDaniel (1989) in using diameters when discussing the kinetic theory of gases. Some of the expressions introduced in Chapter 9 do in fact require the atomic diameter rather than its square. It is also a historical convenience to use σ rather than σ^2, since it was probably usually remembered that atomic diameters were typically a few Ångström units (Å) (1 Å = 10^{-10} m) or are now a few hundred picometres (pm).

The most detailed compilations of viscosity are given in four volumes by Yaws (1995b, 1997). The first three volumes cover more than 1000 organic compounds. The first covers those with up to four carbon atoms, the second from five to seven and the third from eight to 28. The fourth volume gives data for 330 inorganic compounds and elements. In addition to a graphical presentation for each substance with temperatures in Celsius, Yaws gives the coefficients of the equation $\eta = A + BT + CT^2$ with T in Kelvin and η the gaseous viscosity in micropoise. Note that his values must be divided by 10 to obtain values in μPa s. For many common substances, experimental values are fitted to this equation, but in the majority of cases, theoretical estimates are given. The temperature range for vapours is unfortunately higher than that usually required for forming a beam, but the near-linear relationship between η and T makes extrapolation fairly reliable. For gases, values from below laboratory temperature are always given. Unfortunately we have been unable to obtain any information on the computer programs and data files available from the author for a 'nominal fee' as mentioned in the books, so we assume that they are no longer available. Hydrogen and deuterium atoms are not included in viscosity compilations and so the values to use for atomic diameters are discussed when considering the beams formed by these atoms in Chapter 11.

As mentioned earlier, η increases with temperature. In fact it generally does so slightly faster than $T^{1/2}$, so σ decreases slightly with increasing temperature. For example, for molecular nitrogen at 1500 K, $\sigma = 320$ pm, which is a 15% reduction over the value of 373 pm at 295 K. Hence, for accurate design of atomic beams from vapours, knowledge of their viscosity at the temperature of evaporation is required. We give in the Table 2.1 values of σ for common gases at 295 K, using the viscosity

TABLE 2.1
Atomic or Molecular Diameters for Some Common Gases

Gas	Atomic or Molecular Diameter (pm)
Acetylene	481.2
Ammonia	434.7
Argon	360.8
Bromine	615.8
Carbon monoxide	374.6
Carbon dioxide	451.6
Chlorine	539.5
Deuterium	271.3
Fluorine	350.9
Helium	215.6
Hydrogen	272.9
Krypton	409.0
Neon	257.3
Nitrogen	373.2
Oxygen	359.9
Xenon	481.8

data of Yaws (1997), which are derived from experiment in these cases. Note that other values for atomic diameters are often quoted that differ somewhat from these. These may result from one or more combinations of superseded viscosity data, use of other transport properties or other theoretical models or using NTP rather than laboratory temperature.

2.5   ATOMIC BEAM TERMINOLOGY

We use this section to define the terminology we intend to use in the remainder of the discussion of atomic beams and compare it with other expressions that have been used in order to simplify the understanding of other work referred to.

We use input pressure p to mean the pressure in the source before the impedance including the vapour pressure in an oven or furnace. Other terms in use are driving pressure, backing pressure and stagnation pressure. We avoid these since 'driving' suggests that an atomic beam is always influenced by interatomic collisions. Backing pressure is in use for the pressure at the output of a diffusion pump and stagnation pressure suggests that there is no gas flow, which is far from the case.

When we wish to state whether we are considering input pressures greater than, equal to or less than the mean free path λ compared with a tube diameter or length, we will state this specifically. Many authors use Knudsen numbers, which are the quotient of λ and either the radius or diameter of an orifice or tube or the length of a tube. Inverse Knudsen numbers are also sometimes used, which are also Reynolds numbers. We find that using Knudsen numbers adds unnecessary complexity. Another expression used when flow is through a tube or channel is 'transparent,' meaning that no collisions take place in the tube, and 'opaque' when they do. Since an atom entering a tube eventually either leaves it into the vacuum or returns to the input side, and the presence of collisions in the tube does not change this basic fact, we find these terms misleading. Gases at pressures in the range from where viscous flow ceases to take place through the transition region to where atomic flow takes place are also referred to as rarefied gases.

The axial intensity $I(0)$ is frequently also described as the centreline intensity. We avoid the use of the term 'flux' since definitions vary. For example, as well as having the same units as intensity, it has been used for atoms s^{-1}, atoms $cm^{-2} s^{-1}$ and atoms $sr^{-1} s^{-1}$.

A diagram of the shape of a beam formed by a tube is shown in Figure 2.1. It is the cross section of that depicted in Figure 1.3. It is convenient to refer to the full width of the symmetrical beam where the intensity is one half the axial intensity, namely, where $I(\theta) = I(0)/2$ as the halfwidth H. Thus, as already mentioned in Section 2.3, for a beam with a cosine distribution, $H = 120°$. To avoid confusion, we will convert all references to data that refer to the semi-angle of the beam to FWHH, the full width at half height, H as defined above and always quote angles in degrees rather than radians when numerical values are being compared. Angles in mathematical formulae are, of course, expressed in radians.

For gas flow through a tube, the ratio of its length l to its diameter d is important. Although $\gamma = d/l$ is often used to express this ratio, it is psychologically the wrong way up, since a small value of γ is desirable. Hence we use $\Gamma = l/d$ to express the ratio, which is often referred to as the 'aspect ratio.' It was introduced in Section 1.1 in connection with transmission probability. A tube with a large Γ is often referred to as a 'long tube,' which is of course incorrect but convenient providing that its meaning is understood.

Another term that we avoid is the effective length of a tube. Giordmaine and Wang (1960) call it the critical source length, but Becker (1961a) refers to it as the

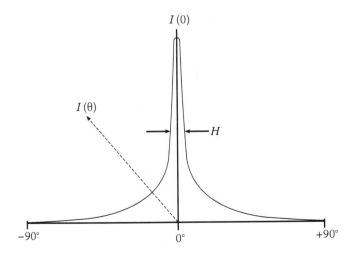

FIGURE 2.1 A cross section of the beam formed by a tube.

effective length, and this is the term that has also been used subsequently. It is discussed in Section 13.2.

Finally in this section we note that, for consistency throughout, we have converted all temperatures in the referenced papers to Kelvin, all Imperial units to SI units, and all magnetic fields to Tesla ($1\ G = 10^{-4}\ T$), making approximations where these seem sensible. Also unless crystalline quartz is referred to explicitly, we have used 'fused silica' in place of 'quartz.'

3 Design of Ovens

3.1 INTRODUCTION

If a beam is to be produced from a material that is solid or liquid at laboratory temperature, then it must be converted to a vapour in an oven before it passes through the beam-forming impedance. In the reviews mentioned in Section 1.2, the approach has usually been to give details of numerous designs. In this chapter, we discuss the general design principles, and then the next two chapters become more of a reference source for the evaporation of different substances. We thus include the experience of a large number of experimenters who have needed to heat a liquid or solid substance to form it into an atomic or molecular beam. If the material can be evaporated and raised to a temperature to give the required input pressure at the input to the impedance at a temperature less than about 1500 K, it is usual to describe the heating device as an oven. At higher temperatures, the device is usually called a furnace, but the distinction is not really necessary. Hence we will avoid the term except for thermal dissociation of a molecular species, discussed in Chapter 6, when very high temperatures are usually needed, so a furnace is a more appropriate term. For convenience we will also term a completely different approach, namely, the evaporation of the material by direct electron beam heating of the surface (Section 3.2.5) as an oven. 'Crucible' is used by many authors interchangeably with 'oven,' but it often implies a container for the material to be evaporated that is placed within a heated envelope.

The majority of early published designs of ovens were used with either circular apertures or slits. However, more recently, single collimating tubes have frequently been used. They reduce contamination of surfaces since the distribution that would be produced by a thin aperture is reduced from the broad cosine distribution to something narrower. The use of collimating arrays in conjunction with ovens is still fairly rare, but the general design principles remain the same if a collimating array is used with them. Hence we can consider the design of ovens independently of the design of beam collimators. The techniques for the construction of metallic circular apertures and slits are often quite detailed, so in Section 3.4 we include information on how they can be made. We also discuss details of making narrow slits in glass apparatus, although these are mainly required where molecules are to be dissociated.

The general design considerations discussed in this chapter are followed in Chapter 4 by details of the production of beams of alkali atoms and in Chapter 5 on other specific elements arranged according to the groups of the periodic table of elements. There are fewer references to compounds so these are considered with their principal element. This arrangement means that elements with similar chemical and physical properties are usually discussed together. Although the evaporation temperature of alkali metals is quite low, their instability in air usually requires different design techniques. They are also suitable for use in recirculating ovens in which the material not formed into a beam is recycled. These are a special type of oven used only with orifices as the beam-forming impedance. These ovens cater for the fact that the beam has a cosine distribution when leaving the oven orifice, so a large percentage of the material is cut off by a second aperture. This material is recycled back to the oven so allowing a much longer running time before the oven has to be reloaded. Recirculating ovens will be discussed separately in Section 4.6. Chapter 5 considers all atomic elements and compounds other than the alkali atoms and their compounds. The special techniques needed to dissociate molecular gases are considered in Chapter 6.

3.2   HEATING METHODS

The first stage in the design of an oven must be to decide the intended method of heating. The several methods that have been employed are discussed in the following subsections. Sometimes an oven is required that is moveable within the vacuum system and it should be clear from the context which heating methods are then most suitable. In general more detailed information of specific heater configurations is given in Chapters 4 and 5.

3.2.1   INDIRECT ELECTRIC HEATING

This is the simplest approach in which current is passed either through a heater wire or cylinder which then heats the oven walls and hence the substance to be evaporated. The former method has been very frequently used. The cylinder requires much higher currents, but should be more reliable.

In most cases, the heater is within the vacuum envelope, but in some cases it can be external to it, which has the obvious advantage of its being replaceable without disturbing the vacuum. This method was used for caesium ampoules, both by Klein (1971) and by Tompa et al. (1987) and is described in Section 4.5. DeGraffenreid et al. (2000) used a 115-mm-diameter UHV tee piece to contain sodium, which was heated with heating mantles to a temperature of nearly 700 K. They produced a supersonic beam, but the principle is also applicable to thermal beams. Gerginov and Tanner (2003) obtained a caesium beam from an ampoule in an ultrahigh vacuum tee piece that was heated externally to the vacuum system. Externally heated ovens were used to provide beams of lithium and sodium from ampoules by Stan and Ketterle (2005). Further details of these last two methods are also given in Section 4.5.

Historically, the first ovens used tungsten wire heaters insulated with a paste as described by Gerlach and Stern (1924) for silver and discussed in Section 5.9. They commented that it was not very successful. As mentioned in Section 4.3, Leu (1927) wound his platinum wire heater on a fused silica tube and isolated it with asbestos initially for producing beams of sodium and potassium. A commercial paste insulator has been employed both by Friedman (1955) who used electric-resistor cement for lithium iodide and by Berkowitz et al. (1979) for their oven for the lithium halides. Paste insulators are clearly particularly useful where an irregular shaped oven is necessary. In their case, the tubular ovens were surrounded by thoria tubes on which the tungsten wire heaters were wound noninductively and sealed into position with the high-temperature cement. The ovens operated at temperatures up to 1400 K. Cushing et al. (2011) wound their tungsten heating wire onto an alumina tube and used a commercially available ceramic liquid to cover the wire. They describe in detail the procedure for preparing the assembly, which could be operated at temperatures up to 1000 K.

For many years, heaters were made by passing wires through single- or multihole ceramic tubes that passed through holes in a massive oven. Ross (1993) shows how to construct quite a simple double oven using the same technique that provided beam collimation. Up to temperatures of 1300 K an insulated heating element that is a miniature version of that used in electric cookers can be employed. It is generally available as thermocoax. Versions with cold ends are available, which simplifies the requirements for vacuum lead throughs. In practice a cylindrical oven with a hemispherical two-start spiral groove is usually used on which the heating element is wound and hard soldered to obtain good thermal contact. A two-start groove means that a bifilar winding is produced, which reduces the magnetic field produced by the heating current (Holland et al. 1981) and is in any case useful since then both ends of the winding are conveniently closer to the current lead throughs in the oven mounting flange than the beam. However, twin-core windings are also available which meet both requirements. An oven of this type is both the easiest to construct and likely to be the most reliable of those where the source of heating is within the vacuum envelope.

3.2.2 Passing a Current Directly through Oven

This method requires the handling of very much higher currents than if heater wires were used, but they are comparable to those required for cylindrical heaters that are mentioned in the previous section. These currents are likely to produce a magnetic field that may well disturb the intended experiment with the beam. A water-cooled heavy current lead has usually to be accommodated near the beam aperture, which could be a disadvantage in many experiments. Use of concentric heating cylinders to overcome this problem has been described by Roberts and Via (1967) and further details are given in Section 5.6. Directly heated graphite ovens have been used in many experiments. Bromberg et al.'s (1975) for caesium fluoride, is described in Section 4.4 and Lew's (1949) for aluminium, in Section 5.11. These very high currents require massive vacuum lead throughs and heavy current carrying wires throughout, usually in the form of copper tubes through which cooling water flows. The design should enable heat to be mainly provided where it is needed by having leads of low resistance.

3.2.3 Electromagnetic Induction

The oven is surrounded by a radiofrequency coil for induction heating. Its use is only likely to be successful where stray electromagnetic fields can be tolerated or the RF power is pulsed so that measurements are made when it is off. This approach has been used by Zavitsanos and Carlson (1973) to produce a beam of carbon as described in Section 5.12. More details of the design of an oven heated by the use of radiofrequency power are described by Bulgin et al. (1977). These authors used a crucible made of graphite which was 10 mm in diameter, 30 mm long, and with a 1-mm wall thickness. Up to 2.75 kW of power at a frequency of 1 MHz was delivered to a seven turn water-cooled coil, which was 25 mm in diameter and 30 mm long and made of 3-mm-diameter copper tubing. The two heat shields situated between the oven and coil were made from 150-μm-thick tantalum or molybdenum sheets that were slotted to minimise inductive effects. In addition slotted aluminium shields 1 mm thick surrounded the coils to reduce the effects of stray electrons. Details are given of the electronic circuitry used to modify the commercial generator to obtain pulsed operation. This reduced the power output to 800 W, which enabled temperatures of 2500 K to be reached. Bulgin et al. give references to the use of their oven to obtain the species NS, PN, SiO, CH_3, P_2 and PN, and they quote furnace temperatures for each. Their oven is also referred to in several sections of Chapter 5.

The International Telecommunications Union has designated a number of frequencies for industrial, scientific and medical applications. To avoid interference with broadcast frequencies these should be used whenever possible, especially if power is likely to be radiated. The frequencies are listed by Withers (1999). Those internationally available are centred at 13.56 MHz, 27.12 MHz, 40.68 MHz, 2.45 GHz, 5.8 GHz, and 24.125 GHz. If personal computers are used close to apparatus employing these frequencies it should be borne in mind that any wireless connections to peripheral devices such as keyboards and mice use some of the lower ones.

3.2.4 INDIRECT HEATING BY ELECTRON BOMBARDMENT

This method is so frequently employed that it is impractical to list specific references here, but details are contained in the relevant succeeding chapters. The oven becomes the anode and a heated usually tungsten wire surrounding it is the cathode. It has the advantage that heat is applied directly to the oven.

Another method, which has, however, only found limited use, is heating the oven by means of field emission, as is described by Pribytkov et al. (1987). Their cathodes consisted of between 700 and 1000 carbonized cellulose hydrate fibres each of 7-μm to 8-μm diameter that were formed into a ring of 500-μm thickness. Four such rings were mounted between steel washers in two pairs, as shown in Figure 3.1. These surrounded the 3-mm-external-diameter cylindrical container that was 25 mm long and made from either graphite or tantalum foil. Up to 8 kV could be applied between the rings and the container mounting and measured temperatures of 2700 K were achieved with the tantalum container at 25-W input power. Unfortunately the authors do not discuss the advantages of field emission, but since 60-h operation is mentioned without substantial change in the emission, reliability appears to be satisfactory.

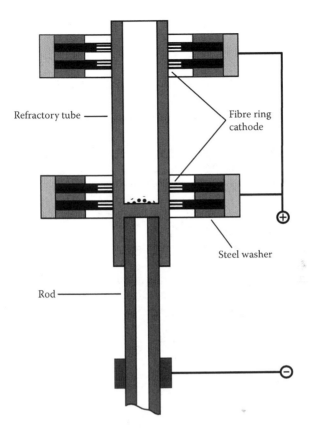

Refractory tube ————

Fibre ring cathode

Steel washer

Rod ——————————

FIGURE 3.1 Pribytkov et al.'s method of heating an oven by means of field emission. (Adapted from Pribytkov VA, Matveev OI and Dibrova AK 1987 A high-temperature furnace with field-emission electron heating *Instr. Exp. Tech.* **30** 746–7 [*Prib. Tekh. Éksp.* **3** 218–9 1987]. With permission from Springer Science & Business Media B V.)

3.2.5 DIRECT HEATING BY ELECTRON BOMBARDMENT

Instead of electron heating of the oven walls, the substance to be evaporated may be heated by the direct impact of a high-power electron beam on its surface. This is usually undertaken either where any possible contamination of the beam by the material of the oven is to be avoided, where creep is serious or if the liquid or vapour of the substance to be formed into a beam is highly reactive. Contamination is minimised and other problems eliminated because evaporation takes place from a pool of molten material surrounded by the solid, which is prevented from becoming molten by cooling around it. Since there is almost no reaction of the evaporating material with the container walls, the beam is of high purity. This method is described in Section 5.9, since the most detailed studies have been made on the production of a copper beam including by Ohba and Shibata (1998). The technique is a development of that used in commercial vacuum evaporation equipment. As mentioned in Section 5.4, a detailed description of the design and operation of a suitable electron gun has been given by Thakur et al. (2001).

The high electron currents can clearly cause problems and the atomic beam tends to be unstable, due mainly to convection currents in the liquid. Stefanov et al. (1988) showed that capillary waves are formed with frequencies of typically a few Hertz. As mentioned in more detail in Section 5.9, Ohba and Shibata (1998) reduced these instabilities by placing a sintered tungsten rod in their beam source. Direct heating by electron bombardment has now been employed to obtain beams of many elements and these are discussed in the relevant sections of Chapter 5. A disadvantage of the method is that it does not appear to have been used to obtain well-collimated beams. However, Graper (1973) made detailed measurements of the angular distribution of aluminium atoms emitted by the electron beam produced by a 12-kW electron gun after deflection through an angle of 270°. He found that the halfwidth was typically 60°, hence one half that of a cosine distribution. Graper fitted his measurements to empirical formulae since the beam shape depended upon the heating power.

Clearly if a sufficiently high-powered electron beam is employed coupled with careful design to reduce radiation and conduction losses, the method should enable beams of refractory materials to be obtained. The cooling by circulating water could be replaced by a coolant operating at a higher temperature, such as mercury, or an oil suitable for high temperatures as mentioned in Section 3.2.7, but the technical problems become formidable.

3.2.6 HEATING BY PHOTONS

We first mention two unusual approaches before turning to the use of external high-power lasers, which have been employed by several authors to heat their ovens.

Although not used to form a molecular beam, Afzal and Giutronich (1974) describe how a solar oven can be used to produce temperatures of 2300 K. The sun's rays are focussed by a parabolic mirror over 3.5 m in diameter with a focal length of 1.2 m. Six triangular roller blinds across its surface were used to control the temperature. The authors consider that design changes would achieve higher temperatures. Feuermann et al. (2002) describe in detail their prototype equipment for transporting solar energy up to a distance of 20 m by a 1-mm-diameter optical fibre. This was also not used for atomic beam production. They used a 200-mm-diameter mirror, which is much smaller than used by Afzal and Giutronich (1974).

Allen et al. (1973) used a 50-W carbon dioxide laser beam directed through a potassium chloride window to reach temperatures up to 1300 K in order to produce beams of numerous elements and compounds. They reported that uneven evolution of the vapour caused problems with their photoelectron spectroscopy. Streets and Berkowitz (1976) evaporated molecular selenium using a 50-W carbon dioxide laser focussed through a germanium window. Cadmium selenide contained in a boron nitride sample holder was heated to a temperature of 1200 K. They reported that many materials have poor absorption at wavelengths of 10.6 μm and that evaporation was very difficult to control. A ruby laser has been employed by Tang et al. (1981) to produce a carbon beam from a thin film.

Though not used to form a thermal atomic beam, Hopkins et al.'s (1983) evaporation of molybdenum by a Q-switched Nd:YAG laser with energies of up to 30 mJ per pulse might form a valuable alternative to direct electron beam evaporation. The laser beam was focussed onto a 3-mm-diameter rotating rod, which the authors consider could be made of any refractory metal. Griffin et al. (2005) showed that by using a laser to heat the alkali metal dispensers that, as we mention in Section 4.2, were initially employed by Fortagh et al. (1998), atom production could be turned on and off in times of less than 1 s. They used up to 4 W of power from a Nd:YAG laser, which was focussed to a spot size of 35 µm on the dispenser.

3.2.7   OTHER METHODS

A novel means of heating an oven to obtain atomic beams of mercury and cadmium is described by Schönhense (1983), namely, by using hot air. This has the advantage that there are no stray magnetic fields near the beam interaction region and stabilising the temperature of the air coupled with the use of a massive 2.5-kg copper oven meant that a long-term temperature stability of about ±2 K could be obtained. The beam-forming nozzle was 30 mm long and 1 mm in internal diameter and had the air inlet close to the beam exit using a double-walled arrangement for both the nozzle and the oven. The beams were condensed on a liquid nitrogen cooled trap with a large surface area. The nozzle could be maintained at a temperature of 20 K to 150 K above that of the oven by adjusting both the rate of the air flow and its temperature.

The air at a pressure of several atmospheres was heated by passing through a 5-m-length, 3-mm-internal-diameter, thin-walled stainless steel tubing that was coiled between two 600-W mains-operated hot plates that were magnetically shielded with mumetal. The hot air was conducted to the apparatus through insulated stainless steel tubing. Temperatures of 900 K could be attained after about 90 min of operation. The nozzle temperature was reached in about 30 min, which meant that it was automatically the hottest part of the vapour system. It remained so on cooling since cooling air flowed in the reverse direction.

Bellman and Raj (1997) heated a copper cup to a temperature of 500 K, by circulating oil, which the authors considered produced better temperature stability than an electric heater. They give further details neither of the oil nor of the method of circulation. It is possible that poly (methyl cyclotrisiloxane) [poly (dimethyl siloxane)] [poly $(C_6H_{18}O_3Si_3)$] silicon transformer oil was used, since this should be usable up to 570 K.

Another evaporation possibility, which would also be important if stray magnetic fields are to be avoided, might be to heat the oven by conduction through the vacuum envelope by burning town or bottled gas.

3.2.8   DISCUSSION OF CHOICE OF HEATING METHOD

Several considerations apply to the choice of heating method. If it is essential that stray magnetic fields are of a very low order, either means other than electric currents could be used or oven heaters could be wound as described by Hertel and Ross

(1968) and discussed further in Section 4.3. In addition, they shielded their oven with a mumetal screen, which indicates that bifilar heater windings are not always sufficient to reduce the magnetic field in particularly sensitive experiments.

If fields at mains frequency can be tolerated, heating current could be supplied from a transformer, possibly fed from a stabilised mains supply. Many experiments require either the modulation of the atomic beam itself, or the beam it interacts with, or both. Hence any power supply for the oven heating that can be pulsed so that current only flows during the time when the beam is interrupted has the advantage that the effect of magnetic fields produced by the heating current can be minimised or eliminated. This was already mentioned with particular reference to RF heating in Section 3.2.3.

In many experiments, the presence of both electrons and their effects, such as excitation and ionisation of the residual gas in their path and the production of metastable atoms and molecules, is particularly disturbing, so direct or indirect heating by electrons in these cases is best avoided unless it is the only method available. Electron emission from the hot oven itself may cause problems as reported by Lemonick et al. (1955) and mentioned in Section 5.9. The effects of electron interactions are clearly most severe when the substance is directly heated by electron bombardment, since the electron currents are highest due to the requirement for the surrounding substance to be cooled sufficiently to remain solid.

The light emitted by a heater winding may also be a problem where the experiment requires photons to be detected. When this is necessary, as mentioned in Section 5.2, Dagdigian et al. (1974) found that for barium, an oven heated by passing current through a heating cylinder was preferable to using a winding.

Heating by lasers has the advantage that the source of heating power is outside the vacuum system and is therefore more easily accessible, but it does not yet appear to offer good beam stability.

Electron beam heating does have the advantage that it is easier to mount the cathode wire so that expansion and contraction is better accommodated compared with a heater in close contact with oven insulation. Heaters may become brittle after use and breakage of heater windings due to expansion and contraction may occur if this is not allowed for. A tungsten heater wire 1 m long will expand about 16 mm on being heated to a temperature of 3000 K, for example. Since tungsten does become brittle, alternative materials with lower melting points have been used, especially if only relatively modest oven temperatures are required. Pure metal wires of tantalum or molybdenum have been employed by many experimenters. Wires of platinum have been used by Leu (1927), whose oven is described further in Section 4.3 and of rhenium by Wang and Wahlbeck (1967) for their oven for caesium chloride, mentioned in Section 4.4. Alloys of platinum–rhodium have been used by Grimley et al. (1972) for the evaporation of potassium chloride and also mentioned in Section 4.4, nichrome by Lambropoulos and Moody (1977) for their alkali–metal recirculating oven described in Section 4.6.2, and kanthal (Ross and Sonntag 1995) have all been used successfully. More information on nichrome and kanthal is given by Rosebury (1965). Nichrome is an alloy of 60% nickel, 16% chromium and 24% iron whereas nichrome V contains 80% nickel and 20% chromium. Kanthal is the name given to a range of alloys of iron, chromium, aluminium and cobalt. Nichrome, kanthal and the platinum–rhodium alloy can be operated in air and the two former alloys have resistances that do not vary greatly with temperature, whereas the hot resistance of pure metal heater wires may be many times the laboratory temperature value. This must be taken into account when selecting heater power supplies. At the highest oven temperatures, evaporation of heaters or oven materials may contaminate the beam, coat insulators with conducting material or result in heaters becoming open circuit.

3.3   OVEN DESIGN

3.3.1   OVEN MATERIAL

The final choice of oven construction material is best left until discussing the production of beams of individual atoms, but we discuss the design principles here. The main one is that not only should the materials used to construct the oven not melt, but also the reaction between the material of the oven and the substance to be evaporated should be negligible. The oven could become porous to the substance to be evaporated or the oven material could form an alloy with it or could even be dissolved. Problems can arise from distortion due to expansion, which can affect the oven causing displacement of its position or leakage of the contents. Reliability is particularly important if long periods of operation are required but it is also difficult to assess other than by trial. Other considerations are the maximum operating temperature and the ease of machinability or fabrication.

It must be remembered that boiling high-purity materials are much more reactive than the same materials in the solid or liquid state and they may be extremely corrosive. There seems to be no reference work that discusses the stability of materials in prolonged contact with substances to be evaporated, particularly their vapours. Some guidance may be obtained from studies of successful vacuum evaporation techniques, for example, to allow a beam to impinge upon a target to produce mirrors or thin films and for molecular beam epitaxy (MBE) and its variants, such as chemical beam epitaxy (CBE) and metal organic vapour phase epitaxy (MOVPE). Olsen et al. (1945), Holland (1956, 1965) and Burden and Walley (1969) consider vacuum coating of many materials and the compatibility of molten materials with possible heaters and crucibles. However, it is usually an advantage in these studies to evaporate the substance quickly to reduce the effects of the background gas in the vacuum system impinging upon the target. Hence materials that might be suitable for periods of seconds or minutes might fail when required to operate reliably for the many hours of typical crossed-beam experiments. However, this is not always the case since, for example, Hahn et al. (1998) mention deposition times of 60 min for a sulphur film. Studies such as those mentioned are therefore more of negative assistance, in indicating what is likely to be unsuitable, rather than what is suitable.

It is again impracticable to reference each utilization of construction material, so we just summarise the possibilities here and give further details in the following chapters. Some possible oven materials are magnetic. For example, early ovens were made of iron, as reported by Gerlach and Stern (1924) for a silver beam, discussed in Section 5.9. Since stainless steel was only developed in 1913, it was not a common laboratory workshop material until much later. Other alloys containing nickel, such as monel metal, inconel and nichrome have also been used for ovens as well as phosphor bronze. Many pure metals have been employed, such as copper, gold, molybdenum, platinum, silver, tantalum, titanium and tungsten. Ovens have also been constructed of glass, fused silica, graphite and many ceramic materials such as alumina, boron nitride, tantalum carbide, thoria and zirconium carbide.

If a nonmagnetic stainless steel oven is required, the choice of stainless steel grade is important. The magnetic properties of austenitic stainless steels were investigated by Post and Eberly (1947). The chemical composition of stainless steels is given in BS ISO 15510 (2010). For simplicity, we use the designations given by the second three-digit subgroup. Types 304 and 321, the most common austenitic stainless steels, with about 18% chromium and 10% nickel, have a residual relative magnetic permeability that depends upon their mechanical history. Less magnetic austenitic stainless steels are the 316 types, which contain slightly more nickel and about 2.5% of molybdenum. In extremely critical applications, ovens can be constructed of type 310, which contains about 25% chromium and 20% nickel. Methods of permeability measurement suitable for these materials are described in BS 5884 (1989).

When measuring velocity distributions of atoms in a beam, it is important that the temperature of the oven is uniform, so that there is little uncertainty in the temperature of the atoms. Hence, as mentioned in Section 4.3, Miller and Kusch (1955) constructed an oven of oxygen-free high-conductivity copper for producing beams of potassium and thallium, which reduced temperature differences to 3.5 K when a massive oven constructed from a block of material was heated to a temperature of 900 K.

Insulating materials for use at high temperatures are usually ceramics and these are also used when a crucible is required which is not wetted by the material to be evaporated. In addition to their use in atomic beams, ceramics are important for the melting of metals, so studies of the resistance of ceramics to molten metals have been made. Fisher (1977) lists alumina [Al_2O_3], beryllia [BeO] boron carbide [B_4C], boron nitride [BN], magnesia [MgO], silicon nitride [Si_3N_4], thoria [ThO_2] and zirconia [ZrO_2] as having good resistance to corrosion by molten metals. He states that thoria is the most refractory oxide ceramic available, and it and magnesia are resistant to attack by molten alkali metals. Silicon nitride is not wetted by the majority of metals but is unsuitable for the alkali metals and titanium. Silicon nitride and silicon carbide [SiC] are suitable for use with sulphur vapour. Titanium nitride [TiN] can be used for cerium and beryllium and a TiN–TiC composite is suitable for melting titanium and zirconium.

Archer (1977) states that pyrolytic boron nitride is not attacked by molten aluminium, antimony, arsenic, bismuth, cadmium, copper, germanium, gold, indium, iron, silicon, silver and tin when in contact for several hours and no wetting occurs at the temperatures of their melting points. However, molten alkali metals do attack boron nitride and he also describes the rate of its attack by aluminium to be acceptably low. Pyrolytic boron nitride is produced commercially by chemical vapour deposition.

Alumina [Al_2O_3] for electrical and thermal insulation of ovens is readily available in the form of ground rods and is suitable up to temperatures of about 2000 K. Under the name of sapphire it is available in the form of small spheres, which aid accurate relocation of insulated components. Boron nitride is a common machinable ceramic that is usable up to a similar temperature. Further details of the mechanical and physical properties of ceramics are given by Rosebury (1965).

3.3.2 Oven Design Considerations

As mentioned in Section 3.1, we follow the general terminology in referring to the container of the material to be formed into a beam as a crucible when it is heated indirectly. It may be used to prevent the beam-forming substance coming into contact with the surrounding oven and reacting with it. A crucible may be used which is not wetted by the material being evaporated from it, thus preventing the thermo-capillary effects discussed in Section 1.1. This is particularly important for metals that have a high surface tension in their liquid state. One solution is to construct the oven of a material such as graphite or some ceramics that are not wetted by the material to be evaporated. As described in Section 4.4, this approach has been adopted by Bromberg et al. (1975), who used an oven constructed entirely of graphite for producing a beam of caesium fluoride. Prescher et al. (1987) used a graphite crucible for titanium as mentioned in Section 5.4. An alternative approach was used by Hubbs et al. (1958), who used a tungsten crucible for plutonium beam production but constructed a razor edge lip to it, as described in Section 5.16. When these techniques have not been employed, it is likely that the causes of an unstable beam, for example, as reported by McCartney et al. (1998) in the case of lead and mentioned in Section 5.12, are due to partial clogging of the beam-forming impedance. This is followed by a build up of pressure, which temporarily clears the obstruction.

Another problem that is reported by Preuss et al. (1979) and discussed in Section 5.9 when evaporating copper is the formation of aerosols. Instead of a pure vapour beam emerging from the beam-forming impedance, it can be mixed with solid or liquid material.

It is generally agreed that the beam-forming impedance must be kept at a higher temperature than the evaporating material, to prevent its being clogged by condensation, either when the oven is being heated, or cooled after a series of measurements has concluded. The temperature difference required is usually not greater than 100 K. An exception to this temperature difference is possibly when the temperature of the material is required to be known accurately, for example, when making measurements of velocity distributions. Simple ways can be used to achieve the higher temperature, for example, by providing more heater windings nearer the impedance as employed by Holland et al. (1981) for barium and mentioned in Section 5.2. If a tubular heater is used through which current is passed and which surrounds the oven, it can be made thinner nearer the impedance. This method was used by Bromberg et al. (1975) for caesium fluoride and detailed in Section 4.4. Another approach that has been used successfully is to provide more radiation shielding around the impedance, as used by Pasternack and Dagdigian (1976) for the alkali earth elements and mentioned briefly in Section 5.2. However, a double oven is frequently employed, with separate heaters for each. The oven containing the beam-forming impedance can then be heated first and cooled last. The main oven contains the material to be evaporated, the vapour pressure of which is controlled by its temperature.

In a double oven design the connecting tube between the two ovens is usually cryogenic thin-walled tubing, to minimise heat conduction. It is important that it does not enable the beam-forming impedance to have a direct line of sight to the surface of the evaporating material. When this occurs, Ward et al. (1967) report an 'inverse pinhole camera effect' for gold, by which they mean that the beam is nonuniform if atoms are evaporated from a pool of liquid or subliming solid that is itself irregular.

It is interesting to note that although it is now noted for the problems of beam instability, silver was used in some of the earliest beam experiments by Gerlach and Stern (1924) as mentioned previously and in Section 5.9. In their case however, the effects of the beam were integrated over time, since atoms were deposited on a glass plate. Presumably any instability in the beam was either present only for a small fraction of this time, or did not disturb the observations.

Early ovens tended to be formed by starting with a massive rectangular block and drilling holes for the heater wiring and the container for the material to be evaporated. This is evident from the chronological order of discussion in Chapters 4 and 5. Massive ovens would have a long thermal time constant, which means that their temperature would remain unaffected by normal short-term fluctuations in the supply voltage. However, some care is needed in bringing an oven up to the correct operating temperature for the first time after filling. For example, as discussed in Section 5.9, Preuss et al. (1979) report that copper emitted dissolved gases as it was brought up to its melting point. As mentioned in Section 5.12, Zavitsanos and Carlson (1973) report that carbon initially evolved hydrogen and hydrocarbons on being brought up to the working temperature of 3000 K. When working with sodium to form a supersonic beam, Gole et al. (1982) heated the stainless steel oven slowly to the melting point of sodium and then at only 2 K min^{-1} to 550 K during which time large quantities of hydrogen were released. Nagata et al. (1986) observed the emission of carbon dioxide when forming a beam of strontium. The resulting violent motion of the materials can result in blocking of the impedance and hence careful temperature control, which is aided by a rapid response of the oven temperature to the power being applied, is advantageous. Clearly ovens constructed of thin-walled materials are one solution, but choice of a construction material with a higher thermal conductivity may also assist when this is compatible with the material to be evaporated. For example, the thermal conductivity of copper is about 18 times greater than for stainless steel. In general, it should be anticipated that most solids and liquids evolve dissolved gases on first being heated to evolve a vapour in vacuum, so accurate knowledge of the oven temperature at the melting point of the solid employed is important so that this temperature can be approached cautiously.

Provision has of course to be made for filling and cleaning the oven and possible replacement of the beam-forming impedance. The mounting design may need realignment following any work upon it to be a minimum. Special treatment is needed for substances such as caesium that are highly unstable in air, and these techniques are detailed under the relevant material. For materials that react more slowly with air several approaches can be adopted. The preparation of the material, transfer to the oven and placing the oven in the vacuum system all under argon have been successfully used by Cvejanovic and Murray (2002) for calcium and described in Section 5.2. Any gas that does not react with the substance such as dry nitrogen could be used. In Ross and Sonntag (1995), Ross describes his use of an inflatable glove box for loading highly reactive substances into ovens and then covering them with diethyl ether, which he emphasises must be dry. This evaporated during the normal evacuation of the vacuum vessel into which the oven was placed. In the case of caesium, which is provided in ampoules, he broke the top of the ampoule and added ether before transferring the metal to the oven. For less reactive materials Ross found that transfer in an atmosphere of argon was sufficient to prevent oxidation.

Oven design should adopt the general principle that no threads terminate in the oven interior, since the mobility of many molten materials means that they will creep into the threads and cause seizure. Some other design points have been offered by Gerlach and Stern (1924) whose oven that is described in Section 5.9 and used to produce a beam of silver was initially operated above its normal operating temperature in order to form a vapour-tight seal in those parts of the oven that operated at temperatures below the melting point of silver. In addition Mais (1934) suggests a tapered plug made from a material with a higher thermal expansion coefficient than the material of the oven body. Crumley et al. (1986) recommend in the case of their oven for copper that interference fits rather than clamps should be used to reduce distortion due to expansion and contraction of the components. For some applications the vacuum chamber must be insulated from the oven body so that an electric potential can be applied to it.

3.3.3 TEMPERATURE MEASUREMENT

For ovens, temperature measurement is usually made with one or more thermocouples. The latter applies if the uniformity of temperature within the oven is important. The vapour pressure of the substance being evaporated is governed by the coolest internal surface. A detailed description of the selection and use of thermocouples is given by Rosebury (1965). In descriptions of ovens that follow, it can be assumed that temperatures are measured by means of appropriate thermocouples in good thermal contact with the oven wall. For high-temperature ovens an optical pyrometer is usually used, so provision has to be made for a window in the vacuum envelope that is unlikely to be contaminated by evaporated material. The advantage of thermocouples is their compactness but Shah et al. (1985) used a platinum resistance thermometer for their oven used to produce a beam of lithium atoms as mentioned in Section 4.3.

3.3.4 RADIATION SHIELDS

Since the main loss of heat from an oven in vacuum is by radiation, shielding by means of a concentric series of preferably polished metal foils surrounding the oven reduces both the power input required and the overheating of vacuum walls and other components of the experiment. Radiation shielding presents something of a compromise. The more shields that are present, the less power required to evaporate the source, yet each shield means that the beam source becomes further away from the place where it is to be used so reducing the usable beam density. The shields also increase the area of material within the vacuum system that will outgas, and they have to be dismantled and cleaned from time to time. Nonmetallic shields can also be used. Roulet and Alexandre (1981) used an alumina tube as a radiation shield for their experiments with copper, silver, gold and nickel, as did Kobrin et al. (1982) for cadmium.

The number of radiation shields that should be used has been calculated by Lawson and Fano (1947). They give general formulae, which also enable the temperature of intermediate shields to be calculated, but show that one thin shield of typical emissivity 0.5 reduces the heating power required compared to the unshielded case to 25%. In a practical case of equal thin shields of identical material of emissivity 0.5, three shields reduce the heater power to 10% of the unshielded case, whereas a further four shields are needed to reduce the power to under 5%. Hence there is normally no need to use more than three shields. Crawford (1972) who was apparently unaware of the work of Lawson and Fano (1947) deduced that for minimum heat losses, the maximum number of shields should be fitted into the smallest space. He made shields from a continuous thin strip of tantalum 25 μm thick to which two rows of narrow spacer strips were spot welded along its length. These buckled on rolling the strip into a spiral to produce a corrugated shield with about 100-μm minimum spacing between windings. End losses were minimised by using a stack of circular disks, which were spaced by dimples.

It is common for the outermost heat shield to be water cooled, to prevent undesirable heating of other parts of the apparatus. This presents no particular technical problems if a continuous copper pipe is used which passes through hollow lead-throughs through the oven mounting flange of the vacuum envelope. An alternative is to cool that part of the vacuum system walls surrounding the oven with cooling coils external to it. Crumley et al. (1986) recommend water cooling the mounted ends of tantalum heat shields in order to increase their working life by reducing distortion, cracking and their becoming very brittle when used with ovens heated to a temperature of 2800 K for the production of a supersonic beam of copper.

3.4 PRODUCTION OF CIRCULAR APERTURES AND SLITS

3.4.1 INTRODUCTION

We conclude this chapter on construction techniques for ovens by describing the special methods that have been developed for making apertures and slits for use in producing atomic beams. Chapter 10 is concerned with the much more technically involved procedures for making collimating arrays of long parallel apertures.

3.4.2 CIRCULAR APERTURES

Hedley et al. (1977) review techniques for producing single small-diameter holes in a range of materials. These include mechanical drilling and drilling by energetic particle impact, such as photons (using lasers), electrons, or nuclear particles. Mechanical twist drills are available that have a diameter of 2.5 μm. Lasers have drilled 125-μm-diameter holes and pulsed electron beams have been used to produce 40-μm-diameter holes.

A detailed description of the method of making single pores in polycarbonate membranes having thicknesses between 2 μm and 10 μm is given by Packard et al. (1986). Basically, the foils are irradiated using a ^{252}Cf radioactive source followed by etching in sodium hydroxide solution to yield near circular single pores having diameters in the range from less than 1 μm to several micrometres. Since they were tested with viscous flow of gas, they are clearly suitable for producing atomic beams. Stinespring et al.'s (1986) fluorine atomic beam mentioned in Section 6.8 employed a 25-μm-diameter orifice made by laser drilling in a stainless steel disk of comparable thickness. Sudraud et al. (1987) give a detailed description of their processes of making submicron diameter apertures in thin nickel foils by ion-beam milling. Ready (1998) details the production of circular apertures down to about 4-μm diameter with lasers.

Grams et al. (2006) suggest the use of commercial electron microscope apertures. Their suggested supplier offers 5-μm-diameter apertures in 100-μm-thick platinum–iridium foil. Larger orifices are available in this and other metals. Patrick (2006) discusses methods of drilling holes having diameters in the range 25 μm to 100 μm in 1-mm-thick silicon carbide ceramic. He found electron discharge machining, ultrasonic abrasion and mechanical drilling and grinding with diamond tools unsatisfactory. Laser machining proved satisfactory, but no further details are given in the paper other than commercial companies undertaking this in Maryland, USA. In Section 10.6.4, we describe Tunna et al.'s (2006) method of producing 50-μm-diameter apertures in 100-μm-thick tungsten foil by lasers. The holes were cut by trepanning, that is to say the laser was used to cut round the perimeter of the holes. Clearly their method is suitable for preparing both circular apertures and also slits, though the latter were not attempted.

3.4.3 SLITS

As mentioned by King and Zacharias (1956), who reviewed a number of possible ways of making slits, the traditional method of producing fixed narrow metal slits is to use a pair of razor blades. They have the required characteristics of being quite thin at their edge yet robust and reasonably straight, as well as being inexpensive and easy to mount and adjust. King and Zacharias suggest that slitting saws may be used to produce slits as narrow as 127 μm in machinable materials. Fortunately for atomic beam work, the complication of adjustment of the slit width under vacuum is rarely required.

Phipps and Taylor (1927) whose Wood's tube source of atomic hydrogen is included in Section 6.3.2, made a glass slit that was 75 μm wide and 3 mm long by pinching a glass tube onto a steel ribbon, which, after annealing the glass, was dissolved in hot hydrochloric acid. Johnson (1928), whose similar source is also described in Section 6.3.2, describes his means of making a slit that was 130 μm wide and 6 mm long in glass by pressing a hot glass tube on to a thin copper strip, which was then dissolved with nitric and hydrofluoric acids. Jackson and Broadway (1930), who obtained a beam of nitrogen atoms as described in Section 6.7, constructed their glass slit by sealing platinum foil to the end of Pyrex glass tubing, attaching a second tube to this so the slit area could be blown out flat, then dissolving the foil in aqua regia. It was 50 μm wide and had a length and thickness of about 1.5 mm. Miller and Kusch (1955) used a slit made from 25-μm- or 38-μm-thick steel shims held in place with a copper strip for their oven used to produce beams of potassium or thallium.

Horton and Young (1970) briefly review the possibilities of obtaining extremely narrow slits by spark erosion, electron beam cutting and diamond scribing. They give details of the production of slits in gold foils that were from 6 μm to 10 μm wide and 300 μm long by photolithograph etching. The foils were produced by electroplating gold on 127-μm-thick nickel foil. More details are given in their paper, but briefly, etching of the substrate enabled the gold slit to be exposed in a hole in the nickel foil. Sudraud et al. (1987), whose method of making apertures is described in the previous section also made slits of less than 1 μm width in gold and nickel foils by ion-beam milling.

The production of glass slits made from edge-polished microscope cover slips is described by Ballard and Bonin (2001). For their purposes an optically transparent assembly was necessary that would withstand an intense laser beam. Various materials including poly (vinyl chloride) [poly (chloro ethene)] [poly (C_2H_3Cl)], poly (methyl methacrylate) [poly ($C_5H_8O_2$)], mica and various other plastics were found less satisfactory than glass. The rough edges of batches of the rectangular cover slips were first removed with successively finer sanding papers from 600 to 2500 grits then the slips were polished with cerium oxide. The best were then selected by examination under a low-power microscope and cleaved with a diamond-tipped pen. They were fixed with epoxy resin against a carefully ground precision slit wall on the slit holder which was about 32 mm wide, 25 mm long and 3 mm thick to form slits from 200 μm to 260 μm wide. The assembly allowed light to pass through a 16-mm-diameter circle.

4 Ovens for Evaporation of Alkali Metals and Their Salts

4.1 INTRODUCTION

Generally lower temperatures of evaporation make most aspects of oven design easier than those discussed in Chapter 5 and included in this group are ovens for the production of beams of alkali metals and their salts. Some special techniques are, however, necessary to handle the alkali metals because of their reactivity with moist air, which increases with atomic number Z. One solution is to produce the beam by a reaction of stable substances in the oven and we begin the discussion of the alkali metals with those cases in Section 4.2. We then consider in Section 4.3 more conventional ovens where the material to be evaporated is loaded directly into the oven. In Section 4.4 we turn to the alkali halides, which have problems due to their corrosiveness in the liquid and vapour state. Ovens where an ampoule of material has to be broken are discussed in Section 4.5. Finally in this chapter we discuss recycling ovens, where material that is not formed into a useful beam is returned to the oven.

Within each section we discuss details in date order of publication, beginning with the earliest, independent of the material, since the techniques are similar. We only include papers where constructional details or operating experience are mentioned unless no detailed information has been located, in which case we refer to whatever information is available. Our aim is to concentrate on operating experience since general design principles have been discussed in Chapter 3. We therefore only give constructional details of ovens where these are innovatory or are discussed in considerable detail.

4.2 OVENS IN WHICH A REACTION TAKES PLACE

Methods of preparing alkali metals in a vacuum by heating a compound with another metal appear to be quite old, since de Boer et al. (1930) state without references that this is the usual method. This is probably why Kasabov (1986) and DeMarco et al. (1999) are the only authors of those included below who have given references to the sources of their information, which led to the use of a precursor for producing elements from materials that are unstable in air. More recent information may possibly be commercially sensitive due to the convenience of precursors for producing photocathodes. If there is a disadvantage compared with starting from the source material to be evaporated, it is that the temperatures required to produce the same intensity of atoms are much higher. Precursors have occasionally been employed to produce beams of substances other than the alkali atoms but these are so few that they do not justify a separate section. They are included in relevant sections of Chapter 5.

According to de Boer et al., the success of the use of precursors depends on the volatility of the alkali metal. The authors reviewed a number of possible combinations of alkali compounds and metals that could be used to produce pure alkali metals in vacuum. In their view, a mixture that is stable in air and produces the alkali metal at temperatures no higher than 800 K is desirable. It is also important that no other material is evaporated other than the desired alkali metal. Hence a mixture of an alkali chloride with calcium was considered unsatisfactory, since calcium oxidises in air and the alkali hydroxides and carbonates are too hydroscopic. This led the authors to investigate zirconium as a suitable metal, in view of its stability in air and its highly stable compounds. Unfortunately it was unsuitable for use with the alkali halides since the zirconium tetrahalide compounds formed in the reaction are too volatile. Hence alkali compounds needed to be selected that produced zirconium oxide from the reaction.

Among the compounds they considered were caesium sulphate and bisulphate, caesium and rubidium chromate and bichromates, and potassium and sodium molybdates and tungstates. It was found necessary to have an excess of zirconium present in the mixtures of powders; otherwise the reaction became uncontrollable on heating and could become explosive. A successful mixture was 10 parts by weight of zirconium to either 1 part of caesium sulphate or to 1 part of caesium bichromate. These produced caesium at respective temperatures of 800 K, and 650 K. Rubidium bichromate and zirconium in the same 1:10 ratio produced rubidium when heated to 640 K, whereas a 1:4 mixture of rubidium chromate and zirconium required a temperature of about 1000 K.

Potassium compounds tended to react explosively with zirconium, but the chromate and tungstate were satisfactory, producing the alkali metal at temperatures of about 1100 K and 840 K respectively with typically only twice as much weight of zirconium to that of the compound. Sodium molybdate in the proportion of 1:4 was usable at about 820 K and sodium tungstate in the same proportion at about 720 K. No satisfactory production of lithium was found by heating a compound with zirconium.

Although not used to form an atomic beam, Hackspill (1928) prepared sodium by heating a mixture of sodium chloride and calcium carbide to a temperature of about 1100 K in vacuum. The reaction taking place is represented by

$$2NaCl + CaC_2 \rightarrow CaCl_2 + 2Na + 2C.$$

For the preparation of potassium, he used a mixture of potassium chloride and calcium. DeMarco et al. (1999) used the same compounds. They found that it was necessary to employ highly purified calcium which they obtained by baking at a temperature of 700 K for four days under vacuum. The potassium salt and metal were also finely powdered, with the latter sieved through a 150-μm mesh. The 5:1 molar mixture of calcium and potassium chloride was evaporated from an electropolished nichrome boat. The authors give further details of the performance of their source, from which about 20% of the potassium was emitted.

With this background we can now study the experience of those who have used precursors to form beams of alkali metals.

Davis et al. (1949a) describe a simple oven made of monel metal for the evaporation of isotopes of sodium and caesium. The authors state that stainless steel, nickel, fused silica and boron carbide were tried unsuccessfully as oven materials because either physical absorption or chemical reaction occurred. The principle of the construction is shown in Figure 4.1. It consisted of two blocks about 6 mm thick. The lower contained a blind hole about 3 mm in diameter and 5 mm deep in which the sample in the form of the alkali azide [NaN_3, CsN_3] solution or sodium chloride was placed together with calcium chips. The upper block contained a milled slot about 28 mm long and 250 μm wide and deep, which formed a single channel beam collimator. The mating surfaces of the two blocks were lapped so that no gasket was required when they were mounted in contact in a holder that contained the heating elements and thermocouples. When the vapour pressure of the evaporated materials was such that the mean free path was longer than the channel, a measured halfwidth of the beam of 3° was obtained.

Bellamy and Smith (1953) used an oven similar to that used by Davis et al. (1949a) for producing beams of isotopes of rubidium and caesium, whereas for isotopes of sodium and potassium they used a larger monel metal oven which could contain 2 g of material. Lew (1953) briefly describes an iron oven containing sodium and caesium chloride that produced a caesium beam when heated to a temperature of about 500 K. Zacharias and Haun (1954) briefly report that they used a mixture of caesium carbonate and potassium to obtain a caesium beam from their crinkly foil oven. Their measurements of the beam properties at temperatures of 341 K and 417 K are discussed in Sections 11.3.2 and 11.3.3. A beam of rubidium atoms was produced by Senitzky and Rabi (1956) by heating rubidium chloride with calcium shavings to a temperature of about 600 K. Hobson et al. (1956) report the use of rubidium bromide with an excess of calcium for the same purpose, which required higher temperatures.

In order to obtain a beam of caesium, Stroke et al. (1957) placed caesium chloride in a weak solution of hydrochloric acid in a cup that fitted into their oven, which is shown in Figure 4.2. They then neutralised the acid with sodium carbonate. Following evaporation of the liquid, potassium was added. A plated copper seal was used to seal the oven. It was made of monel metal and heated with molybdenum

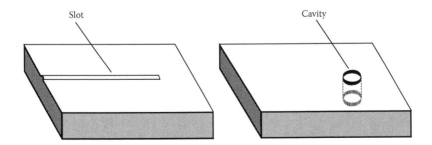

Slot   Cavity

FIGURE 4.1   The oven described by Davis et al. to produce a beam of sodium and caesium. (Adapted from Davis L, Nagle DE and Zacharias JR, 1949. Atomic beam magnetic resonance experiments with radioactive elements Na[23], K[40], Cs[135] and Cs[137]. *Phys. Rev.* **76** 1068–75. Copyright 1949 American Physical Society.)

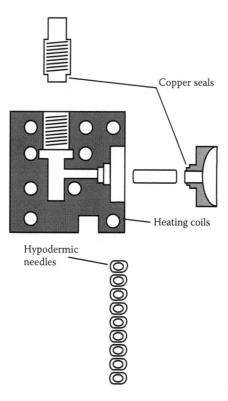

Copper seals

Heating coils

Hypodermic
needles

FIGURE 4.2　The oven used by Stroke et al. to produce a beam of caesium. (Adapted from Stroke HH, Jaccarino V, Edmonds DS and Weiss R, 1957. Magnetic moments and hyperfine-structure anomalies of Cs^{133}, Cs^{134}, Cs^{135} and Cs^{137}. *Phys. Rev.* **105** 590–603. Copyright 1957 American Physical Society.)

wires to a temperature of about 470 K. The wires were insulated with fused silica tubing. Baffles in the oven that are not shown in Figure 4.2 prevented source material from entering the collimator. The beam emerged through nine hypodermic needles that were 12.5 mm long and which were flattened to give a slit-shaped assembly that was about 3 mm high and 460 μm wide containing channels that were 250 μm in diameter. This leads to an open area of 31%, but the authors claim it is 56%.

Essen and Parry (1958) used equal weights of caesium chloride and sodium, which produced a beam of caesium atoms when heated to a temperature of about 500 K. This was loaded into the oven in air. Boutry et al. (1964) prepared caesium from both caesium chloride reduced by calcium and caesium dichromate reduced by silicon. The production of caesium atoms from a mixture of one part of caesium chromate and two parts of silicon powder by weight was investigated by Eichenbaum and Moi (1964). They found that caesium was produced at a temperature of 1000 K. To obtain a constant beam over a period of some hours as the caesium became depleted, a gradual increase in temperature to about 1300 K was necessary. No caesium was produced at temperatures of 800 K, which allowed the vacuum system to be baked with the source in place. It was estimated that the total yield was about 90%

of the caesium atoms contained in the chromate. The authors note that impurities are also emitted which can be trapped with a tungsten wire pad followed by silica wool heated to a temperature of about 600 K. Apparently unaware of the work of Eichenbaum and Moi (1964), Schaeffer (1970) used three parts by weight of silicon to one of caesium chromate to form a caesium beam which was then ionised by passage through 20% porosity tungsten.

Kasabov (1986) reduced sodium tungstate dihydrate [$Na_2WO_4 \cdot 2H_2O$] and sodium molybdate dihydrate [$Na_2MoO_4 \cdot 2 H_2O$] with zirconium. Initially the tungsten compound was heated in air at a temperature of 483 K and the molybdenum compound at 513 K to remove the water of crystallisation. Final dehydration was continued in vacuum. One part by weight of each salt was then mixed with four parts by weight of zirconium powder with a grain size of 40 µm and formed into pellets. These contained an excess of zirconium for reactions of the following type:

$$2Na_2MoO_4 + Zr = 4Na + Zr(MoO_4)_2.$$

Pellets were placed in a 150-µm-thick tantalum foil oven, which was heated by a coil of 200-µm-diameter tungsten wire. Sodium atoms effused through an 800-µm-diameter orifice. The sodium tungstate pellets first emitted sodium atoms when heated to a temperature of about 700 K and the sodium molybdate at about 800 K but operating temperatures were about 300 K higher. Mass spectrometric studies gave an upper limit for impurities of 0.01%, with no ions present. Kasabov preferred the higher temperature emitter because it enabled baking of the ultrahigh vacuum system at a higher temperature. He noted that the emission properties of the pellets were unchanged after exposure to air.

Another means of producing caesium atoms is given by Bañares and Ureña (1989), who reacted equal weights of caesium chloride and barium at temperatures up to 900 K. Their stainless steel oven was heated by passing a current of 200 A through it, which enabled it to reach a temperature of nearly 1300 K. The current leads were 10-mm-diameter copper tubes through which cooling water passed. The stainless steel oven was 20 mm in diameter and 80 mm long, and the wall thickness was 250 µm. A filling of 2.5 g of each material produced a caesium beam through a 500-µm-diameter orifice for 15 h. A diagram of their oven is given in Figure 4.3.

Dinklage et al. (1998) briefly mention that they obtained a beam of lithium from an alloy consisting of aluminium as well as lithium. It operated at temperatures between 835 K and 870 K. A rubidium beam was obtained by heating rubidium chloride with barium to a temperature of 800 K by Aguilar et al. (2001).

We collect together several papers that describe the use of dispenser sources of alkali atoms. Fortagh et al. (1998) state that resistively heated alkali–metal dispensers are available commercially. The one they employed, which was compatible with ultrahigh vacuum, contained rubidium chromate and a reducing agent which produced rubidium atoms when heated with a current from 2.8 A. Since their requirements were for only small numbers of atoms, it is unclear how suitable the source would be for producing an intense atomic beam. Further details of the use of the dispenser source of rubidium atoms are given by Rapol et al. (2001). They found that it was necessary to operate the source below the threshold current for the emission of rubidium

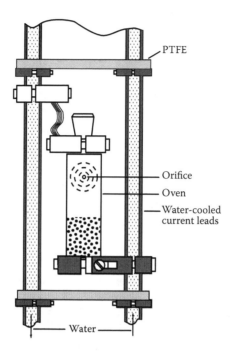

FIGURE 4.3 Bañares and Ureña's oven used to produce a caesium beam. (Adapted from Bañares L and Ureña AG, 1989. Simple oven design for highly reactive metal beam applications. *J. Phys. E: Sci. Instrum.* **22** 1046–7. Copyright IoP Publishing 1989.)

for some hours when their ultrahigh vacuum system was being baked, in order to achieve the lowest pressure in it. However, it was then necessary to heat the source with a current of 8 A for periods of a few seconds at a time before stable emission of atoms with a current of 2.7 A was obtained. The use of a dispenser source to produce an atomic beam is also described by Roach and Henclewood (2004), who valued the short time required to turn it on and off, namely, about 100 s. They report that each dispenser only releases about 4.5 mg of rubidium. Additional details of their operating experience with the dispenser are given in their paper. As mentioned in Section 3.2.6, Griffin et al. (2005) employed laser heating of their source. McDowall et al. (2012) describe heating their rubidium dispenser initially with a current of 15 A for 3 s followed by 9 A for 1 s before the 5-A operating current was applied.

The technical literature indicates that alkali metal dispensers are available for the five commonest alkali metals, the chromates of which are mixed with an alloy consisting of 84% zirconium and 16% aluminium. Their original purpose was to facilitate the production of photocathodes. At currents of up to 7.5 A which produce temperatures up to 1120 K, atoms are emitted from a slit source free from loose particles and chemically active (noninert) gases.

Fantz et al. (2012) report on the use of a source of caesium atoms obtained from a compound with bismuth that is available from another supplier. The one they used was operated at currents of up to 10 A and contained 100 mg of caesium. Dispensers are available for the alkali and alkali earth metals as well as ytterbium that can contain much larger amounts of material according to technical literature. These have potential to form atomic beams.

In conclusion, the absence of comparative studies, both of alternative precursors and also of the methods described in Sections 4.3 and 4.5, unfortunately makes the drawing of meaningful conclusions to this section inappropriate.

4.3   CONVENTIONAL OVENS FOR PRODUCTION OF BEAMS OF ALKALI METALS

We describe as conventional those ovens in which the same material as that to be evaporated is directly placed (without mention of an ampoule being broken in place by means external to the oven) and no recycling of material that is not formed into a beam is used. Unless stated otherwise, no discussion of transfer of the alkali to the oven is given.

Leu (1927) employed a copper oven, shown in Figure 4.4, for the evaporation of sodium and potassium. The oven was mounted on the side of a cylindrical copper block heated with a 200-μm-diameter platinum wire 400 mm long, wound on a fused silica tube, insulated with asbestos and contained in a constantan tube. Hence the material to be evaporated was cooler than the intermediate chamber, the exit tube from the oven and the beam-forming slit, which had no direct line of sight with the evaporating material. A water-cooled copper tube surrounded the oven. The sodium and potassium were stored under xylol and each washed in pure petroleum ether before being transferred to the oven. The film protected the metals from the formation of a hydroxide layer and evaporated rapidly when the apparatus was evacuated.

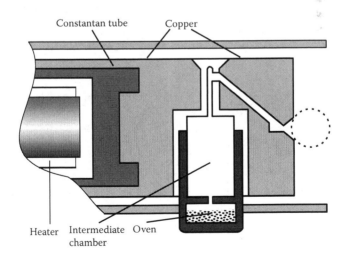

FIGURE 4.4   The oven Leu used for the evaporation of sodium and potassium. (Adapted from Leu A, 1927. Versuche über die Ablenkung von Molekularstrahlen im Magnetfeld. *Z. Physik* **41** 551–62. With permission from Springer Science & Business Media B V.)

Taylor's (1929) apparatus was similar to that described by Leu (1927), but with an oven made from tungsten steel. It was supported on a constantan rod because of its poor thermal conductivity. The oven was radiatively heated to a temperature of 580 K with a tungsten coil to obtain a beam of potassium atoms. Heating in addition by electron impact was employed to obtain a beam of lithium atoms at 1020 K. He noted that lithium above its melting point attacked glass.

The above oven designs were improved by Lewis (1931) for use with lithium, sodium and potassium in order to determine the molecular components in the beam. The solids were heated carefully to their melting points in order to prevent contamination of the beam. The oven was designed both to be at a uniform temperature and to prevent solid material being emitted in the beam or clogging the beam-forming slit. It was constructed from a solid block of metal that was bored to take a container for the alkali metal. The beam emerged through a 1-mm-diameter channel and a slit that was 10 μm distant from its exit. This assembly was surrounded by a thick gold-plated copper shield that contained two tungsten heating coils. A power of 25 W produced temperatures of 870 K with temperature differences not exceeding 5 K.

Mais (1934) employed a nickel oven with a monel metal lid to obtain a beam of potassium. Since the coefficient of expansion of the lid was greater than that of the body, the lid formed a 'very tight' joint. Heating was provided by means of 250-μm-diameter tungsten wires insulated with fused silica tubing and contained in a copper cylinder surrounding the oven. The potassium beam emerged through a slit that was 10 μm wide and 2 mm high. The metal was cleaned under petroleum ether by cutting off the hydroxide coating and was then transferred to the oven under petroleum ether.

An oven constructed of nickel was used by Rabi and Cohen (1934) to evaporate sodium, which left the oven through a 13-μm-wide exit slit via a channel in a tapered monel metal plug, which reduced creep of metal to the slit. The tungsten heaters were close to the slit so this was maintained at a higher temperature than the rest of the oven. Sodium was transferred to the oven under petroleum ether, which, together with occluded gases, was removed by heating the oven to a temperature of about 470 K. Following overnight storage in the vacuum system, the oven was heated at a rate of about 3 K min^{-1} to about 570 K and then more slowly to the operating temperature of 633 K when the vapour pressure of sodium was about 27 Pa.

Miller and Kusch (1955) constructed their oven from a rectangular block of oxygen-free high-conductivity copper in order to minimise temperature differences within the oven. This was particularly important for their measurements of the velocity distributions of beams of potassium, which were made at temperatures between 466 K and 544 K. The oven appears identical to that described as a single-chamber oven by Miller and Kusch (1956) so we include further details from that paper. A diagram of the oven is shown in Figure 4.5. The oven was heated by coils of 250-μm-diameter tungsten wire, which were insulated by ceramic tubing, which was inserted in holes drilled through the block. The beam emerged through a slit made from 25-μm- or 38-μm-thick steel shims held in place with a copper strip.

A double oven made of nickel was used by Taylor and Datz (1955) for the production of a beam of potassium. It was evaporated from the first at a temperature of 540 K and the vapour passed through a thin-walled tube into the second, which was up to 300 K hotter. The beam was collimated by slits to be of rectangular cross section.

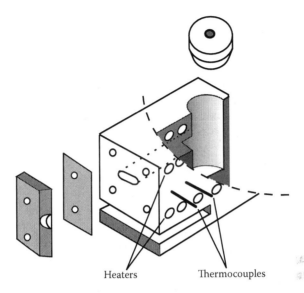

Heaters   Thermocouples

FIGURE 4.5   The oven used by Miller and Kusch to study the velocity distribution of a potassium beam. (Adapted with permission from Miller RC and Kusch P, 1956. Molecular composition of alkali halide vapors. *J. Chem. Phys.* **25** 860–76. Copyright 1955, American Institute of Physics.)

Martin and Kinsey (1967) briefly describe a two-chamber oven for the production of a beam of potassium. This was evaporated from the lower oven at a temperature of about 600 K. The beam was emitted from the upper oven at about 800 K through a collimating plate that was 10 mm long and 3 mm thick containing a row of 150-μm-diameter tubes. No details of the beam properties are given.

A double oven with a crinkly copper foil collimator for use with the alkali metals at temperatures up to about 1000 K is described by Hertel and Ross (1968) and introduced in Section 3.2.8. In order to reduce the magnetic fields introduced by the 250-μm-diameter tantalum heating wire it was wound in 12 twin bore ceramic tubes in each oven. The tubes were so arranged that the wire links between the bores were closer to the interaction region than the wires linking tubes. The heater for the lower oven was arranged in the same way but with direct current flowing in the reverse direction. In addition, a mumetal cover surrounded the oven.

Rassi et al. (1977) describe the evaporation of lithium rods in their main oven at a temperature of just over 800 K with the front oven maintained 120 K hotter in order to reduce, but not eliminate, dimer formation. The stainless-steel double oven briefly described by Shuttleworth et al. (1977) that was used to obtain a beam of sodium atoms is a little unusual in that the main oven contained molten, rather than boiling sodium at a temperature of 520 K. The front oven was at a temperature of 600 K and liquid sodium transferred to this through a multistrand stainless-steel wire wick. The sodium beam was collimated by a collimated holes structure that contained 50 channels mm^{-2} with $\Gamma = 14$.

Shuttleworth et al. (1979) briefly describe their oven for the production of a beam of lithium atoms. They used a two-stage stainless steel oven. The first stage was heated to a temperature of 1100 K and the second was 100 K hotter, which resulted in an estimated dimer concentration of about 2%. The lithium beam was collimated by an array of hypodermic needles having an internal diameter of 150 μm and length of 12 mm; hence $\Gamma = 80$.

Touchard et al. (1981) give brief details of a directly heated cylindrical oven 92 mm long and 8 mm in diameter, and made of 200-μm-thick tantalum, which they used to form beams of the isotopes of lithium, sodium, and potassium. It operated at a temperature of over 2000 K when heated with a current of 500 A. A 100-μm-thick sheet of molybdenum lined the oven to prevent the tantalum being attacked by graphite that was used in other parts of the apparatus.

Aushev et al. (1982) describe an oven for producing a beam of lithium atoms, which enabled the beam to be produced vertically downward. A diagram of the oven is shown in Figure 4.6. The oven was made of a heat-resistant stainless steel because of the reactivity of liquid lithium. Unfortunately a more exact specification of the alloy is not given, but it is presumably type 310, or similar, since this is a high-temperature austenitic stainless steel. The gasket to seal the filling aperture was made of nickel. Tungsten heater windings enabled temperatures of 1300 K to be reached with molybdenum radiation shields, the outer one of which was water cooled. The lithium beam emerged through a funnel-shaped orifice, with a cylindrical part 400 μm in internal diameter and 1.6 mm long.

Shah et al. (1985) constructed an oven 20 mm in internal diameter and 40 mm deep from a soft iron block that was electrically heated to a temperature of over 840 K from a supply of constant current to heating coils. The block was surrounded by a water-cooled radiation shield. A beam was produced from a side tube from about 2 g of lithium, which provided about 50 h of operation at 763 K as measured with a platinum resistance thermometer. The oven walls were about 6 mm thick, so it took about 5 h to attain a stable beam that remained within 10% of its intensity for up to 4 h.

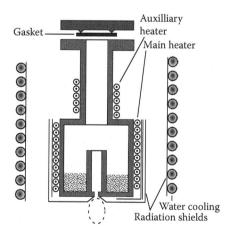

FIGURE 4.6 Aushev et al.'s double oven used to produce a beam of lithium. (Adapted from Aushev VE, Zaika NI and Mokhnach AV, 1982. Properties of a lithium atomic beam. *Sov. Phys. Tech. Phys.* **27** 878–80 [*Zh. Tekh. Fiz.* **52** 1438–41 1982]. With permission from Springer Science & Business Media B V.)

Borovik et al. (1995) describe a single oven for the evaporation of lithium and potassium at temperatures up to 900 K. The stainless steel oven was 72 mm long, 13 mm in diameter with a wall thickness of 1 mm. Heating was provided by 250-μm-diameter tantalum wires wound noninductively in nine four-bore ceramic tubes 40 mm long which were placed closest to the beam exit, in order to ensure that it was maintained at a higher temperature. A polished stainless steel 500-μm-diameter tube was used as a radiation shield, which was enclosed in a water jacket. Particular care was taken in the construction to avoid heat conduction through the use of minimum contact areas. Some collimation of the vertical beam was produced by a channel 700 μm in diameter and 2 mm long, which was attached with a copper gasket between a double knife-edge seal after loading the oven with up to 5 ml of material, which provided continuous operation for over a week.

Egorov et al. (2002) produced a beam of rubidium through a 3.9-mm-diameter orifice from an oven heated to temperatures up to 630 K. The beam passed into a cryogenic cell that was cooled with helium buffer gas. Details are given of the cooling technique and measurement on the properties of the resulting slow atom beam. The authors consider that the cryogenic cooling technique could be used with any atom or molecule that can be formed into a beam.

An oven for the production of a lithium beam is briefly described by Wehlitz et al. (2002). No construction details are given but it is stated that the oven was surrounded by a water-cooled jacket. A graph of electrical power against oven temperature showed that temperatures of over 1200 K were reached when power of just over 140 W was applied. The heating arrangement was designed to produce zero magnetic fields. The oven was mounted on a ceramic bolt so that it could be electrically biased to suppress the emission of thermal ions. Alignment of the oven under vacuum was possible.

Murray (2005) used the double oven that was developed from a design by Cvejanovic and Murray (2002) for calcium to produce beams of sodium and potassium. It is therefore described in Section 5.2, which also refers to Murray and Cvejanovic (2003) and Murray et al. (2006). The important design change was that the sealing plug was made from oxygen-free copper, which was found to react 'only slowly' with the metallic liquids. The oven was filled in an atmosphere of argon inside a glove bag. The main oven was heated to a temperature of about 600 K to produce a beam of sodium and 520 K for potassium, with the second oven 20 K hotter in the case of sodium but only 10 K hotter for potassium.

Saleem et al. (2006) used a stainless steel oven that was 9 mm in internal diameter and 75 mm long to evaporate lithium. The vertical beam was emitted through a 2-mm-diameter aperture in the 500-μm top cap of the oven, which was mounted in a ceramic cylinder that provided thermal and electrical insulation. A thermocoax heater winding enabled a beam to be produced at a temperature of 650 K.

4.4 OVENS FOR ALKALI HALIDES

Experiments on beams of the alkali halides also have a long history. Wrede (1927) used a glass oven within the glass vacuum envelope on which the heater was wound. However, the evaporation temperatures of nearly 1000 K were close to the softening temperature of the glass. Ochs et al. (1954) provided separate heating for the slit housing of their oven, but the housing was still in good thermal contact with that part of the oven containing the sodium chloride to be evaporated.

Miller and Kusch (1956) made a detailed study of the velocity distributions of nine of the alkali halides using the double oven shown in Figure 4.7. The upper oven was made of copper as mentioned for their single oven design for the evaporation of potassium discussed in Section 4.3, but since the copper was close to its temperature limit, the lower oven and connecting components were made of iron. However, for sodium fluoride, because of its higher operating temperature, the single oven design but made of molybdenum was used. The heaters were 250-μm-diameter tungsten wire threaded through ceramic tubes inserted into holes in the oven blocks. The lower oven temperature was kept constant to 0.25 K. Their results confirm the presence of dimers and trimers in most cases.

The alkali halides need evaporation temperatures of about 1000 K, which is somewhat higher than that required for the alkali metals. A major problem in the study of beams of alkali halides, namely, that dimers, trimers and tetramers are formed, was investigated in detail by Berkowitz and Chupka (1958) using a mass spectrometer technique, although not in conjunction with beams. They volatilized all the halides of lithium, sodium and potassium, from a tungsten crucible heated by a tantalum filament with the exception of lithium iodide, which had been studied

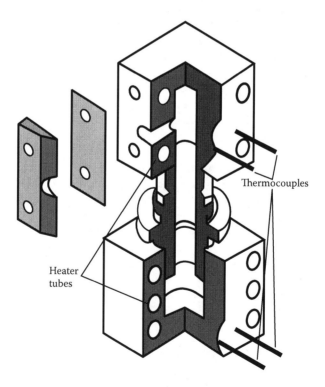

FIGURE 4.7   The double oven used by Miller and Kusch to produce alkali halide beams. (Adapted with permission from Miller RC and Kusch P, 1956. Molecular composition of alkali halide vapors. *J. Chem. Phys.* **25** 860–76. Copyright 1955, American Institute of Physics.)

by Friedman (1955), who used a nichrome oven, and they also studied rubidium and caesium chlorides.

The single oven used by Eisenstadt et al. (1958) for studying the velocity distribution of the alkali fluorides was constructed of molybdenum and was operated at temperatures up to 1170 K, where molybdenum is mechanically more stable than copper. It was similar in design to that shown in Figure 4.7. Whereas the velocity distribution of lithium fluoride is clearly made up of considerable dimer and trimer components, their abundance decreases with increasing alkali atomic number.

Lew et al. (1958) used a single oven heated to a temperature of 1030 K by molybdenum coils enclosed in ceramic tubes to produce a beam of rubidium fluoride from a slit. The oven was surrounded by two radiation shields, the outer one of which was water cooled. The rubidium fluoride is highly deliquescent, so it was dried by heating to temperatures between 800 K and 900 K under vacuum before being transferred to the oven. Problems were encountered with clogging of the oven slit unless only about 200 mg of material were used.

For their spectroscopic study of the alkali bromides and iodides, Rusk and Gordy (1962) used a double oven constructed of stainless steel and heated with molybdenum wire about 760 μm in diameter, which was insulated by either high-temperature glass, fused silica or thoria ceramic for the highest temperatures of nearly 1100 K. The molecular beams were collimated with about 50 stainless steel tubes, which were about 6 mm long and had a 400-μm bore size and a 500-μm outer diameter.

We have already mentioned Martin and Kinsey's (1967) oven for the production of a beam of potassium in Section 4.3. Though not strictly an alkali halide, we include here mention that their beam was crossed with a beam of tritium bromide, which was collimated by a glass array that was 1 mm wide, 10 mm long and 750 μm thick, containing 5-μm-diameter capillaries. Hence $\Gamma = 150$. The tritium bromide was prepared in their laboratory and frozen in a gold-plated stainless steel container before being transferred to the scattering apparatus.

Wang and Wahlbeck (1967) briefly describe a double oven made of nickel for the production of a beam of caesium chloride, which emerged through a 500-μm-diameter orifice in a 127-μm-thick platinum sheet. The oven was heated by 250-μm-diameter rhenium wire that was insulated with porcelain tubes. Evaporation temperatures were from 776 K to 993 K with front oven and orifice temperatures some tens of degrees hotter.

Adams et al. (1968) made a single oven from oxygen-free high-conductivity copper for the evaporation of 1 g caesium chloride. It was heated with tungsten heating coils insulated with alundum (aluminium oxide) tubes contained in bores both in the oven and in its massive lid, which contained the cylindrical channel through which the beam emerged and which was maintained between 1 K and 1.5 K hotter. Experiments were made with channel diameters of about 280 μm, with lengths of 140 μm and 390 μm at temperatures from 820 K to 1017 K.

Grimley et al. (1972) constructed an oven of inconel for the evaporation of potassium chloride at temperatures up to nearly 900 K. The oven was heated with 500-μm-diameter wire of platinum–rhodium alloy about 810 mm long, which was bifilarly wound in seven turns on four alumina rods. The assembly was surrounded by five radiation shields made of tantalum. A direct current power supply regulated to 0.01% enabled temperature stabilities of ±0.1 K to be obtained.

The use of a graphite oven to produce a beam of caesium fluoride is described by Bromberg et al. (1975). Graphite was found to be far superior to stainless steel, molybdenum and tantalum, which all corroded in the presence of caesium fluoride. However, graphite has the disadvantage of being permeable to caesium fluoride vapour. The oven was made from a hollowed out graphite rod about 25 mm in diameter and 120 mm long, bored out to be about 16 mm in diameter. An upper graphite supporting rod was initially a screwed and, in a later design, a taper fit into this. The beam emerged through a 250-μm-diameter orifice in the side of the rod as shown in Figure 4.8a. The oven was surrounded by a graphite tube, which was heated by a current of about 200 A. The wall of the tube was thinner opposite the orifice, which resulted in the temperature being hotter by at least 150 K than the rest of the oven, which could be heated up to a temperature of 1200 K. The oven was surrounded by two radiation shields and a water cooled jacket as shown in Figure 4.8b.

A two-stage oven described by Berkowitz et al. (1979) is shown in Figure 4.9. The lower oven was commercially available. It was heated by radiation from a noninductively would tungsten spiral, which was surrounded by radiation shields. The current leads were water cooled. The evaporated lithium halide under study passed into the tubular upper oven. As mentioned in Section 3.2.1, high-temperature cement was used to seal its noninductively wound tungsten wire heating element to a thoria tube that surrounded this oven. This in turn was surrounded by a radiation shield. In addition both ovens were surrounded by water-cooled copper cylinders. The apparatus was used to produce beams of the lithium halides. By altering the heating power to the ovens, the dimer concentration in the beams could be varied.

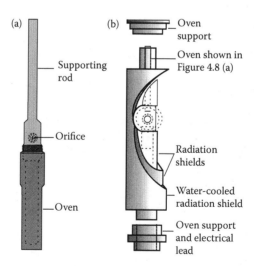

FIGURE 4.8 (a) The graphite oven for caesium fluoride used by Bromberg et al. (b) Bromberg et al.'s oven assembly. (Adapted with permission from Bromberg EEA, Proctor AE and Bernstein RB, 1975. Pure-state molecular beams: production of rotationally, vibrationally, and translationally selected CsF beams. *J. Chem. Phys.* **63** 3287–94. Copyright 1975, American Institute of Physics.)

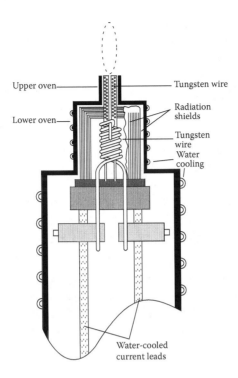

FIGURE 4.9 Berkowitz et al.'s oven. (Adapted with permission from Berkowitz J, Batson CH and Goodman GL, 1979. PES of higher temperature vapors: Lithium halide monomers and dimers. *J. Chem. Phys.* **71** 2624–36. Copyright 1979, American Institute of Physics.)

4.5 EVAPORATION FROM AMPOULES

Caesium is too unstable in air to be provided in other than sealed ampoules and some of the other alkali metals may be provided in these as well, or transferred to them, as described in the next paragraph. Various techniques have been used for breaking ampoules under vacuum and these are discussed in this section. Other experiments where there is a specific mention of ampoules are also included here, since breaking them within the vacuum system of the beam apparatus is not the only method that has been used.

Kratzenstein (1935) describes an oven for the production of a beam of potassium that he prepared in a sealed glass ampoule, which contained 2 g to 3 g and which could be broken under vacuum in the oven. It was 100 mm long and 30 mm in diameter and was heated with a tungsten heater capable of delivering a power of 100 W. The beam emerged through a 950-μm-diameter aperture in a 70-μm-thick steel foil, which was maintained from 17 K to 57 K hotter than the body of the oven, which operated at temperatures between 456 K and 548 K.

Estermann et al. (1947) used glass capsules containing 500 mg of caesium and broke these in nitrogen at atmospheric pressure before sealing the oven's filling plug. The approach used by Stickney et al. (1967) was to cool an ampoule containing 5 g of caesium in liquid nitrogen (77 K), then to break the capsule before transferring the caesium to their stainless steel double oven. The heating elements were molybdenum wires insulated with fused silica tubes. The oven containing the beam-forming impedances was maintained 10 K hotter than that containing the caesium.

The glass vial of caesium used by Klein (1971) was placed in a 10-mm-diameter oxygen-free high-conductivity copper tube. After the apparatus had cooled following baking to a temperature of over 500 K, the vial was broken by crushing the copper tube with a pinch-off tool. Evaporation of the alkali was accomplished by using a heating tape wrapped around the copper tube. A similar approach to breaking the vial is described by Yasunaga (1976). Tompa et al. (1987) used a high-temperature valve to break a caesium ampoule, which fitted into its 9.5-mm bore and was broken by closing the valve. It was heated with heating tapes up to a temperature of 500 K to form the oven. Timp et al. (1992) evaporated sodium at 500 K from a glass ampoule that contained 5 g. It had a 70-mm-long stem with an internal diameter of 3.5 mm, which therefore acted as a collimator with a Γ of 20. Unfortunately they do not mention how the ampoule was broken.

As already mentioned in Section 3.2.1, Gerginov and Tanner (2003) give a brief description of their oven for producing a caesium beam. The element was contained in an ampoule in an ultrahigh vacuum tee piece that was heated externally to the vacuum system at a temperature of 383 K. A second chamber containing the beam-forming array that is described in Section 10.6.2 was maintained 60 K hotter than the oven, again with external heaters. A vacuum valve isolated the rest of the apparatus from that part of the vacuum system containing the ampoule. The means of opening it is not discussed.

Stan and Ketterle (2005) discuss the particular problems of obtaining an atomic beam of a mixed species of atoms, since placing both in the same oven is in general unsuitable because of their different vapour pressures. Hence a mixing chamber for the vapours produced by separate ovens is required before the beam-forming impedance. As already mentioned in Section 3.2.1, Stan and Ketterle used externally heated ovens for both ^{23}Na and ^{6}Li. The sodium oven was operated at a temperature of about 630 K and the lithium oven 20 K hotter. Sodium was obtained in sealed glass ampoules containing 25 g of metal and the isotopically enriched lithium was transferred by a commercial organization into ampoules containing 10 g. These were broken open just before being transferred to the apparatus. The sodium provided about 1250 h of continuous operation and the lithium was estimated to last for about eight times longer than this. The sodium beam emerged into the mixing chamber through a 2-mm-diameter nozzle that was 32 mm long and was mounted in a vacuum flange, which was heated to a temperature of 720 K. The nozzle effectively prevented lithium from entering the sodium oven. The lithium vapour passed directly into the mixing chamber. After mixing, the beam emerged through a 4-mm-diameter orifice. All heaters were commercially available and incorporated thermocouples. Insulation was provided by wrapping a 10-mm layer of ceramic fibre tape around each of the heaters, which were then wrapped with aluminium foil. Since 720 K is above the safe operating temperature of commercial ultrahigh vacuum fittings used to construct the apparatus, modifications were made as described in their paper. The authors consider that their apparatus design should be suitable for all other alkali metals as well as calcium, strontium and ytterbium.

So that their 5 g rubidium ampoule could be broken after evacuating and baking their apparatus, Bell et al. (2010) placed it in its bellows section. A heating tape external to the vacuum system enabled an evaporation temperature of about 350 K to be maintained with 12 W of power. The beam collimator design was unusual in that it consisted of a 10-mm-diameter stainless steel tube that was 200 mm long and heated to 390 K.

In conclusion, the choice of means of using ampoules to introduce an alkali metal into a vacuum system depends largely on its complexity and overall purpose.

4.6 RECIRCULATING OVENS

4.6.1 INTRODUCTION

Several authors have described the means of returning alkali metals that have not passed through a collimating orifice or slit in front of the oven back to it, using either direct return of condensed material or by capillary action. Some designs were used for continuous operation, but in others, measurement had to be interrupted during recirculation. We consider the designs using direct return in Section 4.6.2, followed by those using capillary action in Section 4.6.3. A design not falling into either category is discussed in Section 4.6.4.

4.6.2 DIRECT RETURN OF METAL

Lambropoulos and Moody (1977) describe a three-stage device used with sodium and potassium. Its design, shown in Figure 4.10, also incorporated the means of reducing dimers in the beam. The first stage was a 50-mm-diameter cylindrical oven containing initially 20 g of material. It was heated with six heaters of nichrome wire wound on ceramic cores and inserted into ceramic tubes. Heating power of

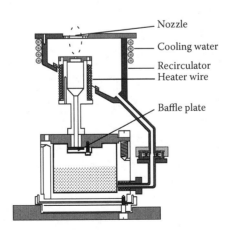

FIGURE 4.10 The three-stage recirculating oven described by Lambropoulos and Moody. (Adapted with permission from Lambropoulos M and Moody SE, 1977. Design of a three-stage alkali beam source. *Rev. Sci. Instrum.* **48** 131–4. Copyright 1977, American Institute of Physics.)

75 W produced operating temperatures of over 600 K. A baffle below the exit tube allowed only vapour to pass into a smaller superheater, which was heated separately to a temperature about 100 K higher. This was estimated to reduce the dimer concentration by about a factor of 5. The third stage was the recirculator, which can be both heated with thermocoax and water cooled. The return tube was also fitted with a thermocoax heater so that it could be unblocked before recirculation began. The authors give additional details of the construction in their paper.

Witteveen (1977) describes a double oven for the continuous recirculation of sodium, as shown in Figure 4.11. Sodium was evaporated at a temperature of 750 K from a stainless steel oven clad in copper to provide uniform heating. The sodium beam emerged through a 3-mm orifice into a 36-mm-diameter aperture into the copper collimator chamber, which was 3 mm away from it and maintained at 400 K, at which temperature the sodium was liquid. The useful beam emerged through a 6-mm-diameter aperture. Sodium that did not pass through this aperture was collected by a funnel in the collimator chamber and returned to the oven via a 6-mm-diameter stainless steel tube. Since the pressure in the oven was higher than in the collimator chamber, the liquid in the latter was about 1 mm higher for each 9-Pa difference in pressure. The oven was heated with 350 W of power by passing a current of up to 17 A through 900 mm of 500-μm-diameter tungsten wire. The collimator was heated with 600 mm of the same wire. The sodium charge was 12.5 g, and it was estimated that 96.5% of the atoms leaving the oven were recycled.

Carter and Pritchard (1978) describe a stainless steel recirculating oven that was resistively heated. It was used successfully for producing vertical beams of potassium, rubidium and caesium. Unlike some other designs, operation was not continuous since the metals were solidified in an upper chamber, which was cooled by circulating liquid nitrogen through copper tubing. A 2-mm constriction in the

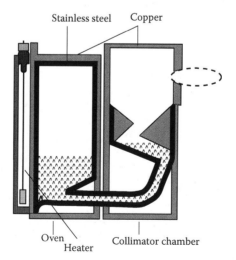

FIGURE 4.11   Witteveen's design for recirculating sodium. (Adapted with permission from Witteveen GJ, 1977. Low-consumption atomic beam source *Rev. Sci. Instrum.* **48** 1131–2. Copyright 1977, American Institute of Physics.)

5-mm-outside-diameter thin wall stainless steel return tube to the oven was also cooled to form a seal so preventing alkali atoms from the oven leaving via this route. When materials were to be returned to the oven, either the liquid nitrogen was shut off so that the heated oven melted them, or steam was passed through the copper tubing. A diagram of this oven is shown in Figure 4.12. With a filling of 3 g of metal, recirculation took about 10 min.

Another non-continuous recirculating oven is described by Haberland and Weber (1980) who used it for sodium. The oven was constructed of stainless steel with nickel gaskets. In their case only water cooling was used to condense the sodium on a cold shield. Initially the oven was heated slowly in order to condense sodium in the return tube and block it after about 15 min. In operation the oven was heated to a temperature of 950 K and the horizontal nozzle to 1150 K since a supersonic beam was obtained, using a heated skimmer. Recycling was effected by reducing the temperatures of the oven and nozzle and warming up the shield.

A recycling oven for caesium is described by Bacal et al. (1982). They produced a supersonic beam, but the principle applies to a thermal beam, though with probably much less than the 2 l of caesium used in their apparatus and a much lower temperature than 800 K, which required heater powers of about 3 kW. Their oven was built from inconel 600 and was resistively heated and insulated with a 30-mm thickness of a 'fibrous refractory'. The beam was condensed on two condensers, one that intercepted the centre of the beam and one the periphery. They were cooled by oil, which was cooled by water in a heat exchanger in order that water could not come into contact with caesium under fault conditions. The oil was maintained at a

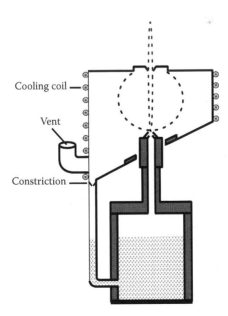

FIGURE 4.12 Carter and Pritchard's recirculating oven. (Adapted with permission from Carter GM and Pritchard DE, 1978. Recirculating atomic beam oven. *Rev. Sci. Instrum.* **49** 120–1. Copyright 1978, American Institute of Physics.)

temperature of 408 K, using heating coils during startup, to avoid caesium solidifying on the condensers. Pyrolytic graphite lined the walls of the interaction chamber to reduce the vapour pressure of caesium away from the target area. Bacal et al. give a schematic diagram of the ingenious recycling system that incorporated pumps and automatic control of caesium levels. It also allowed the alkali to be transferred from its original container to the reservoir. They also describe how the apparatus could be safely cleaned by the introduction of water vapour.

A diagram of the recirculating oven used by McClelland et al. (1989) is shown in Figure 4.13. The reservoir, which was surrounded by radiation shields, was loaded with about 20 g of sodium, which was evaporated at temperatures up to about 700 K. The vapour exit from this oven through a 90° bend terminated in a 1-mm-diameter aperture. This part of the device was maintained at about 100 K hotter than the sodium to prevent clogging of the aperture and to reduce dimer formation. Most of the beam leaving this aperture was trapped by a recirculator that was maintained at a temperature of about 400 K to allow sodium to condense to a liquid and be returned to the reservoir. The useful beam passed through an aperture 3.2 mm in diameter.

A recirculating oven on the same principle for use with caesium is described by Baum et al. (1991) and is shown in Figure 4.14. The recycling funnel was maintained at a temperature of about 300 K, just above the melting point of caesium, but the lower part of the stem had to be water cooled to prevent re-evaporation of caesium back into the funnel. The vapour passed through a 1-mm-diameter nozzle, and this and the pipe leading to it were heated by a separate heater (not shown) to about 400 K to prevent blockage and to reduce dimers in the beam. Caesium was loaded into the oven from an auxiliary vacuum system by breaking the ampoules under vacuum

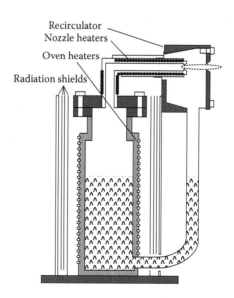

FIGURE 4.13 The recirculating oven for sodium described by McClelland et al. (Adapted from McClelland JJ, Kelley MH and Celotta RJ, 1989. Superelastic scattering of spin-polarized electrons from sodium. *Phys. Rev. A* **40** 2321–9. Copyright 1989 American Physical Society.)

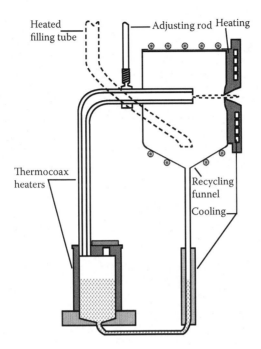

FIGURE 4.14 The recycling system for caesium used by Baum et al. (Adapted from Baum G, Granitza B, Hesse S, Leuer B, Raith W, Rott K, Tondera M and Witthuhn B, 1991. An optically pumped, highly polarized cesium beam for the study of spin-dependent electron scattering *Z. Physik D* **22** 431–6. With permission from Springer Science & Business Media B V.)

with a bellows assembly and heating this part of the apparatus to over 600 K so that the caesium vaporised and condensed on the cold surface of the recycling funnel.

4.6.3 Recycling by Capillary Action

We now consider designs where the alkali metal is returned to the reservoir by capillary flow. These fall into those where the flow is horizontal and the so-called candlestick designs, where it is vertical.

A horizontal design to produce a caesium beam by this means is described by Swenumson and Even (1981). A diagram of their apparatus is shown in Figure 4.15a. The body of the oven was heated with a thermocoax element to temperatures from about 400 K to 500 K with a power of up to 100 W. From 30 to 40 turns of glass fibre cloth insulated the oven. It was lined with fine stainless steel mesh, which caused the caesium to be distributed over its surface and also provided better thermal contact. The beam emerged from the oven through a 3-mm aperture and was then collimated further by a series of baffles with apertures 4 mm in diameter. These had to be maintained above the melting point of caesium but sufficiently cool that caesium was not appreciably evaporated to form a beam. This was achieved by a coil containing compressed air at laboratory temperature. The baffle structure was also lined with stainless steel mesh, which consisted of 40-μm wires spaced 80 μm apart. The oven was

(a)

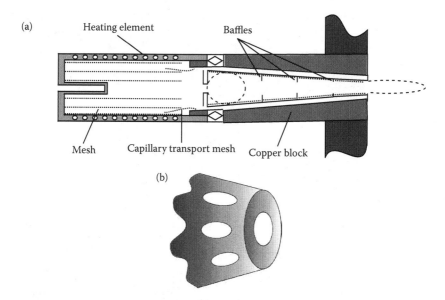

(b)

FIGURE 4.15 (a) The caesium recycling oven by Swenumson and Even. (b) A baffle in Swenumson and Even's oven. (Adapted with permission from Swenumson RD and Even U, 1981. Continuous flow reflux oven as the source of an effusive molecular Cs beam. *Rev. Sci. Instrum.* **52** 559–61. Copyright 1981, American Institute of Physics.)

loaded in an argon atmosphere with two 5-g ampoules of caesium. In use, caesium was returned by capillarity effects from the baffle structure to the oven. Figure 4.15b shows an example of the baffle construction. The authors demonstrated that the recycling process improved the lifetime of the oven by a factor 70. Some further details of construction of the oven, which was ultrahigh vacuum compatible, are given in the authors' paper.

Drullinger et al. (1985) describe a simpler means of constructing a caesium recirculating oven by using a porous matrix of tungsten instead of a fine mesh. The principle of the design is shown in Figure 4.16. Their caesium ampoule was contained in a thin-walled stainless steel tube, which was crushed to release the caesium. This was drawn completely into the 50% dense tungsten porous matrix by capillary action. The collimating tube was also made of porous tungsten but of 80% density. It had an internal diameter of 2 mm and an external diameter of 2.5 mm with a length of 45 mm. The oven was operated at temperatures from about 325 K to 460 K whereas the tube at the beam exit was close to laboratory

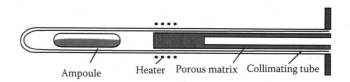

FIGURE 4.16 The porous tungsten recirculating oven described by Drullinger et al. (From Drullinger, RE, Heat pipe oven molecular beam source, US Patent 4 789 779, 1988.)

temperature. Atoms that collided with the tube walls were absorbed by the porous wall and the temperature gradient caused their capillary flow back to the oven. The authors made some measurements of the atomic beam produced by their device, which are discussed in Sections 11.2.3 and 11.2.4. Drullinger (1988) discusses the advantages of the above design over that of Swenumson and Even (1981), described above, emphasising the simplicity of his design. He suggests it is also useful for forming beams of organic compounds such as formaldehyde. Some alternative designs are presented for obtaining the required temperature gradient in the collimating tube. In addition to sintered tungsten, other sintered metals, including molybdenum, and stainless steel had been used, and nickel, copper and alumina silicates were suggested.

The following designs used a vertical recycling system by capillary action. Hau et al. (1994) developed the principle for sodium and termed it a candlestick source, because the metal rises up the gold-plated stainless steel wire cloth inside the oven by capillary action like molten paraffin wax in a candle wick. The design of the oven is shown in Figure 4.17. A photograph of it is shown in Hau (2001). The sodium was maintained above its melting point with one heater and a second heater heated the orifice in the molybdenum candlestick to a temperature of 625 K. The authors give further details of their design and operating conditions in their paper. Although it had not then been tested with other alkali metals, the authors present data that suggest it should be suitable.

Indeed, Walkiewicz et al. (2000) further developed this source for producing a beam of rubidium. This was evaporated at a temperature of about 350 K and the candlestick was operated at up to 500 K. The rubidium was placed in the oven by breaking ampoules in an argon atmosphere and heating them until the metal was liquid. The authors also give further details of their design and operating conditions in their paper. The appearance of their oven is not sufficiently different from the previous one to make a diagram worthwhile. Some additional details of a rubidium candlestick source are given by Slowe et al. (2005). In particular, details of loading the source are given.

Pailloux et al.'s (2007) design of a candlestick oven was used to obtain a beam of caesium. The reservoir contained 10 g of the metal, which was loaded in an atmosphere of dry argon. The temperature of the liquid was maintained at 308 K and the evaporation temperature from the wick was between 370 K and 540 K. The design is very similar to those published since 1994 and already described above with the exception of the material of the wick. The authors discuss the properties of suitable materials for this and investigated the wetting of some of them by caesium from which they concluded that high-purity silica braid fulfilled their requirements. These included a more economic solution than the gold-plated stainless steel mesh used in previous experiments. The braid consisted of about 1-µm-diameter 98% silicon fibres formed into a thread that was about 1 mm in diameter.

Bell et al. (2010) report that they found candlestick ovens relatively complicated to construct and maintain, difficult to load with rubidium and unreliable in establishing the correct wicking and recirculating action. Hence they used the apparatus mentioned in Section 4.5.

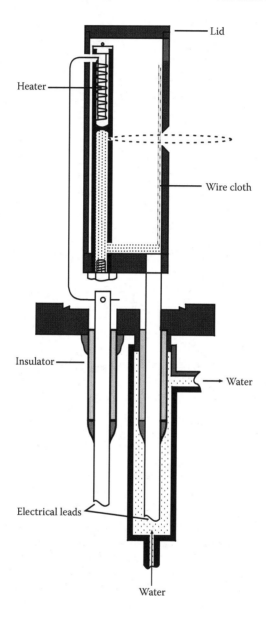

FIGURE 4.17 The candlestick source of Hau et al. for sodium. (Adapted with permission from Hau LV, Golovchenko JA and Burns MM, 1994. A new atomic beam source: The 'candlestick'. *Rev. Sci. Instrum.* **65** 3746–50. Copyright 1994, American Institute of Physics.)

4.6.4 RECIRCULATION BY ION FORMATION

Dinneen et al.'s (1996) method of obtaining a beam of francium, which has no stable or long-lived isotopes, is a novel means of obtaining recirculation without the complications of the designs already discussed. The oven is shown in Figure 4.18. It

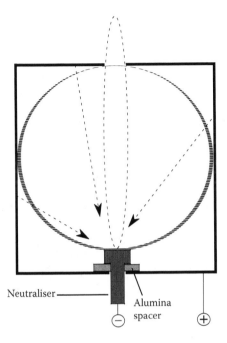

Neutraliser ——

Alumina
spacer

FIGURE 4.18   Dinneen et al.'s method of producing a beam of francium. (Adapted with permission from Dinneen T, Ghiorso A and Gould H, 1996. An orthotropic source of thermal atoms. *Rev. Sci. Instrum.* **67** 752–5. Copyright 1996, American Institute of Physics.)

consisted of a stainless steel cube of 40-mm sides which had platinum sputtered on to its inside surface and which could be heated to a temperature of 1220 K. A 3-mm-diameter yttrium plug in the centre of the base of the cube was insulated from it with an alumina spacer and was biased 100 V negative with respect to the cube. Positive ions formed on the hot platinum surface with a work function of 5.65 eV were therefore attracted to the plug with a work function of 3.4 eV and were neutralised. Atoms emitted in the forward direction formed the beam by passing through a 5-mm-diameter aperture in the top of the oven that was 23 mm from the neutraliser. All other atoms struck the platinum surface and were ionized, and the ions were therefore attracted to the neutraliser. A measured beam halfwidth of 12.4° was consistent with the geometry of the oven. The authors describe how they prepared the francium from actinium.

Dinneen et al. suggest that the source would be suitable for preparing beams of many other elements, some of which would need a tungsten oxide ionising surface. It is perhaps worth mentioning that the authors refer to their source as 'orthotropic,' which is not normally associated with atomic beams. According to the Oxford English Dictionary Online it is a botanical term implying vertical growth and an engineering term for elastic symmetry. We assume in the context of an atomic beam that the authors refer to it as being vertically upward, like many others we discuss.

4.6.5 DISCUSSION OF RECIRCULATING SYSTEMS

In conclusion, it is disappointing that none of the authors quoted in Section 4.6 critically compare the advantages and disadvantages of recirculating ovens, in which the first aperture produces a cosine or near-cosine distribution, with ovens that use a collimating array to produce a beam. When an array is used, much more of the substance is utilised, so no recirculating system is required. Various collimators have been used for the alkali metals for many years, including in atomic clocks, which are designed to produce a caesium beam for many years without refilling. As already mentioned in Section 4.3, Hertel and Ross (1968) describe such an oven for use with the alkali metals as well as other elements. If necessary, a collimator could be used in conjunction with a recirculating oven, but this does not appear to have been tried. The design of recirculating ovens is clearly much more complex than even a sophisticated double oven, and they contain amounts of alkali metal, which, under fault conditions, could be quite dangerous. Hence we conclude that the decision on whether to use a recirculating oven should be made with great care. The method to be employed depends on the required purity of the beam including freedom from compounds and any problems from introducing fluids into the apparatus.

5 Ovens for Higher Temperatures

5.1 INTRODUCTION

There is a large number of atoms and molecules that generally need higher temperatures for evaporation than those of the alkali metals and their salts. Usually elements that are chemically similar have related problems on evaporation, such as whether polymers are formed, whether the choice of oven materials is governed by the material's corrosiveness and whether thermocapillary effects cause an unstable beam. Hence we order our discussion of the non-gaseous elements in the groups 2 to 16 of the periodic table and discuss compounds with their constituent element. We group the lanthanides and actinides in Sections 5.15 and 5.16, respectively. An exception is made for the various means of producing beams of refractory elements, since the extremely high temperatures needed for their evaporation to form a beam require special techniques that are the same for the entire group. They are therefore grouped together in Section 5.17. The selection and ordering approach mentioned in Section 4.1 also applies to this chapter, although this means that sometimes the same technique is described in more than one section.

5.2 BERYLLIUM, MAGNESIUM, CALCIUM, STRONTIUM AND BARIUM BEAMS

The first Group 2 substances we mention that have produced beams are those of barium and calcium. Ottinger and Zare (1970) report briefly on the production of beams of these elements from a molybdenum crucible heated to a temperature of 1400 K by a graphite cylinder. These experiments were extended by Jonah et al. (1972) to include magnesium and strontium. They used a three-phase molybdenum wire winding to provide about 1 kW of power to heat their rather large molybdenum oven, which was 100 mm long and 40 mm in diameter. It was surrounded by a single molybdenum radiation shield followed by a water-cooled copper shield. The authors noted that magnesium, calcium and strontium could clog the 3-mm-diameter exit aperture of the oven, but this did not occur with barium, which was evaporated from the liquid state. Aleksakhin and Zayats (1974) obtained a beam of beryllium from a graphite crucible heated by a graphite heater to a temperature of 1800 K.

Although it was not used to form a beam, we mention Afzal and Giutronich's (1974) furnace, used to study magnesium oxide, since it could be adapted to form a beam. Currents of up to 1200 A were passed from water-cooled copper conductors through a graphite tube, which was surrounded by a water jacket. The absence of radiation shielding enabled the specimen to reach stable temperatures up to 3300 K in about 3 minutes. The graphite tube was 112.5 mm long, 12 mm in internal diameter and with a wall thickness of 2.9 mm, reducing to 2.2 mm at its ends so that more heating power was applied where the conduction losses were higher. Thermal expansion was accommodated by the graphite tube being a sliding fit into the lower conductor. Tubes were made with their axes parallel to the extrusion direction of the graphite since the resistivity in the perpendicular direction was approximately double.

Dagdigian et al. (1974) obtained a barium beam from an orifice at a temperature of 1200 K. Their molybdenum oven was heated to 1050 K by a molybdenum tube about twice its length, as already mentioned in Section 3.2.8. The design enabled the light emitted by the oven to be much less than that emitted by an oven heated with heater wires. Bernhardt (1976) obtained a beam of barium from a crucible that was about 12 mm in outer diameter, 1.6 mm in wall thickness and 12 mm long and made of molybdenum. This was placed in a cylinder that was 19 mm in diameter and the same length, which had a wall thickness of 3.1 mm. The purpose of this double-walled oven is not explained, but it is stated that the crucible had a sharp upper edge to prevent migration of liquid barium. The assembly was radiatively heated by a molybdenum heating jacket, which was surrounded by five tungsten radiation shields that were 25 μm thick. The atomic beam emerged through a 500-μm aperture in a tightly fitting lid over the outer crucible.

Pasternack and Dagdigian (1976) used a stainless steel crucible heated to a temperature of about 1300 K by a stainless steel heating tube to produce beams of calcium, strontium and barium. Radiation shielding enabled the exit aperture to be maintained 100 K hotter. Barium was evaporated from a molybdenum oven by Holland et al. (1981). The tungsten heater wire was wound in grooves in a machinable glass former 15 mm in diameter and 42 mm long. The use of a two-start groove on the former enabled a non-inductive winding to be produced and the doubling of the grooves near the oven exit prevented blockage of the 8-mm-long, 600-μm-diameter beam-forming nozzle. The winding was insulated by a tube of the same glass and surrounded further by water-cooled radiation shields.

The direct electron-beam heating method of producing a beam of iron used by Dembczyński et al. (1980) and described in Section 5.8 was also used by Aydin et al. (1982) to obtain a beam of calcium.

Although they give no details of their oven construction, Nagata et al. (1986) report that their oven containing strontium, which was operated at a temperature of 850 K, emitted impurities, particularly carbon dioxide. Hence it had to be warmed up very slowly, as mentioned in Section 3.3.2. Krause and Caldwell (1987) used a tantalum oven at a temperature of 1400 K as a source of beryllium for their electron spectroscopy experiments.

Martin et al. (1987) describe briefly how beams of radioactive atoms of barium and strontium can be obtained by firing a heavy ion van de Graaff accelerator at a thin target and collecting reaction products in a tantalum oven. Atoms were emitted when this was heated at temperatures up to 2000 K and were collimated in a tantalum tube 600 μm in internal diameter and 10 mm long.

Czarnetzki et al. (1988) produced a beryllium beam by electron-beam heating of a 6-mm-internal-diameter tantalum crucible. This was surrounded by a water-cooled copper vessel, which also served as the beam-defining aperture. The crucible was surrounded by a 100-μm-diameter tungsten wire spaced 5 mm away from it, through which a current of 6.5 A was passed. The emitted electrons were accelerated to the crucible through a potential of 900 V.

González Ureña et al. (1990) obtained a beam of calcium atoms from a stainless steel oven 25 mm in diameter and 110 mm long, containing about 5 g of metal. It could be heated up to a temperature of 1450 K by passing a current of 300 A through a stainless steel cylinder 36.5 mm in diameter and 175 mm long. The 500-μm beam orifice was maintained at a higher temperature by reducing the tube wall thickness from 250 μm to 180 μm. The copper current-lead pipes and a stainless steel radiation shield were water cooled. Insulation of ceramic enabled either a pulsed or a dc discharge to be run between the oven and heater so that metastable atoms could be produced.

A molybdenum oven was used to produce a beam of magnesium atoms by McCallion et al. (1992). The oven was supported by an aluminium oxide ring within a tantalum tube, which formed the heater. Alternating current passed through water-cooled leads via molybdenum rings to the tube. A heating current of 75 A was used and the oven contained about 5 g of magnesium. The temperature of operation is not stated.

For the production of a calcium beam, Kämmerling et al. (1994) describe a stainless steel oven that was heated to a temperature of about 1000 K by a thermocoax coiled-coil heating element requiring a power of 100 W, which produced a negligible magnetic field. Thermal insulation of the oven from its two stainless steel radiation shields was provided by ruby spheres. The final shield was of water-cooled copper.

Shah et al. (1996) evaporated magnesium from a molybdenum crucible which was heated by passing a current of up to 200 A through a tantalum tube 78 mm long, 16 mm in diameter and 25 μm thick, which was mounted by molybdenum cylinders in water-cooled electrodes, as shown in Figure 5.1. Copper tubes provided both the cooling water and the heating current.

A magnesium beam was obtained by Boivin and Srivastava (1998) by using an electron beam to heat a molybdenum crucible. A tungsten filament was maintained at about 900 V negative relative to the oven to produce heating of its base. The vertical beam was produced at temperatures up to about 800 K and was collimated by a single tube 790 μm long and 123 μm in internal diameter. The authors did not report any problems with instability in the beam.

Ross and West (1998) describe a novel form of oven for the production of beams of calcium and strontium in which the beam is produced by three or more cylindrical channels 1.5 mm in diameter and 9 mm long, which were directed toward the interaction region. A diagram of this oven is shown in Figure 5.2. The heating method was based on the principle described by Ross (1993) in Section 3.2.1, but the beam-forming channels were arranged to be hotter than that part of the oven containing the substance to be evaporated. The oven was of welded construction, made from stainless steel and operated at a temperature of about 1000 K when supplied with 150 W of power to the 500-μm-diameter molybdenum wire. Three layers of tantalum

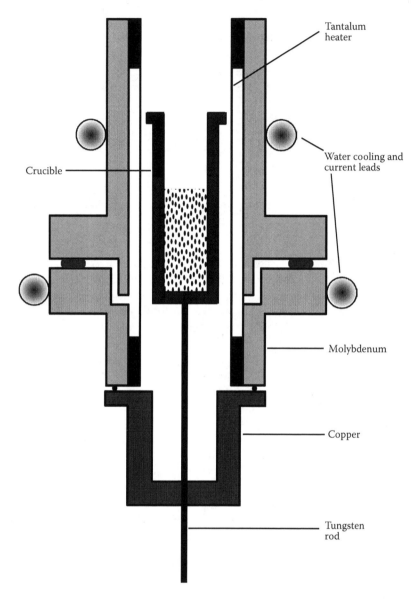

FIGURE 5.1 The oven design used by Shah et al. to produce a magnesium beam. (Adapted from Shah MB, Bolorizadeh MA, Patton CJ and Gilbody HB, 1996. Simple metallic atom source for crossed beam collision studies. *Meas. Sci. Technol.* **7** 709–11. Copyright IoP Publishing 1996.)

foil 100 μm thick were used for radiation shielding. The oven charge was loaded into an annular cavity since the oven was required to surround a fused silica light guide.

Yagi and Nagata (2000) used a tantalum oven heated by an electron beam in order to obtain a beam of barium. The oven was surrounded by one radiation shield of tantalum with two outer shields of stainless steel. Electrons were produced by three suspended 300-μm-diameter tungsten wires and were accelerated with potentials

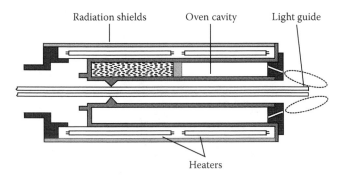

Radiation shields   Oven cavity   Light guide

Heaters

FIGURE 5.2 The oven used by Ross and West for calcium and strontium. (Adapted from Ross KJ and West JB, 1998. A resistively heated vapour beam oven for synchrotron radiation studies. *Meas. Sci. Technol.* **9** 1236–8. Copyright IoP Publishing 1998.)

between 500 V and 2600 V. The oven assembly is shown in Figure 5.3 and was water cooled. The oven used by Wehlitz et al. (2002) and described in Section 4.3 to obtain a lithium beam was also used to obtain a beam of beryllium, at a temperature of 1150 K.

We collect together four papers which essentially describe the same design of double oven from which the beam is emitted through a single capillary tube, but with some modifications. Cvejanovic and Murray (2002) give a detailed description of their oven to produce a beam of calcium. The body of the ovens was made from oxygen-free copper, but they were lined with stainless steel since calcium vapour reacts with copper. The beam was emitted from a single tube 20 mm long and 900 µm in internal diameter and was found to diverge at about 3°. Calcium was evaporated from the main oven at about 1000 K, a temperature that was achieved with a thermocoax heating element. A second element heated the front section of the oven containing the exit tube to a temperature up to 100 K higher. Constant-current power supplies enabled temperatures to be maintained stable to within 1 K. A useful feature of the oven was that it could be refilled without disturbing the alignment of the exit tube. This was done in an argon atmosphere to prevent the calcium oxidising. Double radiation shields for each section of the oven were made from type 316 stainless steel.

Murray and Cvejanovic (2003) describe some of the problems they encountered with the production and use of a calcium beam in conjunction with both electron and visible optical components. Murray (2005) employed a similar oven but constructed of type 310 stainless steel. It was used with both calcium pellets and magnesium turnings. Modifications of the design of Cvejanovic and Murray (2002) were necessary to provide additional heating to prevent sublimation of these elements in otherwise cooler parts of the assembly. In addition, filling of the main oven was facilitated by the addition of a knife edged aperture that was sealed by a tapered stainless steel plug that was held in place by a screw assembly. The main oven was heated to a temperature of about 820 K for magnesium and 875 K for calcium with the front section 30 K hotter.

Murray et al. (2006) give extensive details of the design of their apparatus, which incorporates a beam of calcium. In this design, the metal was placed within the

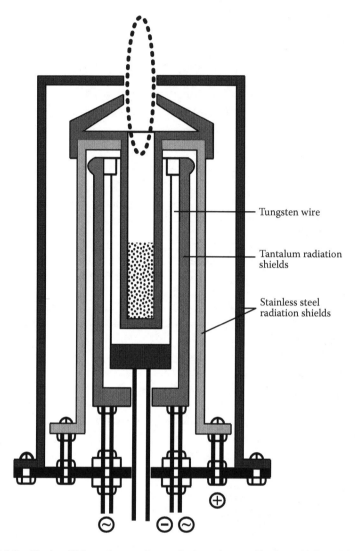

Tungsten wire

Tantalum radiation shields

Stainless steel radiation shields

FIGURE 5.3   Yagi and Nagata's oven for producing a beam of barium. (Adapted from Yagi S and Nagata T, 2000. Absolute total and partial cross-sections for ionization of Ba and Eu atoms by electron impact. *J. Phys. Soc. Jpn.* **69** 1374–83. With permission.)

oven in an ampoule, which was broken in an argon atmosphere. The oven was surrounded by four concentric stainless steel radiation shields that were held in place with 2-mm-diameter ceramic rods. Careful attention was placed on the design of the mounting so that the oven could be accurately replaced after filling. It was heated by a thermocoax spiral wound round it with turns that were closer together where the beam emerged so that this region would be hotter. The heating coil had a cold resistance of 270 Ω and a hot resistance of 275 Ω. It was heated by a constant current power supply, which is described in detail, up to temperatures of about 1100 K when a power of 53 W was required. The calcium beam-forming impedance was a molybdenum tube that was 1.4 mm in internal and 2 mm in external diameter that was at the opposite end of the oven to the screwed filling aperture. The beam was collimated in turn with apertures of 6 mm and 1 mm in diameter leading to a halfwidth of the beam of 2.6°. The beam density in the interaction region was 2.8×10^6 atoms mm^{-3} at a temperature of 1090 K, which was sufficient for the author's purpose.

An oven for the production of a magnesium beam is described by Brown et al. (2003). It had to meet the stringent requirements of both operating reliably over a period of months and having a low residual magnetic field. A diagram of the oven is shown in Figure 5.4. Magnesium was evaporated from a molybdenum oven that was heated to a temperature of 723 K by bifilarly wound inconel wires insulated with magnesium oxide. The beam was emitted through a separately heated single tube that was 10 mm long and 1 mm in internal diameter. More windings near the tube exit prevented condensation of magnesium within the tube, which was heated to 768 K. Temperatures were maintained to within ±2 K by power supplies that were

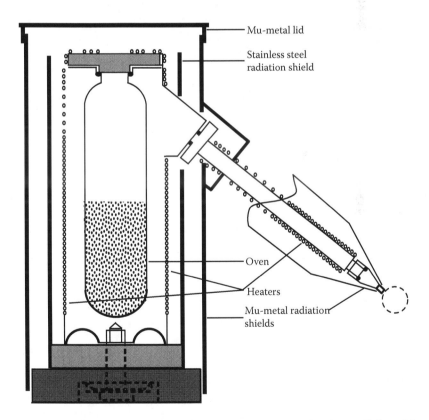

FIGURE 5.4 Brown et al.'s oven for the production of a magnesium beam. (Adapted from Brown DO, Cvejanović D and Crowe A, 2003. The scattering of 40 eV electrons from magnesium: A polarization correlation study for the 3^{1}P state and differential cross sections for elastic scattering and excitation of the 3^{1}P and 3^{3}P states. *J. Phys. B: At. Mol. Opt. Phys.* **36** 3411–23. Copyright IoP Publishing 2003.)

controlled by thermocouples. Concentric stainless steel and mu-metal screens acted as radiation shields, with the latter providing magnetic screening. The assembly was mounted on a machinable ceramic base and filling of the oven was achieved by removing a molybdenum lid.

Continuing with other papers, we note that Milisavljević et al. (2004) used a 20-mm-long single tube that was 1.5 mm in internal diameter to collimate their beam of calcium, hence $\Gamma = 13.3$. It formed the 9.5-mm-diameter stainless-steel end cap of their oven and was heated to a temperature of about 100 K more than the main oven body, which was resistively heated to a temperature of nearly 1000 K. Two radiation shields of stainless steel and one of copper, which was water cooled, surrounded the oven.

Majumder et al.'s (2009) tungsten oven used to produce a beam of barium had a length of 120 mm and a width and height of 18 mm, with a wall thickness of 3 mm. The 5-mm-thick oven lid contained a row of 31 round 2-mm-diameter holes or oval tubes that were 6 mm long in the centre, reducing in steps to 2 mm long at the edges. The authors state that 'the choice of sizes of holes results in relatively larger contribution of inner holes to the extremities of the atomic beam,' but no further discussion of this unusual configuration is made. The oven was heated by passing direct current through 1-mm-diameter tantalum wire, which was supported on 2-mm-diameter insulated tungsten hooks. These were adjustable to allow for deformation of the heater wires following repeated heating cycles. Radiation shielding was provided with tantalum foils, with a final shield of water-cooled copper. About 20 g of the metal was transferred with its protective paraffin to the oven to allow 12 h of operation at temperatures up to about 1000 K.

Schioppo et al. (2012) describe a stainless steel oven for strontium that was 34 mm long and 10 mm in internal diameter. It was heated up to about 750 K with a heating power of 36 W by 300-μm-diameter tantalum wires, which were threaded through an alumina multi-bore insulating tube. A single aluminium radiation shield surrounded the oven. The beam was emitted through about 120 stainless steel tubes that were 200 μm in diameter and 8 mm long, hence $\Gamma = 40$. The fractional open area of the array is not stated. A charge of about 6 g of strontium was estimated to last for 10 years, since the oven was operated at low pressures such that $\lambda > 16 \, l$.

5.3 BEAMS OF SCANDIUM AND YTTRIUM

The lack of detailed descriptions of ovens for the production of beams of Group 3 substances suggests that no particular problems have been encountered.

A directly heated graphite oven for the production of a beam of yttrium is described by Fricke et al. (1959a). It is based on Lew's (1949) design which was used for aluminium and which is described in Section 5.11. The yttrium was contained in a tantalum vessel that was surrounded by corrugated tantalum foil. The beam emerged through a 200-μm-wide slit 8 mm long. The graphite required about 3.5 kW of power to evaporate the beam at a temperature of 2240 K. The shape of the graphite was such that the slit was the hottest point, in order to prevent its clogging. Fricke et al. (1959b) used the same oven for producing a scandium beam, at an evaporation temperature of about 2000 K. They found however that the beam intensity was steadier if the slit width was increased by 50% compared to that used for yttrium. The production of beams of respectively scandium and yttrium by forming powders from the oxides mixed with thorium is described briefly by Chalek and Gole (1976). The powder was placed in either a tantalum or a tungsten crucible, which was surrounded in turn by a directly heated tantalum cylinder, radiation shields and a water-cooled copper jacket. Temperatures of about 2500 K were achieved. Smirnov (2002) obtained a beam of yttrium by direct electron-beam heating in the same way as for lanthanum (Smirnov 2000) and mentioned in Section 5.15 but using a water-cooled copper crucible. The beam was produced at a temperature of 1870 K.

5.4   TITANIUM AND ZIRCONIUM BEAMS

The Group 4 elements are usually classed as refractory, so they are discussed in Section 5.17. Titanium and zirconium are included in this section, since individual experiments were used to produce beams of them.

An oven for the production of a beam of titanium is briefly described by Dubois and Gole (1977). A tantalum cylinder with about 20 mm wall thickness was heated to a temperature of about 2500 K by passing a current through it supplied by water-cooled leads. The titanium was contained in a carbon crucible that was lined with tungsten. The oven was surrounded by tantalum and tungsten radiation shields and finally by a water-cooled copper jacket.

As mentioned in Section 3.3.2, Prescher et al. (1987) used electron bombardment to produce a titanium beam. The metal was contained in a graphite crucible surrounded by a tantalum cylinder. Four heated tungsten wires provided the source of electrons, which were accelerated by up to 1500 V. Power of up to 2.5 kW enabled temperatures over 2000 K to be reached. An outer tantalum cylinder provided radiation shielding and this was surrounded by a water-cooled copper housing.

Benvenuti et al. (2003) report briefly that a titanium dioxide beam was obtained by evaporation of liquid tetraisopropyl titanate (titanium tetraisopropoxide) $[C_{12}H_{28}O_4Ti]$.

Thakur et al. (2005) produced a beam of zirconium from a water-cooled copper crucible that was heated by direct electron bombardment as described in detail by Thakur et al. (2001). A magnetic field deflected a 1.8-A current of electrons at 45 keV through an angle of 270°. In the later experiment, power of 92 kW evaporated zirconium at 3175 K to produce a measured atom density of 10^{12} mm^{-3} at a distance of 110 mm from the crucible.

5.5   BEAMS OF VANADIUM AND COMPOUNDS OF TANTALUM

Apart from vanadium, which can be formed into a beam from a conventional oven, and tantalum compounds, we discuss the Group 5 substances in Section 5.17, since they are refractory. Abernathy (1994) reviewed the precursors that have been used to obtain beams of some Group 5 elements in ultrahigh vacuum. Use of these methods if a pure beam of the element is required is limited because of the mixture of compounds usually found in the beam. His review is also mentioned in several of the following sections.

Bellman and Raj (1997) discuss some of the properties of possible precursors for a range of compounds and used lithium hexaethoxytantalate to produce a beam of lithium tantalate [LiTaO$_3$] by evaporation at temperatures of about 500 K, when the vapour pressure is 27 Pa. The purpose of their beam was to produce a uniform film by chemical beam epitaxy (CBE), rather than a collimated beam, but their methods of production are applicable to the formation of such a beam. The lithium hexa-ethoxytantalate was placed in a stainless steel crucible which itself was contained in an oxygen-free high conductivity copper cup which was heated by circulating oil, as discussed in Section 3.2.7. The beam emerged through a capillary array, which we describe in more detail in Section 10.4.8.

Cochrane et al. (1998) briefly mention that they obtained an atomic beam of vanadium from a tantalum tube that was 1.6 mm in internal diameter. By splitting the tube into four arms that were folded partly back and clamped between two copper rings that carried the 45-A heating current, they overcame the problem of cooling the tube at the exit of the beam where the highest temperatures are required. The base of the tube was pinched and connected to the current return.

5.6 CHROMIUM BEAMS

Chromium is the only non-refractory element in Group 6.

Brix et al. (1953) briefly reported that they obtained a beam of chromium atoms from powder contained in a thoria crucible within a graphite oven heated to a temperature of about 1800 K. A source of a chromium atomic beam is described by Roberts and Via (1967) in which the heating current was passed through concentric tantalum cylinders as shown in Figure 5.5. The inner cylinder was 22 mm in diameter, 102 mm long and 76 μm thick, and the outer was 29 mm in diameter and 127 μm thick. This arrangement allowed both electrical connections to be at the end of the oven away from the beam. The smaller diameter and thinner wall of the inner cylinder meant that its cold resistance would be about twice that of the outer one. The heating cylinders were surrounded by two tantalum radiation shields which were connected to water-cooled flanges. Power requirements were up to 1500 W at 5 V. Dyke et al. (1979) report briefly on the production of a chromium beam at a temperature of 2450 K using the inductively heated oven described in Section 3.2.3 as used by Bulgin et al. (1977) and mentioned in several of the following sections. Dyke et al. used either a zirconia lined carbon crucible or a boron nitride crucible.

Wagner (1983) also mentions producing a chromium beam in connection with the resistively heated oven described in Section 5.15, since it was primarily used for the lanthanides. Schmidt et al. (1984) report briefly on the production of a beam of chromium using an alumina crucible. It was heated by two concentric tantalum cylinders with a current of up to 120 A which produced oven temperatures of up to 2000 K. This heating method compensates for the magnetic field produced by the heater. The atomic beam was collimated by either a niobium or an alumina nozzle for which no dimensions are given. The heater was surrounded by tantalum and stainless-steel radiation shields with a final outer water-cooled copper shield. The authors misleadingly consider that their oven is that described by Wagner (1983) and mentioned above, but his oven was heated with a bifilarly wound spiral filament.

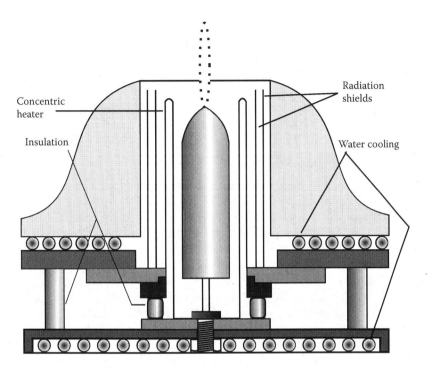

FIGURE 5.5   An outline of Roberts and Via's oven used to produce a beam of chromium. (From Roberts, GC and Via, GG, Monitored evaporant source, US Patent 3 313 914, 1967.)

Schmidt et al.'s oven is similar to that described by Roberts and Via (1967) that is also mentioned above.

5.7   MANGANESE BEAMS

Manganese is the only non-refractory element in Group 7.

Woodgate and Martin (1957) obtained a beam of manganese, referring to an oven described earlier by Woodgate and Hellwarth (1956). However, no details of the oven are given in that paper. Dembczyński et al. (1979) used the direct electron-beam method (Section 3.2.5) to produce a beam of manganese from a tungsten crucible. It was heated by a 13-kV electron gun. The electrons were deflected through 90° and focussed on to the manganese through a 3-mm-diameter aperture in the crucible.

We have already mentioned Dyke et al.'s (1979) oven for producing a beam of chromium in the previous section. They also used the same inductively heated oven of Bulgin et al. (1977) for producing a beam of manganese at a temperature of about 1900 K. They added a boron nitride liner to the graphite oven, since manganese reacts with carbon to form carbides. Schmidt et al.'s (1984) oven which was also mentioned in the previous section was also used to obtain a beam of manganese. Schmidt et al. (1985) produced a manganese atomic beam from a resistively heated furnace at a temperature of 1200 K. This temperature is below the melting point, so the reactive liquid state was avoided. The vapour pressure of up to 1 Pa at this temperature is however too low for many beam experiments. A manganese beam was also obtained by Schmidt (1985) using the same radiofrequency oven described by Bulgin et al. (1977), but with the omission of the single channel collimator, which he describes as having a 'faulty performance'. Schmidt describes manganese as easy to evaporate and free from molecules in the beam. An inductively heated tantalum oven was used for the evaporation of manganese at a temperature of about 1300 K by Ford et al. (1990). No further details are given.

5.8   IRON, COBALT AND NICKEL BEAMS

Since the elements in Groups 8 to 10 are related, we group them into this section. Other elements in these groups that have been formed into atomic beams are refractory and are considered in Section 5.17.

Gerlach (1925) reports that the oven he used for silver and other elements required coating with alumina paste when used to evaporate iron, otherwise it diffused through the material and destroyed the heater winding. Von Ehrenstein (1961) employed a zirconium oxide crucible to contain cobalt. The crucible was surrounded by a molybdenum anode held at a potential of 1000 V relative to the 200-μm-diameter tungsten cathode to produce an electron current of 800 mA which heated the crucible to a temperature of about 2300 K. A radiation shield of tantalum surrounded the oven and this was in turn enclosed in a water-cooled jacket.

A beam of iron was obtained by Dembczyński et al. (1980) by heating it with a 15-kV electron gun that was deflected through 90° and focussed into the top of a carbon crucible containing the iron. This is a similar arrangement to their direct electron-beam heating method described in the previous section for manganese. The authors state that each crucible could be used several times before it was damaged by chemical reactions between the iron and the crucible.

Roulet and Alexandre's (1981) oven described in Section 5.9 is also stated to be suitable for nickel, but no additional details are given for this element. Schmidt et al.'s (1984) oven used for chromium and described in Section 5.6 was also used for producing iron, cobalt and nickel beams. Park et al. (1988) mention that they obtained a nickel beam from an alumina crucible which was placed in a directly heated tantalum oven. Temperatures of 1900 K were reached.

To form a beam of iron Shah et al. (1993) used a 10-mm-diameter crucible that was 25 mm long and made from high-temperature alumina. It was resistively heated by a surrounding tungsten tube to temperatures of about 2000 K. Shah et al.'s (1996) oven described in Section 5.2 was used to produce a beam of iron with an alumina crucible.

5.9   BEAMS OF COPPER, SILVER AND GOLD

We begin our discussion of the Group 11 elements with Gerlach and Stern's (1924) description of two ovens for producing a silver atomic beam. The first was turned from 'purest' iron and was 10 mm long and 4 mm in diameter with a wall thickness of 200 μm. The beam was emitted horizontally through a 1-mm-diameter circular hole in the 100-μm-thick lid of the oven, which was also used to insert a few hundred milligram of silver. The oven was cemented into a thick-walled fused silica capillary tube and mounted in a brass water-cooled tube, which contained a 1-mm-diameter aperture in its front surface, forming the first of three collimating apertures. The oven was heated with a length of between 500 mm and 750 mm of 300-μm-diameter platinum wire which was insulated by fusing on to the oven a paste made from powdered silica, magnesia, kaolin, asbestos fibre and sodium silicate. Since the insulation tended to fail, the heater of the second oven was wound on 'Marquard' (which appears from Gerlach (1925) to be porcelain), which surrounded a similar iron container for the silver, but held in place with a larger fused silica tube. This enabled the oven temperature to be determined optically. By overheating the oven initially, a silver seal was formed on parts which were below the melting point of silver in normal operation. The vacuum envelope was made of glass.

An improved oven arrangement was used by Gerlach and Cilliers (1924) for other metals, including copper and gold, both at temperatures between 1500 K and 1600 K, but no description is given. A further development of the oven is described by Gerlach (1925), using 200-μm-diameter molybdenum wire for the heater, which was wound on a porcelain tube and enabled temperatures up to 2000 K to be reached. Molybdenum was found superior to tungsten, which reacted with other oven materials. The wire was covered with an insulating layer made from baked-on alumina, which however apparently reacted with the heater wire unless alternating current was used.

Wessel and Lew (1953) used the same oven for the production of beams of silver and gold that we describe for Lew's (1953) experiments with praseodymium in Section 5.15 with the exception of the heating jacket, which was constructed of graphite rather than molybdenum. It was kept from direct contact with the thoria crucible with tantalum foil. Lemonick et al. (1955) give brief details of an oven for producing a beam of copper, silver and gold made out of a 32-mm cube of molybdenum which was heated by electron bombardment at 2000 V from a tungsten cathode. The current had to be direct, since with alternating current the filament was burned out due to bombardment with electrons from the molybdenum oven. A current of up to 1 A had to be used to accelerate the electrons.

Woodgate and Hellwarth (1956) briefly mention that a tantalum oven was used for the evaporation of radioactive silver. The oven was heated by electron bombardment to a temperature of about 1300 K. The beam was collimated by a channel having $\Gamma = 50$ since it was about 12 mm long and 250 μm wide. They state that it is probable that a good deal of the silver was lost in the oven by alloying with the tantalum. Reichert (1963) obtained a vertical beam of gold which was evaporated from a molybdenum oven which was heated from below with a tungsten heater spiral. Ward et al. (1967) evaporated gold from a Knudsen cell made from 'high-density Stackpole graphite' about 6 mm in diameter and 13 mm deep and reported that gold did not wet the graphite, but disks of molybdenum and thoria placed in the cell were wetted.

Silver chloride was formed into a beam by Wagner and Grimley (1972). They found that the reaction with inconel was violent at a temperature of 933 K and reaction also took place with the copper in sterling silver and tantalum. Alumina and pure silver were found to be satisfactory materials.

Shen (1978) employed an oven consisting of a 19-mm-internal- and 22-mm-external-diameter graphite tube that was 63 mm long. It was surrounded by two alumina tubes on the first of which a tungsten wire heating element was wound. A tantalum radiation shield surrounded the alumina tubes. A heating power of 100 W enabled temperatures of 1600 K to be reached. The silver beam-forming impedance was an array of 240 channels that were 787 µm in diameter and 7.87 mm long. No further details of their construction are given. An oven made of graphite with wall thicknesses up to 1 mm is reported to be satisfactory for the evaporation of copper, silver, and gold by Dyke et al. (1979), who heated the oven inductively as used by Bulgin et al. 1977 and described in Section 3.2.3.

Although Preuss et al. (1979) produced a supersonic beam of copper, their work indicates some problems likely to be encountered with a thermal copper beam. For example, they reported that a substantial quantity of gas was evolved near the 1356-K melting temperature of copper. Their graphite crucible, which was heated with a woven tungsten heating element, was provided with baffles so that the exit nozzle was shielded from the molten copper, otherwise condensed liquid copper was ejected by vapour to form an aerosol. They felt that the double-oven technique, as described in Section 3.3.2, was difficult to employ at their oven temperatures of 2450 K. A tungsten oven was also tested, with less baffling, which produced intensity fluctuations in the copper beam. It should be noted that diatomic and triatomic copper molecules were present in the beam, so beams of copper produced at lower temperatures may not be completely free from molecules.

Krause (1980) gives brief details of a tantalum oven resistively heated to a temperature of 1300 K with either tantalum or tungsten 250-µm-diameter heater wires for the production of a beam of silver. The oven for the production of beams of silver and copper described by Plekhotkina (1981) was heated by a 15-mm-diameter tantalum cylinder which was 40 mm long. It was 50 µm thick and required about 1 kW of power. The advantage of tantalum was stated to be that it remains elastic, but the lifetime of the heating element was 150 hours when the oven was used at temperatures up to 2800 K. Roulet and Alexandre (1981) describe an ultra-high vacuum compatible oven made of alumina for the production of beams of copper, silver and gold. Heaters were made of tungsten wire, and radiation shields were constructed from an alumina tube and nickel foils. Since the beams were required for deposition studies, any short-term instabilities in the beams were unimportant.

Schmidt et al.'s (1984) oven described in Section 5.6 was also used to produce a beam of copper. Krause et al. (1985) report briefly on the further use of a tantalum oven heated resistively by a constant current source to a temperature of about 1250 K to produce a beam of silver.

Crumley et al. (1986) describe several possible oven designs for producing supersonic beams of copper. They stress the importance of using high-density graphite since otherwise metal vapours will diffuse through it. They attempted both directly heated graphite ovens and those heated by passing current through a graphite cylinder which surrounded a crucible. One design used a slotted 31-mm-diameter tube of 2 mm in wall thickness with pairs of large slots at right angles to increase the electrical resistance. It could be used up to a temperature of 2900 K.

Adam et al. (1989) also briefly report the use of a resistively heated tantalum oven to produce a beam of silver atoms at a temperature of about 1300 K. Shah et al. (1996) evaporated copper from their oven that is initially described in Section 5.2 but with a molybdenum instead of an alumina crucible.

For the evaporation of copper, Ohba and Shibata (1998) used a modification of the basic direct electron-beam evaporation method mentioned in Section 3.2.5, which has been used to produce beams of several elements. The object was to reduce the instabilities due to convection currents in the liquid. A sintered tungsten rod 10 mm in diameter and 24 mm long having a porosity of 40% was placed on a molybdenum liner in a water-cooled copper crucible having a capacity of 4×10^4 mm^3. The rod was surrounded by the copper to be evaporated. When a commercial deflected-beam electron gun able to supply up to 600 mA of electrons at 10 kV heated the top of the tungsten rod, copper evaporation occurred. The metal was replenished as it evaporated in order to obtain long periods of operation. With an electron beam power of 2 kW, the beam density was measured to be over three orders of magnitude greater than when copper was evaporated from a pool of liquid. The authors found no contamination of the copper beam by tungsten from the sintered rod and no dimers in the beam. Ohba and Shibata made detailed measurements of the operating conditions including the temperature across the crucible. The halfwidth of the beam was typically about 90° and decreased with increasing beam intensity. Of particular interest is the peak-to-peak fluctuation in the beam intensity without the rod of about 25% of the mean value at a frequency of 30 Hz. Ohba et al. (1994a) attribute this to mains ripple in the filament power supply but Ohba and Shibata (1998) eliminated this source and then believed the fluctuation to be caused by vibration due to convection in the liquid pool. The fluctuations were eliminated by the use of the sintered rod.

5.10   ZINC, CADMIUM AND MERCURY BEAMS

The earliest report of the formation of a beam of a Group 12 element is with mercury, which is probably the easiest substance to form into a beam. It is stable in air, so filling an oven presents no difficulty, and its large contact angle means that there are no thermocapillary effects. Hence papers which specifically describe mercury beam formation are rare.

The oven used by Knauer and Stern (1926) consisted of a glass tube 13 mm in diameter and 20 mm long coated with multiple platinum films which achieved a resistance of several Ohm. Temperatures of about 400 K produced a vapour pressure of about 130 Pa and the beam was formed by a slit 6 mm by 15 µm, which was arranged to be the warmest part of the oven. It was sealed after a few 100 mg of mercury were added. The apparatus was largely made of glass, but critical metallic parts were made of phosphor bronze, which is insensitive to mercury.

The experiments already mentioned in Section 4.3 for sodium and potassium which were performed by Leu (1927) were also conducted in zinc and cadmium, apparently without any difficulty in forming the beam. Wahlbeck and Phipps (1968) evaporated 3 g of cadmium from a hard-glass oven which was heated by its being immersed in an outer hard-glass vessel containing mercury in an atmosphere of carbon dioxide, the pressure of which could be controlled to adjust the boiling temperature of the mercury and so provide oven temperatures between about 550 K and 700 K. The outer vessel had an embedded tungsten heater. The cadmium beam emerged through a 240-µm-diameter tube about 300 µm long, which was maintained about 10 K hotter than the oven.

A double oven for the evaporation of cadmium is described by Cvejanović et al. (1975) and shown in Figure 5.6. The oven was heated by about 2 m of tantalum wire 260 μm in diameter which was threaded through six-bore ceramic rods. The vertical oven containing the cadmium was heated to a temperature of 500 K. The beam-forming impedance with Γ about 66 was a 500-μm-internal-diameter hypodermic needle about 33 mm long at the exit of a horizontal oven heated to a temperature about 10 K to 20 K hotter. The inner part of the vertical oven was made of stainless steel about 19 mm in internal diameter and 70 mm long, which was surrounded by an oxygen-free high conductivity copper tube containing the heaters, which was also the material of the horizontal oven. Working temperature was achieved in about 90 min with 28 W of heating power, but cooling back to laboratory temperature took 14 hours.

Shannon and Codling (1978) report briefly on a stainless steel oven used to produce beams of mercury and cadmium. Heating to a temperature of about 800 K was by means of a nichrome wire. The beams emerged through a 1-mm-diameter collimating tube which was separately heated to a slightly higher temperature by a machinable glass heating jacket and provided with its own radiation shield. Back et al. (1981) used an oven of titanium heated to a temperature of about 700 K to produce a beam of zinc. They reported that zinc appears to form a low melting point eutectic with stainless steel, which could not therefore be used.

Cadmium was evaporated from a stainless steel oven by Kobrin et al. (1982). The oven containing the cadmium could be removed from the outer structure containing the heating mantle without disturbing the alignment of the 1-mm-diameter

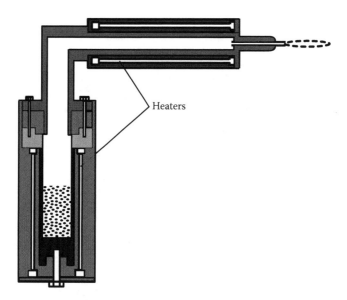

FIGURE 5.6   The oven used by Cvejanović et al. for production of a cadmium beam. (Adapted from Cvejanović D, Adams A, Imhof RE and King GC, 1975. An efficient atomic beam oven for use in low energy electron scattering experiments. *J. Phys. E: Sci. Instrum.* **8** 809–10. Copyright IoP Publishing 1975.)

collimating orifice. Radiation shields were made from alumina as well as stainless steel foil. Myers and Schetzina (1982) evaporated cadmium from a fused silica cell which was placed in the commercial ultrahigh vacuum compatible oven shown in Figure 5.7. The tungsten wire heater required 15 W of power to reach temperatures of 720 K. By having a re-entrant beam-forming impedance in the form of a tube 2 mm in internal diameter and 15 mm long, the need for a double oven was avoided.

Schönhense's (1983) hot air oven described in Section 3.2.7 was used to form beams of mercury and cadmium. Marinković et al. (1991) give brief details of their source of cadmium atoms from an oven maintained at a temperature of 580 K. The beam collimator was a tube 11 mm long and 1.5 mm in diameter that was maintained at a temperature 50 K higher. A molybdenum oven was used by Napier et al. (2008) to produce a beam of zinc from pellets. The beam-forming impedance was a single tube that was 150 mm long and 1 mm in diameter, so $\Gamma = 150$. It was maintained at 30 K hotter than the temperature of about 760 K of the oven. Heating

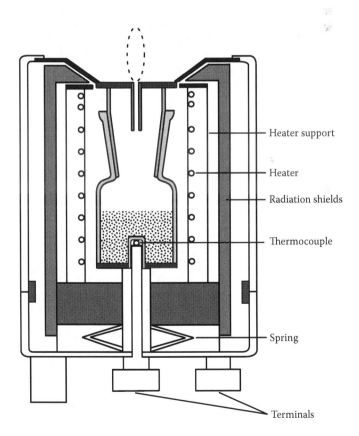

FIGURE 5.7 Myers and Schetzina's oven used with cadmium. (Adapted from Myers TH and Schetzina JF, 1982. Molecular beam source for high vapor pressure materials. *J. Vac. Sci. Technol.* **20** 134–6. With permission from the American Vacuum Society, Copyright 1982, and the Luxel Corporation.)

was provided by twin filament thermocoax heating elements that were embedded in copper cylinders that fitted round the oven and collimating tube.

5.11　BEAMS OF BORON, ALUMINIUM, GALLIUM, INDIUM AND THALLIUM

The Group 13 elements and their compounds were used to produce some of the earliest atomic beams, namely, of thallium. Gerlach and Cilliers (1924) produced a beam of thallium at a temperature of about 1000 K, but no other details are given. We have already referred to Leu's (1927) experiments in sodium, potassium, zinc and cadmium. Thallium proved much more difficult to form into a beam because of creep over the iron components leading to escape of the vapour. Replacing these with phosphor bronze enabled a successful experiment to be performed.

Though aluminium is an easy material to evaporate, for example, to aluminise a mirror, it is quite difficult to form into an atomic beam as Lew (1949) describes. He used a graphite heater, shown in Figure 5.8, through which a current was passed directly and inside which a metal crucible was placed to contain the aluminium. Lew reports that molybdenum is unsuitable, since aluminium creeps up the walls of the tube and clogs the slit. He also rejected graphite since aluminium was absorbed by it and cracked it. Cracking also occurred if a tantalum crucible was used since aluminium crept over its sides and reacted with the graphite. An alumina crucible was satisfactory provided it was isolated from the graphite heater with tantalum foil; otherwise the alumina was reduced by the graphite. Lew's graphite heater was surrounded by a water-cooled radiation shield (not shown) and required 800 W of alternating current to heat it to the required temperature of 1670 K. This is within 30 K of the maximum operating temperature of the oven when the alumina starts to decompose into aluminium and oxygen.

The oxygen free, high conductivity copper oven described in Section 4.3 for the production of a beam of potassium atoms was also used by Miller and Kusch (1955) without modification to obtain a beam of thallium atoms at temperatures up to 950 K. For producing a boron beam, Lew and Title (1960) used the same type of oven as is described by Wessel and Lew (1953) for silver and gold and discussed in Section 5.9. It is a development of that described by Lew (1953) for producing a beam of praseodymium and described in Section 5.15. However, a thoria crucible to contain the boron was found to be impracticable and unnecessary.

The design of Nemirovskii (1968) which is mentioned in Section 5.12 was improved by Nemirovskii and Seidman (1971) to overcome the problem of evaporated material causing failure of the electron gun. By employing 90° magnetic deflection of the electron beam this problem was avoided. Their apparatus is shown in Figure 5.9. Electrons from a tantalum cathode were accelerated to a voltage of 20 kV to give an emission current of up to 25 mA and a focussed spot on the metal to be evaporated of 1 mm in diameter. Their apparatus was used to evaporate aluminium. Its life was limited by that of the cathode but was at least 100 h. Since this was an evaporation source, not used for continuous measurements with the beam, there is no indication of its stability.

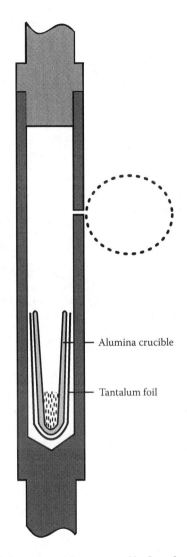

Alumina crucible

Tantalum foil

FIGURE 5.8 The directly heated graphite oven used by Lew for aluminium. (Adapted from Lew H, 1949. The hyperfine structure of the $^2P_{3/2}$ state of Al^{27}. The nuclear electric quadrupole moment. *Phys. Rev.* **76** 1086–92. Copyright 1949 American Physical Society.)

Molecular beams of the chloride, bromide and iodide of thallium were produced by Berkowitz (1972) using a platinum oven with 250-μm-thick walls, heated by nichrome wire up to a temperature of about 700 K. It was insulated with glass fibre and surrounded by a tantalum radiation shield. A diagram of his apparatus is shown in Figure 5.10, from which it can be seen that the beam appears to be emitted vertically downward and collimated by a similarly heated tube which is over 30 diameters long and surrounded by a silver radiation shield, but there is no discussion of this.

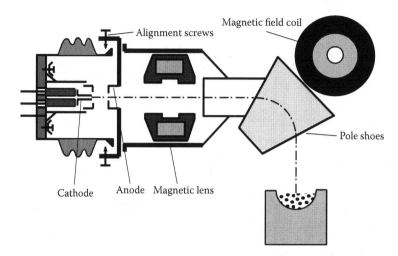

FIGURE 5.9 Nemirovskii and Seidman's electron beam evaporation system. (Adapter from Nemirovskii LN, 1968. Electron gun for floating-zone evaporation of germanium. *Instrum. Exper. Technol.* 6 1482–3 [*Prib. Tekh. Éksp.* **6** 192–3 1968] with permission from Springer Science & Business Media BV.)

Gole and Zare (1972) briefly report that they formed a beam of aluminium by placing filings in a molybdenum oven which was heated to a temperature of 1700 K by a resistance-heated graphite heating element. This appeared satisfactory for their measurement times of about 15 min. Graper's (1973) angular distribution measurements are mentioned in Section 3.2.5. He evaporated aluminium contained in a vitreous carbon liner to a water-cooled container. The electron-beam heating was provided by a 10-kV commercial electron gun that provided up to 12 kW of power to the sample following deflection through an angle of 270°. Crucibles of alumina, tantalum, graphite and boron nitride were tested for the production of aluminium beams by Rosenwaks et al. (1975). They reported that only high-purity boron nitride was satisfactory.

Lindsay and Gole (1977) also found that boron nitride was the only satisfactory crucible material for aluminium. However, the crucible only had an operating life of about 20 hours due both to chemical attack by aluminium and the decomposition of the boron nitride. The heaters had an even shorter life, due to the creeping of aluminium on to them. The crucible was resistively heated with a commercially available tungsten basket heater, which was wrapped with several layers of zirconia cloth to give maximum operating temperatures of over 2000 K. Heating currents of 80 A at 12 V were required.

A high temperature atomic beam source was used to form a beam of indium at 1400 K by Stockdale et al. (1977). The indium was contained in a stainless steel crucible but tantalum crucibles are suggested for higher temperatures. The crucible was heated by passing current through concentric tantalum cylinders 25 µm thick, as shown in Figure 5.11. Measured temperatures of over 3000 K were obtained when the heating current was 120 A requiring power in excess of 900 W. Four concentric

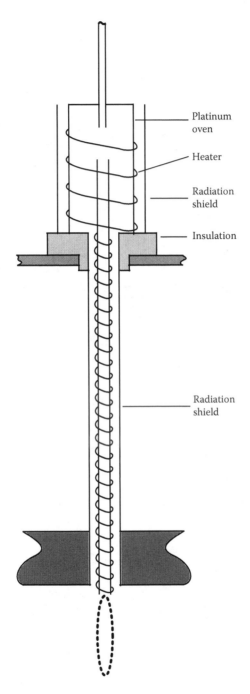

Platinum oven

Heater

Radiation shield

Insulation

Radiation shield

FIGURE 5.10 The apparatus used by Berkowitz to produce beams of some thallium halides. (Adapted with permission from Berkowitz J, 1972. Photoelectron spectroscopy of high-temperature vapors. I. TlCl, TlBr and TlI. *J. Chem. Phys.* **56** 2766–74. Copyright 1972, American Institute of Physics.)

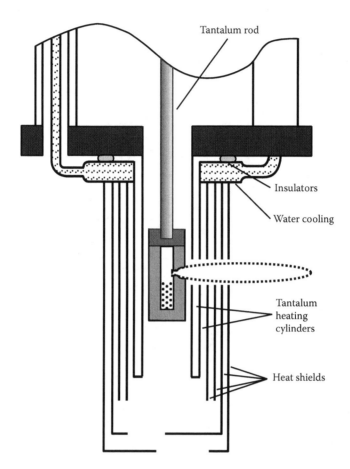

Tantalum rod

Insulators

Water cooling

Tantalum
heating
cylinders

Heat shields

FIGURE 5.11 The indium atomic beam source used by Stockdale et al. (Adapted with permission from Stockdale JA, Schumann L, Brown HH and Bederson B, 1977. High-temperature atomic beam source. *Rev. Sci. Instrum.* **48** 938–9. Copyright 1977, American Institute of Physics.)

radiation shields were attached to water-cooled flanges. Below the oven, staggered shielding enabled good pumping to be obtained. The construction of the oven assembly enabled it to expand freely downward. The horizontal beam passed through 6.35-mm apertures in the heaters and shields.

A graphite crucible heated to temperatures in the range 2175 K to 2675 K was used to obtain a beam of boron by Hanner and Gole (1980). They found that a wall thickness of 3.5 mm to 4.0 mm was necessary because of the corrosive nature of boron. The crucible was heated with a commercial tungsten basket heater.

Oblath and Gole (1980) follow Rosenwaks et al. (1975) in stressing that if the crucible containing aluminium is made from boron nitride, it must be high grade. Their crucible was placed in a carbon sleeve to protect their tantalum radiator from attack by the aluminium. They give no further details of the oven. Riley et al. (1982) used a directly heated graphite tube to produce a beam of a number of metals, which were placed in either a graphite or tungsten crucible. Temperatures of about 1700 K were used to obtain an aluminium beam through an aperture 840 μm in diameter. James et al.'s (1985) approach to producing an aluminium beam was to use an oven fabricated from titanium diboride/boron nitride alloy, which was radiofrequency heated to about 1600 K. Unfortunately no other details of the beam production are given in their paper. A tungsten oven was employed for aluminium by Malutzki et al. (1987), who found the beam to be uncontaminated. The oven was heated by radiofrequency power.

An alternative approach to the formation of a beam of aluminium was employed by Abernathy et al. (1990) who used trimethylaminealane [(CH$_3$)$_3$NAlH$_3$] for MBE. This is a high vapour pressure solid at laboratory temperature which reacts strongly with water vapour. They also review use of other metalorganic compounds for beam formation. Unlike precursors for production of alkali metals discussed in Section 4.2, the problems of producing a pure metallic beam are the other decomposition products. Abernathy (1994) also reviews the precursors that have been used to obtain beams of some Group 13 elements for MBE in ultrahigh vacuum. Use of these methods if a pure beam of the element is required is also limited because of the mixture of compounds usually found in the beam.

An intermetallic composite described as TiB$_2$–BN–AlN, which is a commercially available machinable ceramic material that is not wetted by aluminium, was employed by McGowan et al. (1995) for their oven. The metal was formed into a beam at a temperature of about 1700 K by a 1050-W tantalum resistive heater. The beam-forming orifice was 800 μm in diameter. The authors state that aluminium vapour attacks standard high temperature oven and heater materials, but the life of their oven was apparently only 30 hours.

Gallium was formed into an atomic beam by Patton et al. (1996). Gallium was contained in an alumina crucible, which was heated by a tantalum tube which surrounded it and through which heating current was passed. The authors state that '… in spite of careful control of the oven temperature, the Ga atom beam flux was liable to change over the period of each series of measurements'. Unfortunately, no explanation for this is offered. Shah et al.'s (1996) oven described in Section 5.2 was also used to produce a beam of gallium.

5.12 BEAMS OF CARBON, SILICON, GERMANIUM, TIN AND LEAD

The Group 14 elements and their compounds can all be formed into atomic beams. Production of beams of carbon compounds presents no particular problems. Organic molecules are usually straightforward to evaporate since only low temperatures are required and they do not normally react with any oven, heater or insulating materials.

Gerlach and Cilliers (1924) produced atomic beams of tin at temperatures up to about 1500 K and lead up to about 1100 K, but no details of the oven are given. Gerlach (1925) also used his apparatus described for silver to produce beams of tin, and noted that molten tin reacted with silica.

Nemirovskii (1968) used direct electron-beam heating with a power of 350 W to evaporate germanium.

To produce a beam of carbon, Wolber et al. (1969) used a tantalum carbide crucible heating it by electron bombardment to a temperature of about 3000 K. They found that the beam contained about three times as many C_2 molecules and six times as many C_3 molecules as carbon atoms.

The improvements of Nemirovskii and Seidman (1971) described in the previous section were used to evaporate germanium and silicon.

A beam of carbon was obtained by Zavitsanos and Carlson (1973) using a 25-kW induction heater operating at 450 kHz supplying power to a five turn coil which was 22 mm in diameter and made from 3-mm-bore copper tubing through which cooling water passed. Crucibles were 6.3 mm in internal diameter and 11 mm long with a wall thickness of 1.3 mm made either of pyrolytic graphite or of tantalum carbide. A slotted pyrolytic graphite radiation shield was placed between the crucible and the coil. A slotted tantalum shield also formed the 1.5-mm-diameter aperture that produced the beam. The crucible contained the sample in powdered form. It initially evolved hydrogen and hydrocarbons on being brought up to the working temperature of 3000 K.

Oldenborg et al. (1975) briefly mention that they evaporated lead from a graphite crucible heated to a temperature of 1100 K by a graphite tube. Bulgin et al. (1977) used their radiofrequency heated oven described in Section 3.2.3 to produce a beam of silicon monoxide [SiO] at 2300 K.

The production of a beam of silicon in ultrahigh vacuum for use in MBE is described by Ota (1977). Direct electron-beam evaporation of silicon from a water-cooled copper crucible was employed. The beam was deflected through 270° to prevent contamination from the tungsten filament that produced the electron beam. The commercial source produced an electron beam current of 140 mA by an accelerating voltage of 4.5 kV. A silicon collar was used to protect the copper of the crucible from stray electrons.

Green and Gole (1980) produced beams of both germanium and silicon by evaporation from a carbon crucible by a directly heated tantalum tube. This was lined with a carbon sheath to prevent reaction of the silicon vapour with the tube, but was otherwise stated to be the same as that used by Dubois and Gole (1977) for titanium described in Section 5.4 and Lindsay and Gole (1977) for aluminium described in Section 5.11.

Ota's (1983) review gives useful information on producing silicon and also germanium beams for MBE. In addition to giving further details of the apparatus he used for the direct electron-beam heating of silicon mentioned above, Ota also discusses references to evaporation of silane [SiH_4] and tetraiodosilane [SiI_4]. Silicon can also be evaporated from crucibles made of alumina, boron nitride and silica and in addition Ota describes a fine-grain graphite crucible that was heated to a temperature of 1790 K by a current of 400 A at 5 V to produce a beam of silicon directed vertically downward.

Beams of tin and lead have been produced by Derenbach et al. (1984). Their oven was heated by a radiofrequency generator, which could provide 3 kW at 200 kHz. A diagram of the oven is shown in Figure 5.12. The oven containing the substance to be evaporated was made of molybdenum and supported on a thin-walled tantalum cylinder with some of its wall removed to reduce thermal conductivity. Ceramic rods supported the upper part of the oven. The radiofrequency coil was water cooled. The entire oven was surrounded by a water-cooled stainless steel jacket. A lead beam was obtained at a temperature of about 1140 K and a tin beam at about 1470 K. These emerged through a rectangular 8-mm × 2-mm nozzle 20 mm long.

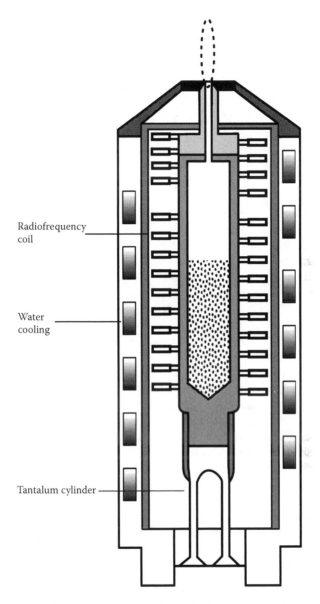

FIGURE 5.12 The radiofrequency oven for tin and lead used by Derenbach et al. (Adapted from Derenbach H, Kossmann H, Malutzki R and Schmidt V, 1984. Photoionisation processes in the 5d, 6s and 6p shells of atomic lead and the 4d shell of atomic tin. *J. Phys. B: Atom. Molec. Phys.* **17** 2781–94. Copyright IoP Publishing 1984.)

Kolodney et al. (1995) used a two-stage oven constructed from 99.8% alumina to obtain a beam of fullerene [C_{60}]. The oven containing it was maintained within 1 K of the desired temperature in the range of 950 K to 1000 K, which yielded a vapour pressure between 13 Pa and 26 Pa. The vapour passed into a second oven which was formed from a 700-μm-diameter tube that was resistively heated up to

2000 K and emerged through an aperture that was about 150 μm in diameter. The authors state that the inert all-ceramic construction prevented 'high-temperature oven chemistry'.

Fullerene was evaporated from a ceramic crucible at a temperature between 950 K and 990 K contained in a ceramic oven by Budrevich et al. (1996). The vapour pressure of the fullerene was about 20 Pa. The ceramic used was recrystallised highly pure ultra-high density impervious alumina usable at temperatures up to 2200 K, in order again to prevent 'high-temperature oven chemistry'. The fullerene beam was collimated by a tube about 11 mm long and 700 μm in diameter but with a short final section about 200 μm in diameter. Unfortunately the rationale behind the design of this capillary is not discussed. This tube was heated by a rhenium ribbon up to 2100 K, since the properties of fullerene at high-temperatures were being studied. Both the oven and the tube were surrounded with single tantalum radiation shields. Vostrikov et al. (1996) report briefly that they obtained a fullerene beam containing 12% C_{70} at temperatures between 800 K and 830 K.

Huels et al. (1998) mention briefly that they obtained a beam of the deoxyribonucleic acid (DNA) nucleobases thymine [$C_5H_6N_2O_2$] and cytosine [$C_4H_5N_3O$] by evaporation at temperatures up to about 450 K.

Keeling et al. (1998) review the use of gaseous germane [GeH_4] and digermane [Ge_2H_6] as precursors for use with chemical vapour deposition of germanium. Because these are hazardous, they used monoethylgermane [GeH_3Et], which is liquid near laboratory temperature and less reactive with air. When deposited on a heated surface, it decomposes into hydrogen and ethylene (ethene) [C_2H_4] as well as germanium, so the precursors are only suitable for surface deposition rather than germanium beam experiments.

The production of a beam of lead is briefly mentioned by McCartney et al. (1998), since it used apparatus previously described by Shah et al. (1993) for iron mentioned in Section 5.8. Concerns about fluctuations in the lead beam are mentioned.

Shimada et al. (1999) describe briefly the evaporation of hydrogen phthalocyanine [H_2Pc, $C_{32}H_{18}N_2$] at temperatures up to about 700 K from a hard glass oven heated with a tantalum wire.

An oxygen-free high conductivity copper oven heated with thermocoax was used by Jaensch and Kamke (2000) to produce a beam of fullerene, since an oven with a uniform temperature was important. The beam was collimated by a cylindrical hole 1.6 mm in diameter in the 5-mm-thick oven. Their paper gives measurements of fullerene vapour pressure and an estimate of 5900 pm for the molecular diameter, both of which are necessary for the design of a fullerene beam system. Lobo and Silva (2001) used a stainless steel oven heated to a temperature of 888 K by a current of up to 63 A at 500 V to produce a beam of fullerene. Radiation shielding was provided by two highly polished concentric stainless steel cylinders. The beam emerged through a 500-μm-diameter aperture.

Molecular beams of carbon tetrachloride (tetrachloromethane) [CCl_4], dichlorodifluoromethane [CCl_2F_2], chloroform [$CHCl_3$] and bromoform (tribromomethane) [$CHBr_3$], which are the refrigerant gases, known respectively as R1, Freon 12, R2 and R4, were produced by Mateječík et al. (2003). Their stainless steel oven was heated by tungsten wires insulated with ceramic tubes to the temperature range 300 K to 600 K. Gas pressures in the range of 100 mPa to 7 Pa were used, at which pressures the authors state that the mean free path of the molecules is much larger than the dimensions of the oven. Some collimation was produced by an exit tube which was 4 mm long and 500 μm in diameter. No problems with beam formation were reported.

Ahmed et al. (2005) give extensive details of their oven assembly for producing a beam of tin from a tantalum crucible, which was 25 mm long, 8 mm in internal diameter and 1-mm wall. The crucible was surrounded by a 50-μm-thick tantalum foil that heated it, a radiation shield of the same foil and a water cooled jacket. The heating current of up to 400 A at 6 V passed in turn through a 50-mm-diameter tantalum and copper central rod and returned through a steel cylinder to the mounting flange, as shown in Figure 5.13. Since the oven was designed for higher temperatures than that used to produce a beam of tin at 1700 K it was capable of attaining temperatures of 2800 K with a heating power of 3.6 kW. At temperatures above 2300 K the lifetime of the filament is stated to be more than 20 h. The tin beam at a temperature of 1700 K was collimated by a single 200-μm-internal-diameter tube that was 6 mm long. Measurements of the beam properties were made with a uranium beam which is referred to in Section 5.16.

McRaven et al. (2007) briefly describe how a beam of lead monofluoride [PbF] was produced using a magnesium fluoride [MgF$_2$] tube which contained a bead of lead near its 200-μm-diameter exit orifice. When this tube was heated by radiation to a temperature between 1100 K and 1200 K, the lead reacted to form lead monofluoride, which was emitted by passing helium through the tube.

Tabet et al. (2010) obtained beams of the deoxyribonucleic acid (DNA) and ribonucleic acid (RNA) nucleobases adenine [C$_6$H$_5$N$_5$] and cytosine [C$_4$H$_5$N$_3$O], the

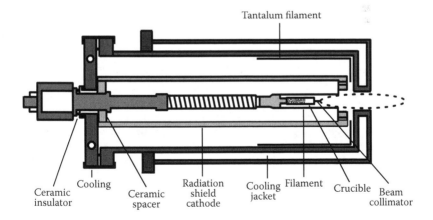

FIGURE 5.13 The tantalum oven employed by Ahmed et al. to obtain a tin beam. (Adapted with permission from Ahmed N, Nadeem A, Nawaz M, Bhatti SA, Iqbal M and Baig MA, 2005. Resistively heated high temperature atomic beam source. *Rev. Sci. Instrum.* **76** 063105-1–4. Copyright 2005, American Institute of Physics.)

DNA nucleobase thymine [$C_5H_6N_2O_2$] and the RNA nucleobase uracil [$C_4H_4N_2O_2$] by heating them in turn in a 10-mm-internal-diameter copper oven that was heated at a temperature of up to 500 K with a resistive filament ribbon. The oven assembly was insulated with an outer casing of a ceramic of unstated composition. The beams were collimated by a single 1-mm-internal-diameter tube that was 35 mm long so $\Gamma = 35$. Graphs are presented showing deposition rates on a cold surface for the four bases as a function of temperature and also angular distribution measurements for uracil.

5.13   BEAMS OF PHOSPHORUS, ARSENIC, ANTIMONY AND BISMUTH

The main problem with obtaining atomic beams of the elements in Group 15 is that the beam obtained by conventional evaporation is molecular and consists of a mixture of species.

Bismuth was formed into an atomic beam at a temperature of about 1050 K by Gerlach and Cilliers (1924), but no details of the oven are given. Bismuth was also formed into an atomic beam by Leu (1928) using a modification of the apparatus used by Leu (1927) for sodium and potassium, that is already described in Section 4.3, since bismuth requires a higher temperature oven. Hence copper was replaced with tungsten steel and the oven was heated by electron bombardment using 200-μm-diameter tungsten wire for the cathode through which about 3.2 A was passed. A heating current of about 30 mA was obtained at a potential difference of 2000 V, which was concentrated on the exit orifice.

Lindgren (1958) used a molybdenum oven for the production of a beam of bismuth, which was collimated by an 8-mm-long channel, described as 500 μm 'wide.' The electron beam from a tungsten cathode was directed at this. As mentioned in a later paper (Lindgren and Johansson, 1959) bismuth vapour at a temperature of about 1000 K consists mainly of polyatomic molecules, but these were dissociated in the channel which was at about 1700 K. They stabilised the 150-W heating power by regulating the current heating the cathode. The author reports that the oven was shielded with two tantalum radiation shields. The temperature of about 1300 K was reached with an emission current of 50 mA to 100 mA at 500 V. Wagner and Grimley (1974) evaporated bismuth at a temperature of 923 K from Knudsen cells made both from sterling silver with an alumina liner and tantalum. They were heated with a platinum–rhodium helical filament bifilarly wound.

McFee et al. (1977) state that they used an aluminium oven at a temperature of 618 K to evaporate phosphorus as P_4 molecules, which passed through a separately heated ceramic beam-forming impedance. Panish (1980) obtained beams of As_2 and P_2 by the high temperature decomposition of arsine [AsH_3] and phosphine [PH_3], respectively, in a resistively heated alumina tube at pressures from about 50 kPa to 300 kPa and temperatures of about 1000 K. This produced As_4 and P_4 molecules which dissociated following their passage through an orifice into an alumina tube.

Chai (1984) produced a molecular beam of the phosphorus dimer by evaporating tin phosphide [Sn_3P_4] at a temperature of up to 623 K from a graphite crucible. Further studies were undertaken by Huet et al. (1985), who investigated the composition of arsenic beams as a function of temperature from arsine [AsH_3] and As_4

molecules as well as phosphorous beams from phosphine [PH_3] and P_4, since these are of importance in MBE. Panish et al. (1985) also studied the production of beams of arsenic and phosphorous dimers from the corresponding hydrides.

Brief details of the production of a beam of antimony are given by Dyke et al. (1986a) who used Bulgin et al.'s (1977) design of induction-heated carbon oven described in Section 3.2.3 heated to a temperature of about 1000 K. This produced Sb_4 and Sb_2 and atomic antimony was produced by passing the vapour through a graphite cylinder heated to about 1600 K. Dyke et al. found that evaporating antimony powder was preferable to heating copper antimonide [Cu_3Sb]. A similar technique was used by Dyke et al. (1986b) to deal with the evaporated species of arsenic. It was evaporated at about 500 K then further heated to a temperature of about 1200 K to produce As_2 in the beam. An alternative means of producing As_2 was by evaporating copper arsenide from a carbon furnace heated to about 1300 K. An atomic beam of arsenic was produced when the graphite cylinder was heated to about 1750 K.

Gericke et al. (1991) studied the effect of filling level of Knudsen cells made of graphite for the evaporation of bismuth at temperatures from 850 K to 1000 K for use in MBE.

Abernathy (1994) also reviews the precursors that have been used to obtain beams of some Group 15 elements in ultrahigh vacuum. Use of these methods if a pure beam of the element is required is again limited because of the mixture of compounds usually found in the beam.

Brewer et al. (1996) present measurements of the composition of an antimony beam as a function of temperature and pressure. Only atoms are present at temperatures above 1400 K, but Sb_2 is present at lower temperatures and Sb_4 below 1000 K.

5.14 SULPHUR, SELENIUM, TELLURIUM AND POLONIUM BEAMS

Olsen et al. (1945) report no problems with evaporating the Group 16 elements selenium or tellurium. Zingaro and Cooper (1974) review the detailed work which has been undertaken on the composition of selenium vapour. Most vapours from Se_2 to Se_9 have been found. The composition of the vapour varies with temperature.

Berkowitz (1975) mentions briefly that he obtained a beam of sulphur molecules [S_2] by evaporation of mercuric sulphide [HgS] from a platinum oven at about 600 K. For his purpose, the presence of mercury vapour in the beam was not critical. He also obtained a beam of diatomic tellurium molecules directly from powdered metal from his platinum oven at a temperature of about 670 K.

Kowalewska et al. (1991) briefly describe the use of crucibles of either molybdenum or graphite that were heated to temperatures between 700 K and 1000 K to produce a beam of polonium. Some of its isotopes were transferred to the oven in silver foil on which polonium was deposited from acidic solution. Others were produced by irradiating lead which was then placed in the crucible.

Gossla et al. (1995) briefly describe a three-stage oven to produce a molecular beam of sulphur for use in MBE. The first stage was maintained at temperatures between 333 K and 393 K so that sulphur sublimed. The vapour passed through an aperture into the second stage, which was maintained at 573 K. The third stage consisted of a ceramic tube maintained at temperatures up to 1273 K at which temperature the predominantly S_8 and S_6 rings are dissociated. Unfortunately no details of the other construction materials are given.

5.15   BEAMS OF THE LANTHANIDES

Elements with atomic numbers from 57 to 71 are known as the lanthanides or more popularly as the rare earth elements.

Lew (1953) formed a beam of praseodymium by placing it inside a thoria crucible which was placed within a molybdenum cylinder that was machined from a solid rod to produce a wall thickness of about 800 μm and internal diameter of about 6 mm. Current was passed through water-cooled copper leads and about 240 A at 10 V passed through this cylinder and raised the temperature to 2000 K. The beam emerged through a slit 100 μm wide by 3 mm long in the centre of the cylinder, which would be the hottest part. Lew mentions that the use of the crucible greatly reduced the chances of the oven slit being clogged by the molten metal. A single water-cooled radiation shield surrounded the assembly. Ting (1957) formed a beam of lanthanum from a tantalum oven heated to a temperature of about 1800 K.

Parr (1971) gives only very brief details of his oven for the production of beams of europium and thulium. The crucible was heated by a tungsten filament to a temperature of about 1000 K. An electrostatic deflector was necessary to remove ions from the beam. A tantalum crucible heated by electron bombardment to about 2000 K was used by Figger and Wolber (1973) to produce a beam of lutetium, which was formed by a slit 100 μm wide and 10 mm long. This is the same design as used by von Ehrenstein (1961) for cobalt and described in Section 5.8. Figger and Wolber reported both an alloying of the lutetium with the tantalum and its creep, causing blockage of the slit. The apparatus already mentioned for Holland et al.'s (1981) barium measurements in Section 5.2 was also used for ytterbium.

Molybdenum was used as the oven material by Neubert and Zmbov (1983) for producing beams of samarium, europium, thulium and ytterbium. It was heated by radiofrequency power. Evaporation temperatures were in the range 770 K to 1200 K. Wagner (1983) describes a resistively heated oven suitable for temperatures up to at least 1700 K for use with the lanthanides and other elements such as chromium. He used a bifilarly wound heater to reduce its magnetic field. Just over 4 kW of power provided by a mains-fed transformer capable of delivering up to 600 A at 20 V were required to reach 1800 K. The current leads internal to the vacuum system were made of niobium. Externally they were of water-cooled copper tubes. The crucible was surrounded by several tantalum radiation shields, within a copper envelope. Electrical insulation was provided by an oxide ceramic. The beam emerged through a conical nozzle.

An oven for the production of a beam of gadolinium is described by Ruster et al. (1989). Temperatures of up to 2800 K could be obtained by electron bombardment of a tungsten, molybdenum or tantalum oven from a tungsten filament. This produced 200 W of heating power. The oven and heater were surrounded by a single tantalum radiation shield, followed by a water-cooled stainless steel housing. This was required to suppress ions or electrons and electrodes in front of the beam exit served the same purpose. The beam exit from the oven was through a 1-mm-diameter, 10-mm-long channel. Ford et al. (1990) produced a beam of samarium from a radiofrequency heated tantalum oven at a temperature of about 1000 K. No details of this are given.

Several research groups produced beams of the lanthanides by the direct electron-beam heating method. A beam of gadolinium was produced by Nishimura et al. (1992) using a commercial axial beam gun which produced up to 30 kW of electron beam heating power at 20 kV. The gadolinium was contained in a water-cooled copper crucible and electrons were incident at 30° to the normal to the surface of the element. Ohba et al. (1994b) obtained a gadolinium atomic beam for which the 500-mA electron beam at 10 kV was deflected through 225°. They describe how the plasma which accompanied the beam could be removed by applying negative potentials to electrodes in the form of parallel plates which were 80 mm long and 50 mm wide on either side of it and separated by 15 mm. The potentials required increased with atom density to just over 60 V.

Direct electron-beam evaporation was also used to produce a beam of praseodymium by Smirnov (1994). It was contained in a tantalum crucible that was placed in a water-cooled copper cup supported on a copper base. Electrons impinged normal to the surface of the lanthanide which reached a temperature of 1600 K. The beam was emitted at an acute angle to the molten surface and was defined by three water-cooled diaphragms that formed part of differential pumping stages. An atomic beam of dysprosium was produced by Tamura et al. (1999a) by the same method. The lanthanide was contained in a water-cooled copper crucible. The maximum electron-gun power was 5 kW at up to 10 kV. Charged particles in the beam were removed using the method described by Ohba et al. (1994b) for gadolinium described above. Tamura et al. used the method described by Ohba and Shibata (1998) for copper in Section 5.9 for the evaporation of cerium. A sintered tungsten rod 10 mm in diameter and 24 mm long having a porosity of 30% was placed on a tantalum liner in the crucible and surrounded by cerium.

Further experiments on gadolinium were conducted by Ohba et al. (2000) with their magnetically deflected electron beam. Smirnov (2000) briefly describes the production of a beam of lanthanum by direct electron-beam heating of a 'bowl-shaped' sample that was mounted on a graphite substrate placed in a crucible. The evaporation temperature was 1950 K. We have already referred to Yagi and Nagata's (2000) oven to produce a beam of barium atoms in Section 5.2. They also used the same apparatus to produce a beam of europium atoms. The same authors (2001) obtained in addition beams of cerium, neodymium, samarium, gadolinium, dysprosium, erbium and ytterbium. They used a tantalum oven to produce beams of cerium and erbium at temperatures of respectively 1700 K and 1400 K. For neodymium, the evaporation temperature was 1500 K; it was 1700 K for gadolinium and 1400 K for dysprosium. For these lanthanides a carbon crucible with a tantalum liner was used. A stainless steel oven was used to evaporate samarium at a temperature of 1100 K, europium at 1000 K and ytterbium at 800 K. Whereas for cerium, neodymium, gadolinium and dysprosium evaporation was from the liquid state, samarium, europium and ytterbium sublimed. Erbium was evaporated either from the solid or liquid state. The authors reported that typically 2 g to 3 g of a sample lasted for 2 days.

Hudson et al. (2002) briefly mentioned that they produced a beam of ytterbium fluoride [YbF] molecules by heating 4 parts by weight of ytterbium with one part of powdered aluminium fluoride [AlF_3] to a temperature of about 1500 K in a molybdenum oven.

5.16　BEAMS OF THE ACTINIDES

Elements with atomic numbers of 89 or more are known as the actinides or transition elements. Olsen et al. (1945) report no problems with the evaporation of thorium from tungsten, which it wetted.

Hubbs et al. (1958) report in detail on the problems of forming a beam of plutonium. It apparently reacts with tantalum and creep is a serious problem. Hence, attempts were made to evaporate plutonium from the decomposition of its compounds with carbon, silicon and oxygen. However, the silicide also reacts with tantalum and decomposition of the others was insufficient at temperatures up to 2300 K. Other crucible materials tried included molybdenum, tungsten, thorium oxide and thorium and cerium sulphides. However, with the exception of tungsten, porosity, creep and chemical attack caused problems. The tungsten crucible had a sharp lip to reduce the effects of creep. This was placed in a tungsten cylinder which was heated by electron bombardment so as to have a large temperature gradient. The crucible was thermally isolated with either cerium, thorium sulphide or its oxide, which enabled it to be cooler than the hottest part of the cylinder. This both reduced creep and also returned plutonium that was not formed into a beam back to the crucible.

Neptunium was formed into a beam by Hubbs and Marrus (1958). It was not found possible to form the beam without separation from the uranium from which it was formed, because of the reaction of the latter with tantalum. However, neptunium oxide was mixed with graphite and placed in a tantalum oven which was initially heated to a temperature of about 1300 K. After the evolution of carbon monoxide ceased, the oven temperature was raised to the range of 2100 K to 2800 K. This same technique to produce neptunium was used by Hubbs et al. (1959) to produce a beam of curium. In this case the first stage required a temperature of about 1500 K and the beam was obtained at 2100 K.

The apparatus Ward et al. (1967) used for obtaining a beam of gold that is described in Section 5.9 was also used for plutonium. Measurements were made at temperatures up to 1800 K. A tantalum oven was tested with crucibles of yttria [Y_2O_3], thoria [ThO_2], magnesia [MgO], zirconium diboride [ZrB_2] or tungsten coated yttria, since otherwise the tantalum was attacked and creep was observed. It was found that the oxides were not wetted by plutonium. Some reaction with the yttria crucible was observed. The thoria and magnesia crucibles were found to be less satisfactory due to reactions and diffusion. Plutonium wetted zirconium diboride, but no reaction appeared to take place. The most satisfactory was a yttria crucible coated with a 127-µm-thick layer of tungsten deposited from tungsten hexafluoride [WF_6]. To prevent creep, the tungsten was ground away at the lip of the crucible to expose the ceramic.

To produce a beam of uranium, Fite et al. (1974) employed an oven that was 50 mm long and 6 mm in diameter. It was constructed from six turns of 25-µm-thick tungsten foil which was mounted in molybdenum end caps that were fixed to water-cooled copper electrodes. A 2-mm-diameter uranium rod was fixed in an alumina sleeve to the upper end cap. A current of about 200 A at 2.5 V enabled furnace temperatures of 2000 K to be achieved which produced a molten ball of uranium on the end of the rod. Uranium evaporated from this ball on to the internal surface of the oven from where it was emitted to enable a uranium beam to be formed by the 2-mm-diameter aperture in the oven. Tungsten wadding reduced uranium migrating to the cooler parts of the assembly. The advantage of this method was that liquid uranium did not normally come into contact with tungsten with which it is extremely reactive. When the uranium ball dripped, the oven was immediately destroyed; otherwise beams were obtained for 2 h to 3 h.

Krikorian (1976) obtained a beam containing over 90% uranium by evaporating uranium rhenide [URe_2] in a rhenium boat heated to a temperature of 2300 K by a current of 100 A at 3.4 V. He found that a tungsten boat could also be used and suggested that other refractory metals than rhenium, namely, iridium, niobium and osmium could be used to form a compound with uranium and used in the boat because of their high vapour pressure. Use of this method overcomes the extreme corrosiveness of uranium. Greenland et al. (1985) briefly describe their method of obtaining a uranium beam by evaporating uranium rhenide from a tungsten foil oven at a temperature of 2300 K, which was heated by passing a current through it. The beam emerged through a 1-mm-wide slit. Allen et al. (1988) report briefly on the production of a beam of uranium atoms, using a tungsten oven. Uranium was placed in the oven in a nitrogen atmosphere to reduce oxidation. An alternative of a mixture of the carbides [UC_2/UC] was also used. In each case oven temperatures in the range of 2250 K to 2330 K were required. An equimolar mixture of uranium with uranium oxide [UO_2] enabled beams containing the oxides UO and UO_2 to be obtained at oven temperatures from 2130 K to 2280 K.

Eichler et al. (1997) made a detailed study of the properties of the actinides, from which they concluded that a sandwich type source is required for elements that are not abundant. The oxide of the element should be deposited on a mechanically and thermally stable substrate which is coated with a strongly reducing thin metallic coating. For the former, tantalum is superior because of its high temperature mechanical stability and because it enriches actinides at the foil surface and forms a barrier against back diffusion into the substrate. They state that titanium has a high reductive potential, low electron work function, low adsorption enthalpy and high diffusion rate for actinides and so is the material of choice for the coating. They give details of the electrolytic deposition of the actinide hydroxides on to a 50-μm-thick tantalum filament which was then coated by evaporation or sputtering with titanium to a thickness of 1 μm. Actual beams were produced from plutonium at an evaporation temperature of 1270 K, curium at 1390 K to 1410 K and berkelium and californium at 1020 K. It was calculated that for other actinides, evaporation temperatures would all lie in the range from 700 K to 2000 K. Since the higher temperatures are close to the melting point of titanium, it is suggested that it should be replaced with zirconium.

Several authors have also used direct electron-beam heating to produce beams of uranium. Tamura et al. (1999b) placed it in a water-cooled crucible with a capacity of 8×10^4 mm³. Electrons impinged on the crucible after acceleration through up to 10 kV and deflection through a large angle with an available power of 16 kW. Bhatia et al. (2000) also used the same method to produce a beam of uranium atoms. Their detailed paper mentions that one source of beam instability was the convection currents in the molten liquid pool causing turbulence leading to temperature fluctuations. Their paper is particularly concerned with the measures necessary to prevent electrothermal feedback which can cause thermal runaway, leading to an uncontrolled increase in furnace temperature. This can happen because an increase in furnace temperature leads to an increase in the filament temperature and increased emission if this is in the temperature-limited region of the emission curve. This was cured by inserting a voltage-dropping element into the high-voltage circuit. An unusual feature was that after initial heating of the oven by the filament, the uranium coated inner radiation shield, which was one of five, became hot enough when at a temperature of 1500 K to emit electrons. This meant that the filament only supplied from 10% to 30% of the total heating current, which reduced the magnetic field produced by the current flowing through it. Another problem that Bhatia et al. had to overcome was the presence of ions in the atom beam. These were successfully suppressed by over a factor of a thousand with two biased grids.

Ohba et al. (2000) produced beams of uranium by the same method. An outline of their apparatus is shown in Figure 5.14. A 16-kW electron beam at 10 kV from a commercial electron gun was deflected by a magnetic field through 270° so that it impinged on the uranium perpendicular to its surface. It was contained in a water-cooled crucible having a capacity of 8×10^4 mm^3. Evaporation was from a 10-mm-diameter disk. In a second experiment a 30-kW commercial electron gun at 20 kV was employed without any magnetic deflection. The electron beam impinged at an angle of 30° to the normal to the surface of a larger crucible with 3.2×10^5 mm^3 capacity. The evaporation area was a 10-mm × 15-mm ellipse. In each case tungsten was added to the crucibles to reduce convection currents. The authors also discuss the fact that the electron beams produce weakly ionised plasma containing multiply charged ions.

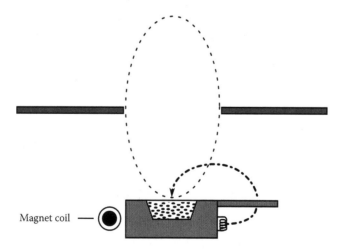

Magnet coil

FIGURE 5.14 The direct electron beam evaporation of uranium employed by Ohba et al. (Adapted from Ohba H, Ogura K, Nishimura A, Tamura K and Shibata T, 2000. Effect of electron beam on velocities of uranium atomic beams produced by electron beam heating. *Jpn. J. Appl. Phys.* **39** 5347–51. With permission from the Japan Society of Applied Physics.)

We have already referred to Ahmed et al.'s (2005) oven in Section 5.12, since it was also used to obtain a beam of tin. They also measured properties of the system when a beam of uranium was obtained with the oven operated at a temperature of 2500 K. They give curves for the current–voltage characteristic of the heater and also the temperature–power curve. Their measurements of the uranium beam angular distribution are discussed in Section 11.2.4.

5.17   BEAM SOURCES FOR REFRACTORY MATERIALS

5.17.1   Introduction

Elements or compounds with evaporation temperatures above about 2400 K are referred to as refractory. The classification is of course somewhat arbitrary, since the actinides discussed in the previous section are among elements requiring high evaporation temperatures. The refractory elements include the metals tungsten, rhenium, tantalum, osmium, molybdenum, iridium, niobium, ruthenium and hafnium listed in decreasing order of melting points. However, we have also included some other elements where the same method of beam formation was used. They are treated as a group since the two techniques described in the next two sub-sections could be applicable to all of them. As discussed in Section 5.17.4, these methods are however now likely to be superseded by the direct electron-beam heating method introduced in Section 3.2.5, since far superior beam densities should be obtained. This method is described in detail in Section 5.9 since it was used for copper by Ohba and Shibata (1998), which appears in that section before the other elements for which it has been employed. The refractory elements cannot be formed into a beam by the use of other ovens since, even if there were a suitable refractory material for the crucible, it might well react with the material to be evaporated in its liquid or vapour state. Hence techniques were developed to enable evaporation by an electron beam from a restricted area of the sample while the remainder remained solid. These include evaporation from a wire and from a rotating cylinder. We will consider them in this order, which is also that of their historical development.

5.17.2   Beams from Wires

Doyle and Marrus (1963) have obtained beams of iridium, rhenium, tantalum and tungsten by evaporation from wires that were either 250 μm or 400 μm in diameter which were heated by electron bombardment from a wire cathode. The authors state that this method produced a more stable beam than that produced by passing a current through the wire. It was not possible to produce a hafnium beam by this method, since insufficient vapour pressure was produced.

Lewis (1964) actually evaporated refractory compounds to deposit thin films, but his technique essentially forms beams. In order to produce a beam of alumina [Al_2O_3], a vertical 2-mm-diameter sapphire rod with a graphite collar was heated by electron bombardment from a tungsten cathode 30 μm in diameter held at a potential of up to 3 kV. The maximum current available was 350 mA. When heated, the rod became conducting and then melted to form a molten blob 5 mm in diameter at its lowest point from which alumina was evaporated. The rod was mounted in a water-cooled tantalum assembly and was fed downward from outside the vacuum envelope by a rack and pinion drive. A spring-loaded tungsten wire stirrup maintained the sapphire rod in close contact with the graphite tip. The apparatus is shown in Figure 5.15. Feed rates were about 500 μm min⁻¹. Lewis reported that both the feed rate and power input required careful hand control. The same technique was used to evaporate silicon dioxide [SiO_2] and magnesium oxide (magnesia) [MgO] from 3-mm-diameter rods without the need for a graphite collar. He notes that magnesia evaporated from the solid rod rather than from the molten state. A wider range of evaporation rates was possible compared with alumina. However, loose deposits of silica and magnesia on electrodes tended to flake off, so Lewis recommends evaporation vertically upward. He also discusses the composition of the beams.

Pendlebury and Ring (1972) describe what they call a molten ball beam source for the production of beams of molybdenum and rhenium. The tip of 300-μm-diameter wire was heated by electron bombardment at 3 kV with a current of 10 mA. It was fed at the rate of about 1 mm min⁻¹. The space charge limited current means that its magnitude depends on the proximity of the resultant liquid ball at the tip of the wire to the filament. Hence there was a balance produced by the heat gained from the heating current and the heat losses of 15 W radiated by the ball of molten metal, which was 'usually' about 1 mm to 2 mm in diameter, and 7 W by the wire. Their

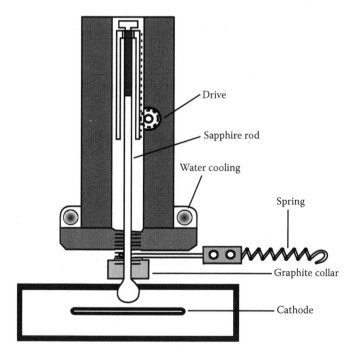

FIGURE 5.15 Lewis' apparatus for producing beams of alumina, silica, and magnesia. (Adapted from Lewis B, 1964. The deposition of alumina, silica and magnesia films by electron bombardment evaporation. *Microelectron. Reliab.* **3** 109–20. With permission from Elsevier. Copyright 1964.)

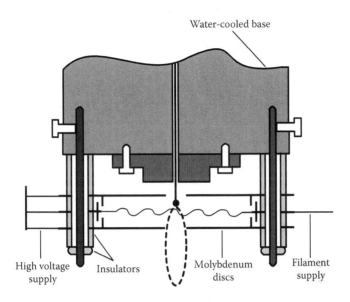

FIGURE 5.16 The apparatus used by Pendlebury and Ring to produce beams of molybdenum and rhenium. (Adapted from Pendlebury JM and Ring DB, 1972. Ground state hyperfine structures and nuclear quadrupole moments of ^{95}Mo and ^{97}Mo. *J. Phys. B: Atom. Molec. Phys.* **5** 386–96. Copyright IoP Publishing 1972.)

apparatus is shown in Figure 5.16. The steel structure was water cooled, and the atomic beam was defined by molybdenum slits to be 200 μm wide with a height defined by the ball diameter. The authors report that slow periodic fluctuations in the beam intensity were common.

Using the same technique, Rubinsztein et al. (1974) employed 1.3-mm-diameter molybdenum wire which formed a ball of about 1 mm larger diameter when melted by electron bombardment. After about 30 min, a stable beam was obtained with a temperature between 10 K and 100 K below the melting point. Their method was used with a neutron-irradiated molybdenum wire, which enabled measurements to be made on radioactive isotopes of both it and technetium. Rubinsztein and Gustafsson (1975) used this method for iridium, molybdenum, niobium, osmium, platinum, tantalum, technetium and rhenium.

In Section 3.2.6 we suggest that Hopkins et al.'s (1983) evaporation of a rotating rod by a laser might form a valuable alternative to direct electron-beam evaporation.

The intensity of the beams formed from wires is clearly not large and hence other techniques have been employed where more intense beams are required.

5.17.3 Beams from Rotating Cylinders

This technique is described by Büttgenbach et al. (1970) and by Büttgenbach and Meisel (1971). A cylinder of the metal to be evaporated, namely, tantalum and tungsten with a diameter of 30 mm and thickness of 10 mm to 15 mm, was rotated at about 1 s^{-1}. In addition it was slowly displaced axially. A diagram of the arrangement

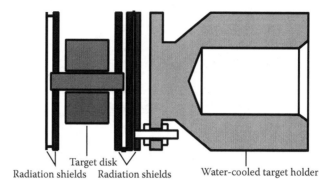

FIGURE 5.17   The rotating beam source for refractory metals described by Büttgenbach and Meisel. (Adapted from Büttgenbach S and Meisel G, 1971. Hyperfine structure measurements in the ground states $^4F_{3/2}$, $^4F_{5/2}$, $^4F_{7/2}$ of Ta^{181} with the atomic beam magnetic resonance method. Z. *Physik* **244** 149–62. With permission from Springer Science & Business Media B V.)

is shown in Figure 5.17. This cylinder was shielded on each side by tantalum radiation shields and this sub-assembly was fixed to a water-cooled copper cylinder. An electron beam current of up to 20 mA from a modified commercial electron beam welder was accelerated up to 100 kV and focussed on to the surface of the cylinder to a slit-shaped area about 300 μm by 6 mm. This shape was achieved by magnetic deflection using a saw-tooth waveform with a repetition rate of 25 Hz. The authors also reported fluctuations in beam intensity, caused both by vibrations of the molten surface and the movement of the molten zone on the cylindrical target. The life of the cylinder was about 10 h after which it needed resurfacing.

Penselin (1978) reviews additional experiments using the rotating target method on atomic beams of carbon, hafnium, iridium, molybdenum, niobium, ruthenium, tantalum, tungsten and zirconium.

5.17.4   DIRECT ELECTRON-BEAM HEATING

The direct electron-beam heating of the surface of a refractory metal in a cooled container is in principle possible. It is discussed in Section 3.2.5 and other sections in this chapter indicate how universal it is becoming. The method should enable much more intense refractory atomic beams to be obtained than those described earlier in this section. Direct electron-beam heating has been used to produce a beam of zirconium as mentioned in Section 5.4 by Thakur et al. (2005), but considerable heating power is required. An alternative possible approach of electromagnetic or electrostatic levitation rather than employing a cooled container is discussed in Section 7.3.

6 Production of Beams of Dissociated Atoms and Other Radicals

6.1 INTRODUCTION

We mainly collect together in this chapter those techniques that have been used for producing atomic beams when the parent substance is a molecule and so has to be dissociated. This process is also sometimes known as 'cracking'. In general these dissociated molecules are called radicals and in addition to the production of beams of atomic hydrogen, oxygen, nitrogen and the halogens we also include any other radicals that we are aware have been formed into a beam. Because of their extreme reactivity, the atomic state of substances that are normally molecular were sometimes referred to as 'active', so, for example, active nitrogen referred to atomic nitrogen. We begin the discussion with hydrogen which occupies several sections and follow it in Section 6.6 with oxygen, since similar techniques are involved. Nitrogen is probably the most difficult molecule to dissociate because of its high dissociation energy, but we consider it next in Section 6.7, because of its importance as the chief atmospheric constituent. The techniques for actually forming a beam of the halogens are not particularly difficult, but working with them is unpopular more because of the problems of handling them in vacuum systems than of forming an atomic beam. They are discussed in Section 6.8. Finally a short Section 6.9 covers beams of other radicals and an even shorter 6.10 exceptionally discusses one production of a molecular beam.

In some cases a beam is required at a lower temperature than that at which it is formed and we do not separate these techniques in the following chronological discussion. Similarly discussion of dissociation mechanisms and the cleaning of discharge tubes are not considered separately. Whether it is required to dissociate a molecule or to transport an atom that prefers to exist as a molecule, surface effects are important. Wise and Wood (1967) provide a detailed review of the recombination of atomic hydrogen, oxygen, nitrogen and the halogens on glass, fused silica and many other metallic and non-metallic surfaces. The temperature dependence of recombination coefficients is discussed.

6.2 PRODUCTION OF ATOMIC HYDROGEN BY THERMAL DISSOCIATION

6.2.1 INTRODUCTION

As mentioned in Section 1.1, the production of beams of atomic hydrogen has received considerable attention. There are two basic methods of producing them.

The molecules can either be dissociated on a tungsten surface when heated to a very high temperature, such as a tungsten wire, foil, or furnace or in any form of electric discharge. The latter method is discussed in Section 6.3. The choice of approach will be clearer after the methods are discussed in detail. The development of furnace and discharge sources started and has continued in parallel. Though the use of dissociation on hot surfaces was slower to develop, we first consider in Section 6.2.2 the extension of the detection of dissociation of hydrogen on a hot filament to the production of atomic hydrogen beams by this means. We next consider furnaces in Section 6.2.3 since they also follow naturally from the previous discussion of ovens. We follow our usual practice of only including papers that contain useful construction or operating details.

In this discussion we do not distinguish between hydrogen and its isotopes deuterium and tritium, since there are no significant differences in producing beams from any of them. The dissociation energy of hydrogen is 4.52 eV and of deuterium 4.60 eV. The recirculation of hydrogen isotopes is discussed in Section 7.2.

6.2.2 THERMAL DISSOCIATION OF HYDROGEN ON TUNGSTEN FILAMENTS AND SURFACES

Langmuir (1912) studied the power consumption of a tungsten wire in an atmosphere of hydrogen and observed that it increased at a greater rate with increasing temperature than was to be expected from the laws of convection and radiation. He correctly deduced that the extra energy was used in dissociating the hydrogen molecules, and so laid the foundations of the means of producing a beam of atomic hydrogen by dissociation on a hot tungsten surface. In this section we consider mainly dissociation on wires, but foils have also been used. These methods have the advantage of simplicity when high beam intensities are not important.

Phipps and Taylor (1927) briefly describe the use of a tungsten filament which consisted of 15 turns of 178-μm-diameter wire which was closely wound on a mandrel about 1 mm in diameter. The wire was within 1 mm of the same slit described in Section 6.3.2. A heating current of 2.5 A raised the temperature of the wire to over 3000 K. The atomic hydrogen beam produced by this source was found to be only about one-half the intensity of that from their Wood's tube used to dissociate hydrogen discussed in Section 6.3.2. It was also less satisfactory for its purpose, which the authors attributed to the higher velocity of both hydrogen atoms and molecules from the heated filament.

A similar hydrogen beam generator, but for use in ultrahigh vacuum, is described by Sugaya and Kawabe (1991). Hydrogen was passed through a 4-mm-internal-diameter and 100-mm-long boron nitride tube which contained a tungsten spiral filament which was maintained at a temperature of about 1800 K. Though this is clearly not an intense source, the authors showed that it was adequate for its intended purpose of cleaning substrates.

Bornscheuer et al. (1993) used a different approach in producing a beam of hydrogen atoms in ultrahigh vacuum. Hydrogen was admitted through a stainless steel 'effusive dosing arrangement' for which no further details are given, but which is claimed to give a factor fifty improvement in atom intensity compared with random gas flow. Dissociation occurred on a 15-mm-diameter flat spiral of tungsten wire which was 250 µm in diameter and 145 mm long and heated to a temperature of about 2000 K. In order to prevent contamination from the wire and to reduce radiation from it reaching the experimental surface, the hydrogen beam was reflected from a pyrex glass plate which was cooled with liquid nitrogen to a temperature of 120 K in order to reduce impurities emitted by hydrogen atom bombardment. It was found that only about 20% of the atoms incident upon it reached the experimental surface.

A design for surface studies where collimation of a hydrogen beam is a disadvantage is described by Bermudez (1996), who directed molecular hydrogen from a molybdenum tube on to a 40-µm-thick tungsten foil heated to a temperature of 1750 K which was supported on 2.5-mm-diameter tantalum rods.

6.2.3 THERMAL DISSOCIATION OF HYDROGEN IN TUNGSTEN AND OTHER FURNACES

Traditionally, molecular hydrogen has been dissociated in a tube made of tungsten. However, as discussed later in this section, a tungsten alloy tube has been employed by Eisenstadt (1965). The tubes have been heated either by passing a large current through them, or by indirect heating from a filament, or by electron bombardment. Since good conductors of electricity are good conductors of heat, the essence of the designs is to enable the atomic beam to be emitted from the hottest part of the tube and some ingenious methods have been employed to do this. Earlier furnaces employed a slit source for the beam, but higher quality beams have been achieved with single tubes and especially with multichannel collimators.

Considerable uncertainty exists over the temperature of the furnace required to obtain highly dissociated hydrogen atoms and we quote the temperatures used in each experiment and the estimated total hydrogen pressure in the furnace when this has been stated. We have collected together in Section 12.5 the available experimental information on the degree of dissociation of hydrogen on hot tungsten surfaces but here we describe each furnace design in its chronological position.

It is important to realise that if the collimator is also the dissociator and the gas pressure is low enough that few collisions occur in the collimator, some molecules will pass through the tube without being dissociated. Tschersich and von Bonin (1998) and Schwarz-Selinger et al. (2000), both of whose apparatus is discussed later in this section, have observed a minimum in the axial intensity of respectively atomic hydrogen and deuterium due to this cause. In the former case, the mean free path at the pressure where the minimum was observed was many times the tube length. In the latter, the more pronounced minimum occurred when dissociation only took place near the tube exit, so more collisions between atoms and molecules would occur than is usual in a dissociation tube.

The first production of an atomic hydrogen beam from a tungsten furnace is attributed to Lamb and Retherford (1950). The principle of their method is still effectively in use, with the change that the tungsten component can now be purchased (Eibl et al. 1998). Lamb and Retherford's first furnace consisted of a tube about 20 mm long, 1.7 mm in external diameter and constructed from 100-µm-thick foil which was mounted in water-cooled molybdenum rods carrying the heating current, one of which contained the hydrogen inlet. When the tube was heated to a temperature of about 2500 K with a voltage-stabilised alternating current of 80 A, the mixed atomic and molecular beam emerged through a 200-μm × 1.5-mm slit in the centre of the tube. The operating temperature was a compromise between the degree of dissociation, which was estimated to be about 64%, and the life of the furnace.

Lamb and Retherford (1951) made some modifications to the furnace to reduce gas leakage and to improve cooling as shown in Figure 6.1. The tungsten dissociator was mounted in more massive 9.5-mm-diameter molybdenum blocks to which water cooled copper blocks were attached. Instead of constructing the tungsten tube from foil it was made from two 1.6-mm-diameter rods which were half ground away and in which 400-μm grooves were ground along their length. The exit beam slit was also ground into the rods before these were assembled to form the tube. This source was found to last for several months at a temperature of 2520 K. It required a heating current of 188 A at 2.31 V. The two halves of the tube were found to weld together in use.

A similar design and method of construction was used by Hendrie (1954), who, however, used tungsten collars in place of the molybdenum since he used the furnace at temperatures up to 3530 K, because the main purpose of his apparatus was to attempt to dissociate nitrogen. Hendrie's furnace is shown in Figure 6.2. It was 25 mm long 1.5 mm in internal and 2.5 mm in external diameter. His slit was 76 μm wide and 3.2 mm long. A current of 300 A was needed to reach a furnace temperature of 3300 K, at which temperature the furnace was found to be dimensionally stable. At temperatures above this, the life was reduced to about 20 min. Gas pressures within the furnace were in the range of about 100 Pa to 500 Pa. The dissociation as a function of furnace temperature was measured at temperatures between about 2000 K and 3000 K. The maximum measured dissociation was 86%. The dissociation percentage is in good agreement with the theoretical curve Hendrie presents.

Fite and Brackmann (1958) constructed their furnace of 25-μm-thick tungsten foil, which was wound into about six layers to form a tube 76 mm long and 5 mm

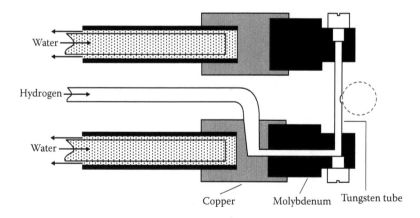

FIGURE 6.1 The furnace used by Lamb and Retherford to produce atomic hydrogen. (Adapted from Lamb WE and Retherford RC, 1951. Fine structure of the hydrogen atom. Part II. *Phys. Rev.* **81** 222–32. Copyright 1951 American Physical Society.)

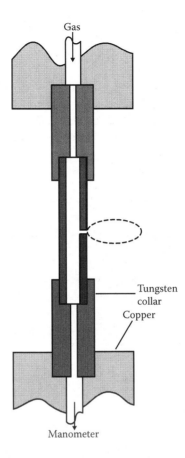

FIGURE 6.2   Hendrie's high-temperature furnace. (Adapted with permission from Hendrie JM, 1954. Dissociation energy of N_2. *J. Chem. Phys.* **22** 1503–7. Copyright 1954, American Institute of Physics.)

in diameter which was mounted in copper current-carrying blocks, one of which contained the hydrogen feed. The tube contained tungsten wadding to ensure that hydrogen emerging from the slit in the middle of the furnace underwent numerous collisions with hot tungsten.

A development of the furnace technique is reported by Kleinpoppen (1961) in which atoms were not emitted through a slit in the side of the tungsten tube, but from its end. The tube was heated to a temperature of 2800 K with electrons from a tungsten cathode at an energy of 2 keV to yield a degree of dissociation above 80%. The tube was mounted in a molybdenum block which was mounted successively in copper and brass blocks, through which hydrogen was admitted. A molybdenum radiation shield surrounded the assembly.

The collimation and intensity of the atom beam were improved by the use of a multichannel furnace heated to a temperature of 2650 K by Kleinpoppen et al. (1962). The construction of this furnace is discussed further in Section 10.2.

Eisenstadt (1965) used a tube of tungsten rhenium alloy containing 26% rhenium to dissociate hydrogen because the alloy is less brittle than pure tungsten, but requires electric arc discharge machining. It was suitable for use up to a temperature of 2400 K. The tube was about 25 mm long, 1.6 mm in diameter and with a wall thickness of 150 μm and was supported by two larger tubes of the same alloy at right angles, to reduce problems caused by thermal expansion. An alternating current of up to 60 A was passed through the assembly when the potential difference across it was less than 1.5 V. The hydrogen beam emerged through a 150-μm-diameter aperture in the centre of the smaller tube. The author gives further details of the construction in his paper, including the brazing of the components of the assembly.

Hils et al. (1966) briefly describe the use of a multichannel furnace to dissociate hydrogen which consisted of about 1000 channels within an area of 3 mm². They were heated to a temperature of 2700 K by electron bombardment to produce dissociation of about 85% to 90%. A diagram of their furnace is shown in Figure 6.3. Further details of the construction of this type of source are also given in Section 10.2.

An improvement in the method of construction of a furnace is given by Fluendy et al. (1967). A hole 3 mm in diameter was first drilled using a tungsten carbide drill through a 50-mm length of tungsten rod 10 mm in diameter. The diameter of the central 40 mm of the tube was then ground down to a diameter of 4.5 mm. Finally a 10-mm-long slit 50 μm wide was made in the tube wall by spark erosion. This tube was mounted in water-cooled copper blocks. Hydrogen was fed through the upper block of the vertically mounted tube. The lower one was mounted between spring-loaded jaws which allowed thermal expansion while maintaining good electrical and thermal contact. Temperatures in the range 2700 K to 3000 K were obtained when a current of about 500 A at 6 V was passed through the tube, giving a calculated dissociation of at least 60% with input pressures of about 67 Pa. The reported lifetime of the furnace tube was in excess of 300 h, the limit being caused by the evaporation of the tungsten.

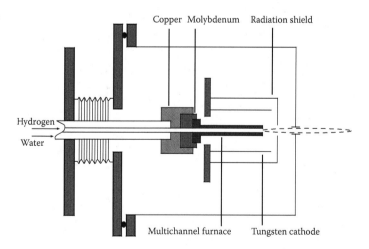

FIGURE 6.3   The multichannel furnace used by Hils et al. (Adapted from Hils D, Kleinpoppen H and Koschmieder H, 1966. Remeasurement of the total cross section for excitation of the hydrogen 2 ²S₁/₂ state by electron impact. *Proc. Phys. Soc.* **89** 35–40. Copyright IoP Publishing 1966.)

Long et al. (1968) rolled four layers of 13-μm-thick tungsten foil on a 4.8-mm-diameter mandrel to construct their tungsten foil furnace, which was heated to a temperature of 2500 K by 425 W of direct current power at 125 A. The hydrogen pressure within the furnace was estimated to be 3 Pa and the dissociation was 95 ± 1%.

Koschmieder et al. (1973) describe the use of a tungsten tube with $\Gamma = 33.3$ since it was 100 mm long, 3 mm in diameter and with a wall thickness of 50 μm to dissociate hydrogen. The tube was heated to a temperature of 2500 K by a tungsten heater which surrounded it. Water cooling both for the supports for the tungsten tube and molybdenum radiation shields was provided. The measured dissociation was 80%. This atomic hydrogen source is described in more detail by Koschmieder and Raible (1975). The furnace tube was only heated near the beam exit. It was constructed of 25-μm-thick tungsten foil to form the 100-mm-long and now 4-mm-diameter tube, so $\Gamma = 25$, as shown in Figure 6.4. The heating loop was a tungsten ribbon which was 80 mm long, 15 mm wide and 150 μm thick which was heat treated as it was bent

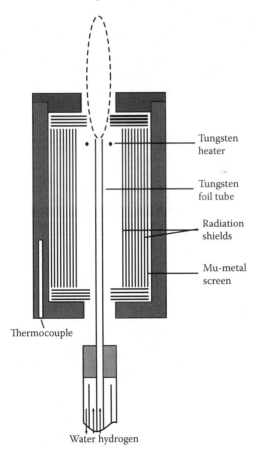

FIGURE 6.4   The single-tube atomic hydrogen source described by Koschmieder and Raible. (Adapted with permission from Koschmieder H and Raible V, 1975. Intense atomic-hydrogen beam source. *Rev. Sci. Instrum.* **46** 536–7. Copyright 1975, American Institute of Physics.)

into shape. It was heated by a current of 150 A at 10 V, which produced temperatures of 2400 K. This assembly was surrounded by seven concentric radiation shields made of tungsten or molybdenum except for a break for the heating current leads. A water-cooled jacket surrounded the shields and the furnace tube mount was also water cooled. Stress relieving of the furnace tube at a temperature of about 1000 K and its single mounting point reduced distortion in use. In order to reduce the magnetic field produced by the heating current to the heater ribbon, the current was supplied through concentric tubes insulated with mica (not shown). Additional mumetal shielding was used to screen the magnetic field even further. The dissociation of the beam was measured to be over 80% at a temperature of just over 2600 K.

The furnace tube employed by Balooch and Olander (1975) was fabricated by chemical vapour deposition. It was 3.7 mm in external and 3.2 mm in internal diameter and was 70 mm long. The beam-forming orifice in the centre of the tube was 1 mm in diameter and was constructed by spark erosion. Hydrogen at a pressure up to 800 Pa was admitted through the open end of the tube which was a press fit in a water-cooled copper block, which provided one of the current leads, as shown in Figure 6.5. The lower end of the tube was closed. It had six 760-μm-diameter tungsten leads wrapped around it through which a current of 150 A could be passed at a power of 750 W. The electrical resistance of the leads was lower than that of the tungsten tube, which at a maximum usable temperature of about 2500 K was hottest closest to the orifice. This method of passing current enabled the tube to slide through the wires when it expanded and contracted. The tungsten tube was surrounded by several radiation shields.

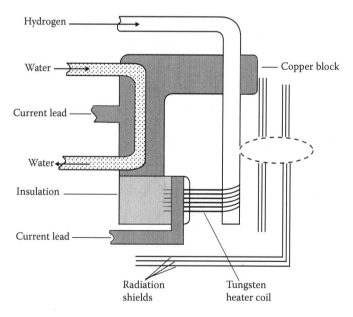

FIGURE 6.5 Balooch and Olander's atomic hydrogen furnace. (Adapted with permission from Balooch M and Olander DR, 1975. Reactions of modulated molecular beams with pyrolytic graphite. III. Hydrogen. *J. Chem. Phys.* **63** 4772–86. Copyright 1975, American Institute of Physics.)

Phaneuf et al. (1978) give brief details of their furnace for the dissociation of hydrogen which was made by rolling 25-μm-thick foil into a 6-mm-diameter tungsten tube. When this was heated with a current of 130 A, a temperature of 2350 K was reached, leading to measured dissociation of over 90%.

Shah and Gilbody (1981) succeeded in producing a directly heated single-tube collimating furnace essentially by enclosing it within a second tube. A diagram of this arrangement is shown in Figure 6.6. The inner tube was 20 mm long and 2 mm in diameter. It fitted at the gas entrance end into a molybdenum cap which was mounted in a water-cooled copper bar which provided both the heating current and the hydrogen gas inlet. The outer tube, which was 40 mm long and 6 mm in diameter, was made from a 25-μm-thick tungsten sheet. It was supported at one end by the same cap and at the other by a larger cap. The beam-exit end of the inner tube was held in the centre of the outer tube by a tungsten button which was 1 mm thick. When a current of 160 A was passed through the assembly, the inner tube was heated both by this current and by radiation from the outer tube.

A resistively heated tungsten tube was also employed to produce a beam of atomic hydrogen by Ludwig and Micklitz (1983). Their tube was 50 mm long and had an internal diameter of about 1 mm and a wall thickness of 500 μm. The beam emerged through a 500-μm-diameter aperture in the centre of the tube, which was mounted in tantalum tubes which were themselves connected to copper electrodes. One of these was at apparatus potential and provided the hydrogen input. The heating current of 150 A at 3 V produced a furnace temperature of about 2400 K, as measured with an optical pyrometer. The tungsten tube was surrounded by a radiation shield and enclosed in a water-cooled stainless steel cylinder.

Hildebrandt et al. (1984) describe a furnace made from a tungsten tube which was 4.5 mm in outside diameter, 700 μm in wall thickness and 38 mm long. The tube was mounted in copper blocks which were water cooled and carried the 500 A of current at 6 V needed to heat the tube. Gas was fed into one of the copper blocks and emerged through three apertures in the 2-mm centre section of the furnace. Their alternative microwave source of atomic hydrogen is discussed in Section 6.3.5.

Van Zyl and Gealy (1986) also recognised that a tube source of atomic hydrogen gave improved beam collimation compared with an orifice. Their furnace tube consisted of

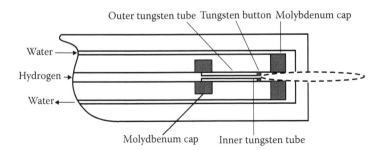

FIGURE 6.6 Shah and Gilbody's furnace design. (Adapted from Shah MB and Gilbody HB, 1981. Experimental study of the ionisation of atomic hydrogen by fast H[+] and He[2+] ions. *J. Phys. B: Atom. Molec. Phys.* **14** 2361–77. Copyright IoP Publishing 1981.)

annealed tungsten and was 2.57 mm in internal and 3.23 mm in external diameter. It was 76 mm long, giving a Γ of nearly 30. The last 35 mm of the tube were heated by electron bombardment using eight iridium loops as cathodes when each carried a current of 6 A with a potential drop of 6 V. The electron emission current was 700 mA at 950 V accelerating potential which provided 665 W of power, leading to a furnace temperature of 2500 K. A small tab inserted in the tube 10 mm from the open end and occupying 60% of the tube area was found to improve the dissociation on the axis of the tube which was 87% at a molecular hydrogen pressure of 13 Pa and a furnace temperature of 2400 K.

Shah and Gilbody's (1981) design was improved by Shah et al. (1987) to reduce the magnetic field in the interaction region produced by the heating current. Their design is shown in Figure 6.7, from which it can be seen that current passed through the inner tube, which was 2 mm in diameter and about 40 mm long and returned through the 4-mm-diameter outer tube of about one-half the length. The tubes were made from 25-μm-thick tungsten sheet. They were mounted in molybdenum blocks and were heated to a temperature of 2600 K by a current of 45 A. The inner tube contained tungsten wool. Although not discussed in the paper, it is presumably serving the same purpose as the wadding described by Fite and Brackmann (1958). However, it would prevent the tube acting as a collimator.

Bischler and Bertel (1993) describe an electron-beam heated tungsten furnace suitable for use in ultrahigh vacuum. A 50-mm-long tungsten tube, which was 1.6 mm in external and 600 μm in internal diameter, was heated to a temperature of about 1800 K by a 30-mA current of electrons at 560 V which was directed at the last 4 mm of the tube. Hydrogen entered this tube through a stainless-steel tube which was isolated from the tungsten tube with a machinable ceramic insulator. The beam produced passed through a water-cooled copper aperture and was estimated to contain

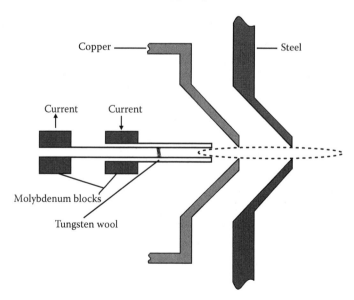

FIGURE 6.7 Shah et al.'s improved design. (Shah, MB et al., *J. Phys. B: Atom. Molec. Phys.* 20, 3501–14, 1987.)

about 45% atomic hydrogen. Eibl et al. (1998) briefly review the importance of atomic hydrogen in surface studies under ultrahigh vacuum conditions. The furnace they used for producing atomic hydrogen is based closely on the design of Bischler and Bertel. Eibl et al. mention that their tungsten tube was purchased. Boh et al. (1998) also used a tungsten furnace heated by an electron beam similar to that described by Bischler and Bertel but their 600-μm-internal-diameter tube was 30 mm long. They found that maximum dissociation was obtained at a furnace temperature of 2000 K when the pressure in the exit region of the tube was 2.5 mPa. A similar design of electron-heated tube is described by Tschersich and von Bonin (1998). Their tungsten tube was 64 mm long and 1 mm in internal diameter, hence $\Gamma = 64$. Electrons were emitted by a coiled tungsten filament surrounding the tube when the coil was heated by alternating current. The tube reached a temperature of 2600 K when the emission current was 60 mA at an accelerating voltage of 1600 V, leading to a degree of dissociation greater than 80%.

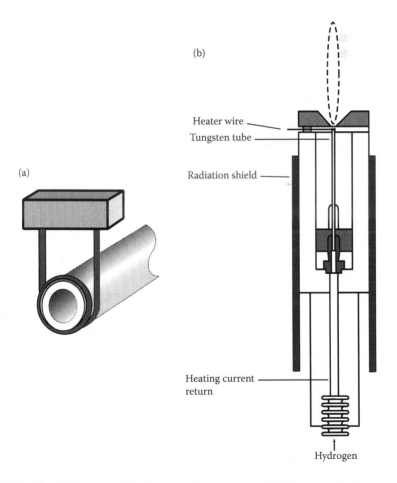

FIGURE 6.8 (a) Horn et al.'s hydrogen furnace tube. (b) The atomic beam system described by Horn et al. (Horn, A. et al., Vorrichtung zur Erzeugung von Radikalen und/oder Reaktionsprodukten German Patent DE 197 57 851 C 1, 1999.)

Horn et al. (1999) give a detailed description of a resistively heated tungsten tube for the production of a beam of atomic hydrogen in ultrahigh vacuum. The method of heating their dissociation tube is shown in Figure 6.8a. One of three similar arrangements used a 50-mm-long tube which was 1.6 mm in external diameter for a length of 10 mm and 1.2 mm in external diameter for the remainder of its length and with a bore of 600 μm. A 300-μm-diameter tungsten wire mounted in a metal block was wrapped one and a half times round the exit of the tube. Since the wire had a higher resistance than the tube, especially when it was hot, a current of 27 A at 1.4 V passed between the wire and the tube mounting enabled the open end of the tube to be heated to a temperature of over 2000 K. Hence the basic difference between this and other indirectly thermally heated tubes is that the tube itself is part of the heater circuit. The authors also suggest that the wire could be made either from rhenium or an alloy of tungsten with rhenium so it could be used in combination with the tungsten tube to form a thermocouple for the measurement of temperature. In this case, alternating current heating would allow the thermocouple electromotive force to be measured. Ceramic tubes or wires, such as tantalum carbide [TaC] or hafnium carbide [HfC] with conducting films, were suggested as alternatives to tungsten for higher temperatures. A copper radiation shield was recommended, as well as cooling of the supporting materials. Other suggestions include a second heater winding close to the first and a thinning of the tube near the windings in order to increase its resistance. An outline of the entire assembly containing the heated tube is shown in Figure 6.8b. Horn et al. only describe possible construction materials in general terms.

Schwarz-Selinger et al. (2000) briefly describe a 50-mm-long tungsten tube which was 1 mm internal diameter, hence $\Gamma = 50$. It was resistively heated at its tip. It was based on the design of Horn et al. (1999), described above, with an additional feature being the ability to vary the heated length of the tube and its temperature gradient. The effect of these variables on the angular distribution of atomic deuterium was studied.

Wnuk et al. (2007) used a commercially available source of atomic hydrogen that was dissociated in an iridium tube with an internal diameter of 2 mm. It was heated by electrons from a thoriated tungsten filament that were accelerated to 2 kV. The temperature of operation is not stated, but the degree of dissociation was estimated to be from 80% to 90%. The assembly was surrounded by a water-cooled copper block.

6.3 DISCHARGE SOURCES FOR ATOMIC HYDROGEN

6.3.1 INTRODUCTION

Several aspects of dissociating molecules in discharge tubes, particularly so for hydrogen, are far from clear cut. Probably the least controversial is that good cooling of the discharge tube leads to less recombination on the walls and so to a higher concentration of atoms in the extracted beam. This is based upon the work of Wood and Wise (1962). Numerous recipes for cleaning the discharge tube either before use or after re-use have been given. These range from simple dust removal by Spence and Steingraber (1988) to complex cleaning procedures involving a number of solutions by Yu-Jahnes et al. (1992) and also to coating the discharge tube which is described by many authors. Only Donnelly et al. (1992) appear to have made a systematic comparison, and conclude that there is no advantage in any one of a number of methods they employed. The other uncertainty is the effects of additives to the hydrogen to improve the dissociation degree. These vary from none (Huber et al. 1983) to being important by most authors. Water is usually favoured, but oxygen and others have been tried. Where authors have given details, we include these in view of their obvious importance.

We divide the techniques for forming beams of atomic hydrogen by dissociation in discharge tubes into four groups. Both Wood's tubes, discussed in Section 6.3.2, and arc sources, which are discussed in Section 6.3.3, have electrodes in the gas. The latter operate at much higher gas pressures than Wood's tubes. Radiofrequency and microwave discharges, which are respectively discussed in Sections 6.3.4 and 6.3.5, mainly differ in their frequency of operation. A general introduction to these is given by Popov (1995). The partially ionised gas is usually referred to as a plasma. With the exception of the arc sources, the discharges are also known as glow discharges.

6.3.2   WOOD'S TUBES

The production of atomic hydrogen in a discharge also has a long history. Wood (1920) was concerned with the observation of the Balmer series of spectral lines from atomic hydrogen since, whereas 32 members of the series had been observed in the solar spectrum, only 12 had been observed in the laboratory. Wood used a discharge tube about 2 m long and a bore of 7 mm which was bent into three lengths for convenience. Spectroscopic observations were only made in the central portion of this tube, where the Balmer lines appeared strongest. A discharge was struck between aluminium foil electrodes at the ends of the tube, which was continuously pumped from one end with gas admitted at the other. An AC potential of 25 kV enabled a current of 200 mA to pass through the discharge. Though Wood did not form an atomic beam, such atomic beam sources are usually called Wood's tubes. He also noted that the presence of water vapour or oxygen also assisted the atomic spectrum. Wood (1921) gives additional details of the construction and operation of the tube. Wood (1922a) explains that the presence of the Balmer series from atomic hydrogen was only in the centre portion of the tube because atomic hydrogen recombined on the aluminium electrodes. Changes in the spectrum and the wall becoming locally hot are attributed by Wood (1922b) to recombination taking place on contaminated parts of the walls of the tube which he called infected spots. He describes the recombination as a catalytic action, so that its prevention was due to poisoning of the catalyst. Wood (1922a) also pumped atomic hydrogen from his discharge tube, thus paving the way for producing a beam of atomic hydrogen.

Wood's studies led Bonhoeffer (1924) to undertake further experiments with atomic hydrogen when pumped outside the discharge region. He used the same design of Wood's tube and undertook numerous studies of the reducing effect of atomic hydrogen. He also investigated its recombination on various metals and compounds by coating a thermometer bulb with them and noting the indicated temperature increase. Of particular interest are the experiments with a Crookes' radiometer which was made of a glass slide silvered on diagonally opposite faces. Its rapid rotation in the path of the atomic hydrogen indicated the presence in the gas flow of atoms which recombined on the silvered surfaces.

Phipps and Taylor (1927) employed a Wood's tube that was 3.5 m long which was constructed of 18-mm pyrex glass tubing. Electrodes were of aluminium cylinders and each was contained in a U-shaped portion of the discharge tube as shown in Figure 6.9 to prevent small metallic particles reaching the centre of the tube. The tube was operated at 25-kV alternating current at a hydrogen pressure of 16 Pa. The beam emerged through a glass slit which was 75 μm wide and 3 mm long. Its construction is described in Section 3.4.3. In fact a differentially pumped triple slit system was used.

Johnson (1928) extracted a beam of hydrogen atoms from the centre of a water-cooled Wood's tube through a slit which was 130 μm wide and 6 mm long. The method of slit construction is also described in Section 3.4.3. The discharge was operated at an alternating voltage of 10 kV when a current of 45 mA flowed when the hydrogen pressure was about 20 Pa. It emerged into a pumped region where the pressure was maintained between 13 mPa and 1 mPa.

Wartenberg and Schultze (1930) investigated various coatings to prevent the recombination of atomic hydrogen and found that washing the glass discharge tube with either orthophosphoric acid [H_3PO_4] or metaphosphoric acid [HPO_3] was successful. They also investigated the effect of water vapour and oxygen on the dissociation and showed that more than 2% water vapour does not increase the dissociation appreciably. Oxygen was believed to be effective because it reacted with hydrogen in the discharge tube to form water vapour.

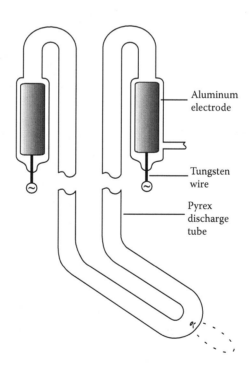

FIGURE 6.9 The Wood's tube used by Phipps and Taylor. (Adapted from Phipps TE and Taylor JB, 1927. The magnetic moment of the hydrogen atom. *Phys. Rev.* **29** 309–20. Copyright 1927 American Physical Society.)

A Wood's tube for obtaining an atomic hydrogen beam was employed by Rabi et al. (1934). Their tube operated at a pressure between 27 Pa and 40 Pa. A 200-mm-long pyrex glass side tube led the atomic hydrogen to a slit which was 20 µm wide and 4 mm long. An improvement to Rabi et al.'s apparatus was made by Kellogg et al. (1936) whose Wood's tube, shown in Figure 6.10, was angled and eccentrically mounted to enable adjustment of the beam position. The tube was arranged so that the exit slit, which was 30 µm wide and 2 mm long, was very close to the water-cooled discharge. The concentration of hydrogen atoms in the beam was increased by this means from the range of from 10% to 20% to between 70% to 90% with a discharge tube gas pressure of about 130 Pa.

Some useful advice on the operation of a Wood's tube is given by Poole (1937), who also reviews earlier work. The correct amount of water was introduced into the discharge tube by bubbling the admitted hydrogen through water at atmospheric temperature and pressure. After the discharge tube was cleaned and coated with meta-phosphoric acid [HPO_3] by melting it on the surface, the most stable production of atomic hydrogen was achieved. A Wood's tube was also used to obtain atomic hydrogen by Wittke and Dicke (1956). Of particular interest is the coating of the pyrex glass tube used to transport the hydrogen from the discharge region. This was a mixture of dichlorodimethylsilane [$C_2H_6Cl_2Si$] and methyltrichlorosilane [CH_3Cl_3Si].

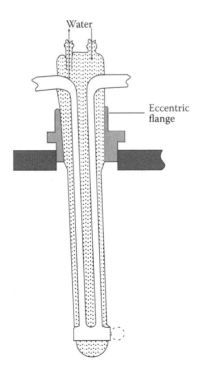

FIGURE 6.10 Kellogg et al.'s Wood's tube. (Adapted from Kellogg JMB, Rabi II and Zacharias JR, 1936. The gyromagnetic properties of the hydrogens. *Phys. Rev.* **50** 472–81. Copyright 1936 American Physical Society.)

What they called a 'Buder' discharge tube was used by Ad'yasevich et al. (1965) to produce atomic hydrogen which was 80% to 90% dissociated. It was constructed of 12-mm-diameter molybdenum glass that was about 3 m long. A current of 200 mA was produced by an alternating voltage of 5 kV at 50 Hz that was applied to aluminium electrodes. The design enabled the discharge tube to be cooled with either air, water or liquid nitrogen. It was cleaned with a 10% solution of potassium hydroxide and either heated in chromic acid or 1% hydrofluoric acid solutions. The atomic hydrogen was collimated by a glass capillary array that is described in Section 10.4.6. Potassium tetraborate was used to coat both the discharge tube and the array.

A detailed description of a Wood's tube used to obtain a beam of atomic hydrogen is given by Walker and St. John (1974). They used a 20-mm-diameter pyrex glass tube which was 2 m long and operated at 13 Pa. The hydrogen contained 2% water vapour obtained from triple-distilled water. An improvement on Wood's original design was to reduce the tube diameters to 6.5 mm for a 100-mm region near each of the electrodes, which were 25 mm in diameter and 76 mm long and made from aluminium as shown in Figure 6.11. These constrictions reduced the hydrogen atom recombination at the electrodes. The tube was operated at a direct current of from 50 mA to 70 mA at up to 10 kV and was forced air cooled. Before use the tube was cleaned with hot chromic acid and rinsed with distilled water followed by ethanol. The beam was extracted from a side tube that was 1.5 m from the gas inlet which was close to one of the electrodes. This side tube was 340 mm long and 20 mm in diameter. The beam emerged through a 10-mm-diameter nozzle that was 35 mm long.

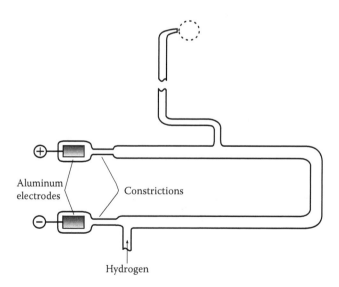

FIGURE 6.11   The Wood's tube developed by Walker and St. John. (Adapted with permission from Walker JD and St. John RM, 1974. Design of a high density atomic hydrogen source and determination of Balmer cross sections. *J. Chem. Phys.* **61** 2394–407. Copyright 1974, American Institute of Physics.)

A development of the Wood's tube used by Walker and St. John (1974) is described by Hood et al. (1978), who used a 1.7-m-long pyrex glass and fused silica discharge tube with a 12-mm bore but which was constricted to 5 mm near the electrodes. Thus it was slightly smaller than that used by Walker and St. John. The hydrogen beam was emitted through a 1-mm-diameter aperture near the centre of the discharge tube. The tube was operated at potentials between 3 kV and 4 kV at currents between 25 mA and 50 mA. It was found that up to 2% water vapour initially added to the discharge improved the atomic hydrogen production.

Huber et al. (1983) describe a means of making a compact Wood's tube by forming the 2.20-m-long pyrex glass tube which was 15 mm in internal diameter into a double helix as shown in Figure 6.12. The exit tube in the centre of the discharge tube was 80 mm long and terminated in a poly (tetrafluoro ethylene) (PTFE) tube 4 mm long and either 1 mm or 1.6 mm in diameter. The discharge tube was surrounded by a cooling jacket and was within the vacuum envelope. It was cooled either to a temperature of 265 K or of 210 K in order to improve the degree of hydrogen dissociation, which was just under 70% at the lower temperature. The authors found no advantage in the usual practice of adding water vapour to the admitted

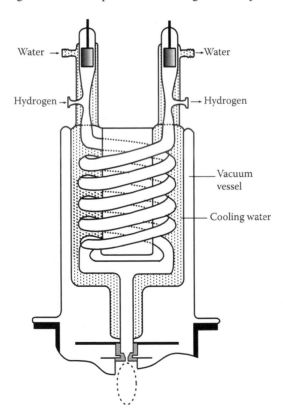

FIGURE 6.12 The double helix Wood's tube of Huber et al. (Adapted from Huber BA, Bumbel A and Wiesemann K, 1983. A high-density effusive target of atomic hydrogen. *J. Phys. E: Sci. Instrum.* **16** 145–50. Copyright IoP Publishing 1983.)

hydrogen, which they believe to be due to its freezing on the cooled surfaces. The discharge was operated at 1.8 kV when its current was 100 mA and the pressure in the tube 10 Pa. Before use, the tube was cleaned with chromic acid, ethanol and distilled water and then rinsed with phosphoric acid.

Thus over sixty years after its development, the Wood's tube was still an option used to produce an intense atomic hydrogen beam. It has the advantage of requiring only inexpensive power supplies, so it may continue to find use where such a source is required.

6.3.3 ARC SOURCES

Before discussing electrodeless discharges, we next consider an entirely different type of discharge, namely, the arc heater. Because of the higher operating pressure compared with other discharges, it is more suitable for producing supersonic beams, but collisionless beams could also be extracted.

An arc source is described by Young et al. (1969) in connection with producing a supersonic beam in helium and argon, but the principle remains the same if hydrogen is to be dissociated, as discussed in the following paragraph. The authors employed a 3.2-mm-diameter 2% thoriated tungsten rod as the anode and the arc was struck between this and a copper anode. Both electrodes and the electrical supply leads were water cooled and the mounting of the cathode on bellows enabled it to be adjusted from outside the vacuum system. An axial magnetic field of 60 mT was produced by a coil external to the vacuum system that surrounded the arc region. This assisted maintaining a stable arc as did injecting the gas tangentially through two inlet ports. The beam emerged through an orifice in the anode and diameters in the range 560 μm to 2.0 mm were tried. The gap between the cathode and the anode was in the range 200 μm to 400 μm. The arc was initiated by bringing the electrodes into contact. Arc currents were from 30 A to 150 A at arc potential differences from 16 V to 45 V.

Way et al. (1976) developed the above arc source to produce a beam of atomic hydrogen. One improvement was that the coil producing the magnetic field was water cooled. Modifications were also made to the construction of the nozzle, which had to be replaced after every use, but which could be operated continuously for over 12 hours. It was not found possible to initiate the arc in hydrogen, so argon at a pressure of about 47 kPa was used initially. The authors give extensive details of the rather difficult procedure for obtaining a stable arc in hydrogen and further references. The best operating conditions were obtained at an arc current of 125 A at 38 V and a hydrogen pressure of 2.4 kPa.

Modifications to the source are reported by Garvey and Kuppermann (1986) to improve its stability and reliability. It is shown in Figure 6.13. They used a 2% thoriated tungsten cathode which was mounted in a water-cooled stainless steel tube which could be adjusted in position by a bellows mounting allowing 19 mm of movement. The anode was made of a 3-mm-thick disk of the same material through which was a 1.2-mm-diameter hole. This disk was mounted in a water-cooled brass anode assembly with screwed copper retaining rings. It required replacement after several periods of 5-h operation. The hydrogen inlets were almost tangential since the resulting swirling motion was considered to improve the stability of the arc. This was also improved by the insulating cylinder that prevented the arc occurring away from the anode disk. The authors give details of the mounting and adjustment of their arc-discharge source as well as the operating conditions which enabled a stable discharge to be obtained in hydrogen. This was when the gap between the cathode tip and the surface of the anode disk was 500 μm. A movable water-cooled coil (not shown) centred on the nozzle produced an axial magnetic field of 100 mT. The discharge had to be initiated in argon at a pressure of 45 kPa by means of a high voltage pulse between the anode and cathode. Once a stable arc had been obtained, hydrogen was gradually mixed in with the argon until its pressure was about 20 kPa, when the argon flow was slowly turned off. The arc operated stably at a pressure of 7.3 kPa at a current of 100 A with a voltage drop of 30 V. The measured dissociation of the hydrogen was 95%.

A beam of deuterium atoms was produced from an arc operating at a direct current of 100 A and voltage of 21 V by Götting et al. (1986). Their arc source is shown in Figure 6.14. The cathode was made of thoriated tungsten which was mounted in a water-cooled shaft. The 1.6-mm-diameter circular anode was made of tungsten which was brazed into a water-cooled copper disk. The distance between the anode and the cathode was adjustable from outside the vacuum system and the arc was usually operated when this was 500 μm. A magnetic field of 100 mT parallel to the axis of the source was provided by permanent magnets which were mounted between soft iron cylindrical yokes. In operation, the arc was initially struck in argon which was changed to deuterium after a few minutes. The discharge could be operated continuously for up to 12 hours. After 50 hours the orifice required replacing. When the gas pressure was 6 kPa the measured deuterium dissociation was between 50% and 99%, with maximum dissociation usually being obtained with the smallest distance between the cathode and anode and the largest orifice diameters.

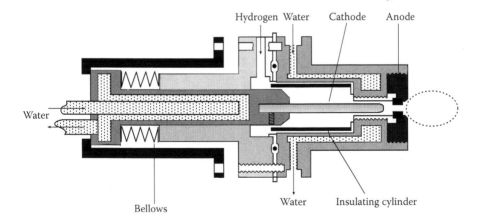

FIGURE 6.13   Garvey and Kuppermann's arc source of atomic hydrogen. (Adapted with permission from Garvey JF and Kuppermann A, 1986. Design and operation of a stable intense high-temperature arc-discharge source of hydrogen atoms and metastable trihydrogen molecules. *Rev. Sci. Instrum.* **57** 1061–5. Copyright 1986, American Institute of Physics.)

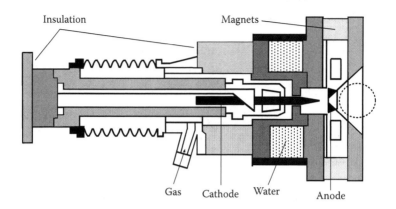

Insulation   Magnets

Gas   Cathode   Water   Anode

FIGURE 6.14   The arc source of deuterium atoms used by Götting et al. (Adapted with permission from Götting R, Mayne HR and Toennies JP, 1986. Molecular beam scattering measurements of differential cross sections for D+H$_2$($\nu = 0$) $\rightarrow$ HD+H at E$_{c.m.}$ = 1.5 eV. *J. Chem. Phys.* **85** 6396–419. Copyright 1986, American Institute of Physics.)

An arc source for the production of a beam of atomic hydrogen is described by Samano et al. (1993). The arc was struck between a water-cooled thoriated tungsten cathode with a sharp tip and the anode which was a water-cooled hollow molybdenum cylinder. The distance between the anode and cathode was adjustable from outside the vacuum system by a micrometer screw adjustment of the cathode, which passed through a vacuum seal, over a range of 8 mm. A short solenoid outside the vacuum system produced an axial magnetic field of 23 mT in the arc region. A diagram of this source is shown in Figure 6.15. A fused silica tube surrounded the discharge region and a pyrex glass tube formed the vacuum envelope, which enabled the arc to be observed. The atomic beam was emitted through an aperture 400 μm in diameter and 1.2 mm long. When running, the arc current was about 15 A at a voltage of about 105 V with a hydrogen pressure of at least 1.6 kPa. The arc temperature was typically 8700 K. The authors give details of how the arc is initiated by first obtaining a glow discharge and adjusting both the hydrogen pressure and applied voltage until an arc discharge is obtained and then turning on the solenoid.

In conclusion, the high powers of several kilowatts needed to run arcs require cooling of both electrodes. Axial magnetic fields of up to 100 mT are necessary and obtaining a stable arc requires careful adjustment, usually following initiation in an inert gas.

6.3.4   Radiofrequency Discharges

Discharge sources of atomic hydrogen operating at continuous radio frequencies have a nearly 60-year history, but they are largely superseded by the microwave discharge sources that are discussed in the next section. We mention any cooling of beams substantially below laboratory temperature, but retain the chronological order.

Kurt and Phipps (1929) describe a ring discharge which they used to dissociate hydrogen as well as oxygen. Since their main studies were with oxygen, we defer discussion to Section 6.6.

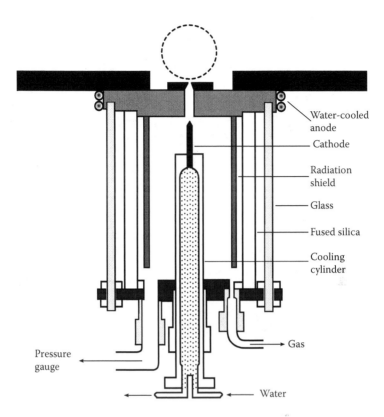

Water-cooled anode

Cathode

Radiation shield

Glass

Fused silica

Cooling cylinder

Gas

Pressure gauge

Water

FIGURE 6.15   The arc discharge source developed by Samano et al. (Adapted with permission from Samano EC, Carr WE, Seidl M and Lee BS, 1993. An arc discharge hydrogen atom source. *Rev. Sci. Instrum.* **64** 2746–52. Copyright 1993, American Institute of Physics.)

Prodell and Kusch (1957) used a radiofrequency discharge operating at about 4 MHz which was coupled into the discharge tube by mercury electrodes which surrounded the vertical water-cooled discharge tube as shown in Figure 6.16. The atomic beam, in this case tritium, was emitted through a slit of unspecified dimensions in the side tube. We refer to this paper again in Section 7.2 for details of the tritium recirculation.

Brackmann and Fite (1961) produced atomic hydrogen using a radiofrequency generator operating at a frequency of 13 MHz. Hydrogen flowed through a 6-mm-external-diameter pyrex glass discharge tube which was surrounded by a cooling jacket which was about 25 mm in diameter. Either liquid nitrogen or carbon tetrachloride (tetrachloromethane) [CCl_4], chosen for its low dielectric constant, flowed through the jacket, since the radiofrequency power was capacitively coupled with electrodes being attached to the outside of the cooling jacket.

A 500-W oscillator operating at 20 MHz was employed by Rudin et al. (1961) to form a deuterium beam. The air-cooled glass discharge tube was about 35 mm in diameter and 160 mm long around which the six-turn RF coil was wound. This part of the apparatus is shown in Figure 6.17. The discharge operated at deuterium

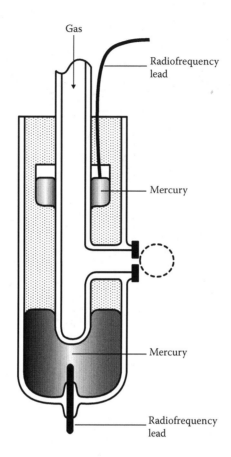

FIGURE 6.16 The discharge tube used by Prodell and Kusch for tritium. (Adapted from Prodell AG and Kusch P, 1957. Hyperfine structure of tritium in the ground state. *Phys. Rev.* **106** 87–9. Copyright 1957 American Physical Society.)

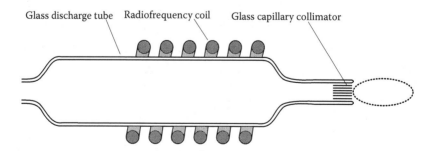

FIGURE 6.17 Rudin et al.'s radiofrequency source. (Adapted from Rudin H, Striebel HR, Baumgartner E, Brown L and Huber P, 1961. Eine Quelle polarisierter Deuteronen und Nachweis der Polarisation durch die (d, T)-Reaktion. *Helv. Phys. Acta* **34** 58–84. With permission from Springer Science & Business Media B V.)

pressures between 67 Pa and 130 Pa. The beam emerged through a bundle of 50 glass tubes 1 mm in internal diameter, 110 μm in wall thickness and 10 mm long, which were packed into the 10-mm-diameter exit of the discharge tube. Dissociation of 50% was achieved.

The following two authors describe the same source. We have reversed the chronological order of presentation in favour of increasing detail. Stafford et al. (1962) dissociated hydrogen at a pressure of about 40 Pa in a 20-MHz radio-frequency discharge. After an operating period of up to five hours, dissociation of over 70% was reported to be reliably obtained with untreated glass surfaces, provided the discharge tube was cooled with an air blast. The hydrogen was collimated by a pyrex glass collimator that consisted of about 1000 tubes that were 100 μm in diameter and 2 mm long, so $\Gamma = 20$. These formed an 8-mm-diameter circular source with a 35% open area. Craddock (1961) describes the source in more detail and indicates that 400 W of RF power were available to dissociate hydrogen in the pyrex glass discharge tube. He gives further details of the construction of the array, which are included in Section 10.4.4. Coating of the collimator with methyltrichlorosilane [CH_3Cl_3Si] and air cooling of the discharge tube produced measured dissociations of hydrogen of typically 90%, depending upon the pressure and radiofrequency power.

Clausnitzer (1963) describes a U-shaped discharge tube, 12 mm in diameter and about 1 m long, which was cooled by a stream of air. Hydrogen was admitted to both arms of this tube and the beam emerged through a short cylindrical aperture 2.6 mm in diameter and about the same length. This part of the apparatus is shown in Figure 6.18. His radiofrequency oscillator was capable of delivering 1 kW of power at 30 MHz which was fed into the discharge at each end of the tube by capacitative coupling through 50-mm-long electrodes, which were not in contact with it. A capacitor in series with these electrodes provided some tuning, but Clausnitzer noted some difficulty in matching the oscillator to the discharge, which was operated at a pressure of about 67 Pa. Before use, the discharge tube was thoroughly cleaned and filled completely with a 30% solution of hydrofluoric acid to remove microscopic rough spots on the glass and rinsed with distilled water. Initially about 10% of water vapour was added to the hydrogen, but this was found not to be necessary after some weeks of operation.

Vályi (1968) used a radiofrequency generator operating at 23 MHz and providing 360 W of power which he coupled into his 12-mm-diameter U-shaped discharge tube using two electrodes spaced 500 mm apart. The hydrogen pressure in the tube was in the range 40 Pa to 53 Pa. It was cooled with an air blast. The atomic beam left the discharge tube through a 4-mm-diameter array of capillary tubes each about 100 μm in diameter and 600 μm long. Their short length was apparently due to Vályi erroneously thinking that the mean free path of the atoms had to be greater than the tube length (Section 1.1). The discharge tube had to be repeatedly cleaned in hydrofluoric acid in order to obtain stable operation. The beam source is shown in Figure 6.19.

A similar type of radiofrequency source was employed by Slabospitskii et al. (1970), in which a 20-mm-diameter discharge tube was operated with 100 W of power at 140 MHz and was also cooled with compressed air. The gas pressure used was 40 Pa. The beam was collimated with an array of 600-μm-diameter glass capillary tubes and lengths of 500 μm, 900 μm and 2 mm were tried. They were cemented in place with a mixture of liquid glass and talcum. Before use the arrays and the

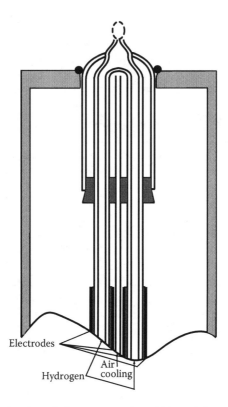

FIGURE 6.18   Clausnitzer's radiofrequency source. (Adapted from Clausnitzer G, 1963. A source of polarized protons. *Nucl. Instrum. Methods* **23** 309–24. With permission from Elsevier. Copyright 1963.)

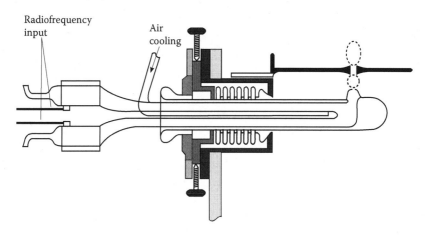

FIGURE 6.19   Válvi's source of atomic hydrogen. (Adapted from Válvi L, 1968. A source of polarized proton and deuteron beams. *Nucl. Instrum. Methods* **58** 21–8. With permission from Elsevier. Copyright 1968.)

discharge tube were coated with dichlorodimethylsilane [$C_2H_6Cl_2Si$] by flushing its vapour through the system. The silane coating reduced recombination of the atoms on surfaces and ensured stable operation, even with dry deuterium.

Wilsch (1972) showed how a cooled beam of atomic hydrogen could be produced from a radiofrequency source operating at 27 MHz by connecting a liquid nitrogen reservoir to a metal shoe surrounding the beam orifice with a thick flexible copper cable. The shoe surrounded the last 20 mm of the discharge tube which was 10 mm in diameter. Wilsch indicated that the addition of 5% to 10% of oxygen produced optimum dissociation. He preferred this to water vapour, since otherwise condensation could occur in the cooled section.

A radiofrequency oscillator operating at 27 MHz with a power output of 650 W which fed power to an eight turn coil wound on a ceramic former is described by Schumacher et al. (1975). Hydrogen was fed into a 50-mm-diameter glass tube that was a seal into a ceramic holder, to enable water cooling as shown in Figure 6.20. The hydrogen beam emerged through a copper nozzle that was 3 mm in diameter and 5 mm long which was cooled by flowing liquid nitrogen. The authors noted that recombination of atomic hydrogen on the copper surface 'does not seem to be a serious problem'.

Sepehrad et al. (1979) describe their cleaning procedure for a fused silica or pyrex glass tube to reduce recombination of atomic hydrogen. The first stage is to leave it for about 15 h in contact with a 10-molar aqueous solution of sodium hydroxide. Distilled water was then used to wash the tube several times after which it was left in contact with 10-molar nitric acid for 15 h. Finally it was again washed several times with distilled water and dried. The tube did not require recleaning following exposure to air.

Toennies et al. (1979) used a radiofrequency generator operating at a frequency of 27.12 MHz to dissociate hydrogen which was fed at a pressure in the range 40 Pa to 67 Pa through the discharge region in a water-cooled pyrex glass tube. The radiofrequency coil consisted of five turns that were 150 mm long which dissipated power of about 50 W. A copper tube surrounding the resonator provided a radiofrequency shield, as shown in Figure 6.21a. The beam left the discharge region through a 5-mm-long aluminium channel that was 500 μm wide and 5 mm high. The channel was surrounded by a copper annulus which was cooled with liquid nitrogen. This so-called moderator is shown in Figure 6.21b. Aluminium was found to be a superior material to pyrex glass, boron nitride and copper for cooling the atoms with minimum recombination. The authors do not indicate that cleaning of the discharge tube before use was necessary, but after 150 to 200 hours use, it was cleaned in a 40% to 50% solution of hydrofluoric acid.

The radiofrequency cavity operating at 35 MHz employed by Slevin and Stirling (1981) to dissociate atomic hydrogen is shown in Figure 6.22. Gas flowed at a pressure of about 20 Pa through a 240-mm-long water-cooled pyrex glass discharge tube which was 18 mm in internal diameter. This was surrounded by a coaxial copper cavity containing a 12-turn coil. Power of about 25 W was coupled into this with a single turn (not shown). The beam, which was 95% dissociated, emerged through a 1-mm-diameter tube of unstated length. The authors state that a stable beam was produced over periods of several thousand hours, provided that the discharge tube was successively cleaned with hot chromic acid, acetone, hydrofluoric acid and distilled water and precautions were taken to prevent its contamination.

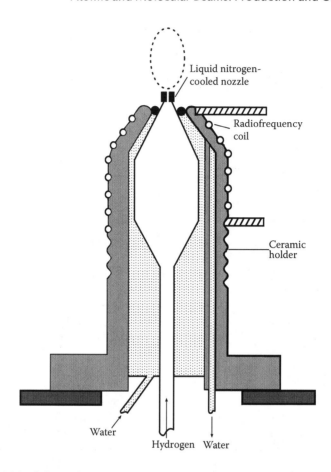

Liquid nitrogen-
cooled nozzle

Radiofrequency
coil

Ceramic
holder

Water

Hydrogen Water

FIGURE 6.20 Schumacher et al.'s atomic hydrogen beam source. (Adapted from Schumacher W, Barz F, Dreesen E, Hammon W, Hansen HH, Penselin S and Scholzen A, 1975. The polarized proton and deuteron source for the Bonn isochronous cyclotron. *Nucl. Instrum. Methods* **127** 157–62. With permission from Elsevier. Copyright 1975.)

Harvey (1982) describes the production of atomic hydrogen in a 40-mm-diameter pyrex glass bulb by excitation at 30 MHz by a 25-W radiofrequency oscillator. Of particular interest is the collimation of the beam by a capillary array 3 mm in diameter containing tubes 50 μm in diameter and 2 mm long, hence with $\Gamma = 40$. The array had a transparency of 50%. The dissociation was estimated to be 50%.

In the design briefly described by Hershcovitch et al. (1987), hydrogen was dissociated in a pulsed radiofrequency discharge and passed through a nozzle cooled to a temperature of 80 K. It then flowed through a 3-mm-internal diameter PTFE tube which was 20 mm long and maintained at a temperature above 80 K to minimise recombination. After passing a thermal isolation gap of 300 μm it was cooled to liquid helium temperatures in a tube of the same dimensions.

One of the problems of using a radiofrequency generator coupled to the discharge coil with a coaxial cable is the difficulty of tuning because the discharge varies the

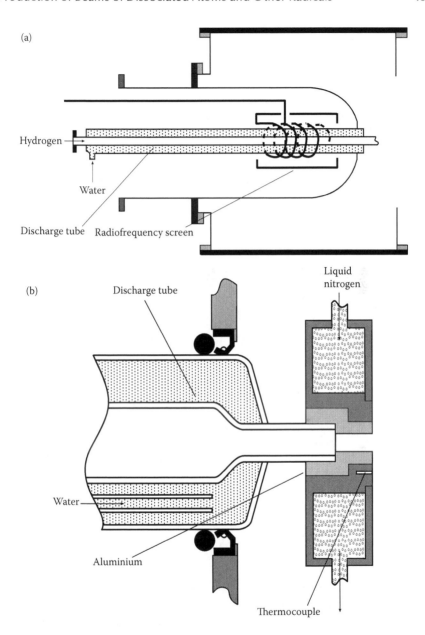

FIGURE 6.21   (a) The hydrogen atomic beam source used by Toennies et al. (b) Toennies et al.'s moderator used to cool the atomic beam. (Adapted with permission from Toennies JP, Welz W and Wolf G, 1979. Molecular beam scattering studies of orbiting resonances and the determination of van der Waals potentials for H-Ne, Ar, Kr, and Xe and for H_2-Ar, Kr, and Xe. *J. Chem. Phys.* **71** 614–42. Copyright 1979, American Institute of Physics.)

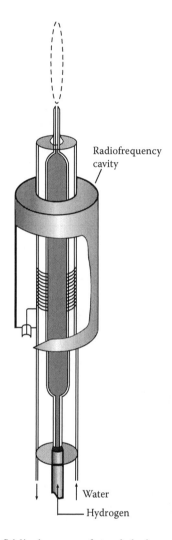

Radiofrequency
cavity

Water

Hydrogen

FIGURE 6.22   Slevin and Stirling's source of atomic hydrogen. (Adapted with permission from Slevin J and Stirling W, 1981. Radio frequency atomic hydrogen beam source. *Rev. Sci. Instrum.* **52** 1780–2. Copyright 1981, American Institute of Physics.)

impedance, as can be seen, for example, in the complicated matching system employed by Sibener et al. (1980) discussed in Section 6.6. Hodgson and Haasz's (1991) approach was to construct a sheet aluminium cavity of dimensions 80 mm × 100 mm × 400 mm which surrounded their pyrex glass discharge tube which was 50 mm in diameter and 240 mm long. The authors preferred a large ratio of tube volume to surface to reduce the effects of recombination on the glass surface. A tetrode power oscillator which was located within the cavity delivered up to 300 W of power at 200 MHz in transverse electromagnetic mode via quarter-wave aluminium strip lines. Feedback to the control grid of the oscillator was provided by a similar strip connected to the wall of the cavity. The oscillator was supplied with a variable high voltage from an unregulated transformer, rectifier and filter combination which was supplied by an autotransformer to regulate the oscillator power. It required no impedance matching or periodic adjustment. The assembly, shown in Figure 6.23, was air cooled. The beam emerged through a 1-mm-diameter tube which was 5 mm long. No additives were used in the hydrogen, which probably explains the relatively low measured dissociation of under 60%. Before use the discharge tube was, however, cleaned with acetone, followed by rinsing in distilled water, dilute hydrofluoric acid and distilled water again.

Donnelly et al. (1992) studied the performance of the source described by Slevin and Stirling (1981), discussed above. They confirmed the importance of coating the discharge tube with orthophosphoric acid as described by Ding et al. (1977). They also found that a palladium filter used to improve the purity of the admitted hydrogen increased the dissociation from about 65% to about 90%, even if 99.999% pure

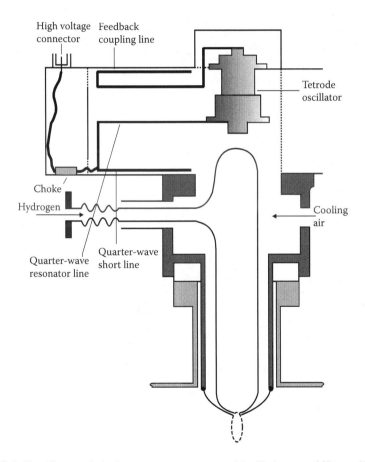

FIGURE 6.23 The atomic hydrogen source constructed by Hodgson and Haasz. (Adapted with permission Hodgson JAB and Haasz AA, 1991. Compact radio-frequency glow-discharge atomic hydrogen beam source. *Rev. Sci. Instrum.* **62** 96–9. Copyright 1991, American Institute of Physics.)

hydrogen was used. The dissociation was fairly constant over more than a factor two change in gas pressure around 84 Pa.

Wise et al. (1993) showed that the recombination of atomic hydrogen on aluminium, when it was treated with concentrated nitric acid, was insignificant. This is to be compared with the aluminium electrodes in Wood's tubes, on which the protective oxide layer is presumably removed by the discharge.

Akulov et al. (1997) returned to the use both of a lower frequency oscillator and to electrodes rather than a coil to produce the discharge. The oscillator was capable of producing 300 W of power at 1 MHz, which was fed directly to cylindrical electrodes surrounding a 500-mm-long discharge tube made of molybdenum glass that had an internal diameter of 10 mm and 1 mm wall thickness. The authors state that this glass has a low recombination coefficient for atomic hydrogen, which emerged through a 100-μm-diameter PTFE tube that was 2 mm long. Hence at this point in their apparatus a beam was formed but the atoms were led to a mass spectrometer through an elbowed PTFE tube that was 250 mm long and 3 mm in internal diameter. Molecular hydrogen was admitted to the discharge tube at pressures in the range from 13 Pa to 130 Pa and power levels of up to 70 W were used, when almost complete dissociation of the hydrogen molecules was believed to be obtained.

6.3.5 MICROWAVE DISCHARGES

A general introduction to the subject of plasmas including those produced by microwaves, though not for atomic beam production, is given by Rosenkranz and Bettmer (2000). The retention of chronological order in this section means that there is some mixture of slotted radiator and surface wave designs with other designs using microwave cavities.

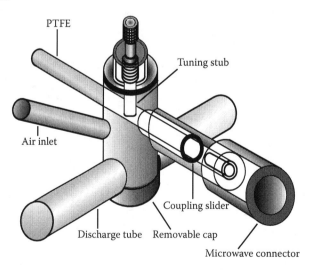

FIGURE 6.24 Fehsenfeld et al.'s microwave discharge. (Adapted with permission from Fehsenfeld FC, Evenson KM and Broida HP, 1965. Microwave discharge cavities operating at 2450 MHz. *Rev. Sci. Instrum.* **36** 294–8. Copyright 1965, American Institute of Physics.)

Davis et al. (1949b) briefly describe the production of a beam of atomic hydrogen which is dissociated in a microwave cavity operating at 3 GHz with 50 W of power input. The gas was admitted through a glass tube placed at a voltage antinode and emerged through a narrow slit in the glass tube at the edge of the cavity. The tube was cooled with an air blast. We refer to this again in Section 6.8 since the same method was used to produce a beam of atomic chlorine.

The most common source of microwave power is now the S-band magnetron operating at 2.45 GHz, since it is reliable, inexpensive and is available for high powers. It is ubiquitous in the household microwave oven, though this does not appear to be readily adaptable to form a dissociator. Fehsenfeld et al. (1965) investigated the use of five microwave cavities to dissociate hydrogen. Although they did not extract an atomic beam in their work, microwave dissociators are probably now the most common means of generating atomic hydrogen beams. They state that among the advantages of these discharges is that highly dissociated hydrogen is produced without undue heating of the background gas, with little electrical interference and with no internal electrodes. However, they emphasise that both tuning and matching adjustments must be provided in order to obtain efficient operation over a wide range of discharge conditions. The one-quarter wavelength cavities were designed to operate when connected by a coaxial cable to a 2.45 GHz medical diathermy unit, which provides 125 W of power, and to dissociate hydrogen in a 13-mm-outside-diameter fused silica tube through which gas flowed slowly. Fehsenfeld et al. give details of all five cavities, but since the fifth was found to be the most successful we only show that one in Figure 6.24. Its advantages are that it has both a tuning stub and a coupling slider which enabled it to operate efficiently over a wide range of gas pressures. The discharge tube was cooled by air. The cavity had a removable cap, which enabled it to be positioned around the discharge tube without disturbing the vacuum system.

Since the correct tuning depends upon the gas pressure, the authors recommend first placing the slider in its position of minimum penetration then adjusting the tuning stub for minimum reflected power as measured with a bi-directional power meter located at the output of the diathermy unit. The slider and stub are then adjusted successively. Some additional operating procedures are given in their paper. They note in particular that hot spots can occur at higher operating pressures. These were believed to be caused by recombination on the discharge tube walls.

This design has been frequently employed to dissociate gases other than hydrogen, as we shall see in later sections. It is usually referred to as an 'Evenson cavity,' for reasons which are not apparent. As pointed out by Murphy and Brophy (1979), the cavity may be extended by an integral number of half wavelengths beyond a one-quarter wavelength, or by increments of about 61.2 mm at 2.45 GHz, and these extended cavities have often been employed.

The use of this design for producing an atomic beam is described by Brink et al. (1968), whose preliminary experiments with a microwave cavity within their vacuum system were unsatisfactory due to problems caused by the heat generated. Gas entered the cavity through 12-mm-external-diameter fused silica tubing and left through an orifice in the end of the tube. The cavity, shown in Figure 6.25, was 325 mm long, namely, 11 quarter wavelengths at 2.45 GHz and made from a 25-mm-internal-diameter copper tube and a 13-mm-internal-diameter copper tube that surrounded

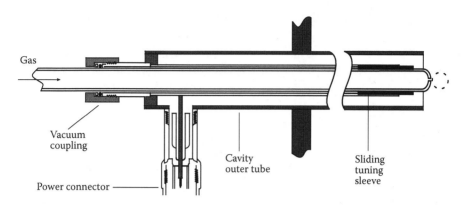

FIGURE 6.25 The microwave beam source of Brink et al. (Adapted with permission from Brink GO, Fluegge RA and Hull RJ, 1968. Microwave discharge source for atomic and molecular beam production. *Rev. Sci. Instrum.* **39** 1171–2. Copyright 1968, American Institute of Physics.)

the discharge tube. Power from a 100-W magnetron was delivered through a coaxial cable and a triple stub tuner to a tap situated 12.5 mm from the input end of the cavity that forms a short circuit. The smaller copper tube was fitted with a sliding sleeve for tuning purposes. The sleeve and tuner were adjusted for minimum reflected power before the cavity was placed in the vacuum system and without any discharge in the tube. The discharge was then initiated with a 'sparker wire,' placed near the orifice and then the tuner was used to reduce the reflected power to less than 1%. No further tuning was required over a range of discharge pressures from about 2.7 Pa to several hundred Pa in what they describe as a variety of gases, which included helium, oxygen, carbon dioxide, nitrous oxide [N_2O] and nitrogen peroxide [NO_2–N_2O_4].

Although not used to form an atomic beam, Lisitano et al. (1968) describe a slotted tube cavity in which axial slots are spaced uniformly around its circumference. The length of each corresponds to one half wavelength of the exciting frequency. Short azimuthal slots connect alternate ends of the axial slots. The authors state that the configuration can be visualised as a long slot antenna which is folded back and forth on a cylinder. An advantage of their design is that the dimensions could be scaled to the chosen frequency, which was from 500 MHz to 10 GHz at a power of 50 W. An axial magnetic field of up to 230 mT could be applied. The reflected power was much reduced when plasma was present and could be minimised by adjusting the magnetic field strength. The device was tested in argon, helium and hydrogen and operated at pressures as low as 1 mPa. We mention out of chronological order that detailed studies of slot antenna sources of this type have been made by Werner et al. (1994). They describe an annular waveguide that was designed to operate at 2.45 GHz so that the five slots on the inner cylinder were 61.2 mm long. Their experiments showed that their design and matching system for coupling the magnetron to the plasma were an improvement over Lisitano et al.'s cable coupling. However, their apparatus was also not designed to extract an atomic beam.

A microwave dissociator was used by Aquilanti et al. (1972) to obtain a beam of atomic hydrogen. The source arrangement is shown in Figure 6.26. The discharge

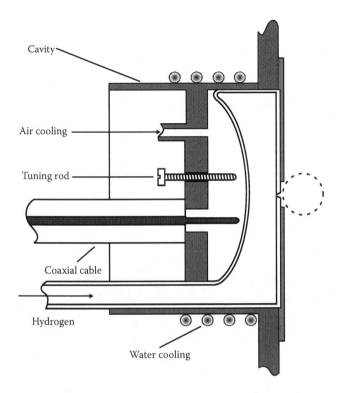

FIGURE 6.26   The microwave discharge source of atomic hydrogen used by Aquilanti et al. (Adapted from Aquilanti V, Liuti G, Luzzatti E, Vecchio-Cattivi F and Volpi GG, 1972. Production and detection of an energy selected beam of hydrogen atoms in the range 0.01–0.20 eV. *Z. phys. Chem.* **79** 200–8. Copyright 1972 Oldenbourg Wissenschaftsverlag GmbH.)

tube was made of fused silica and was both air and water cooled. The cavity was tuned so that its linear dimension was an odd multiple of a quarter wavelength at the operating frequency of 2.45 GHz. The atoms were emitted through a 1.5-mm-diameter aperture. The discharge could be operated at pressures between about 100 Pa and 400 Pa and 3% of water vapour was found to improve the dissociation.

A coaxial tee microwave source is described by Miller (1974). The outer conductor was 22.2 mm in outside diameter through which helium gas was passed for cooling at pressures from 100 kPa to 200 kPa. The inner conductor was 9.5 mm in outside diameter to which power at 3 GHz was fed through a coaxial cable. One end of the tee was short circuited as shown in Figure 6.27 whereas the other end of the inner conductor was 3 mm inside the end cap of the outer conductor. Hydrogen at a pressure of 400 Pa containing about 27 Pa water vapour passed though the centre tube and emerged through a slit that was 250 µm wide and 4.56 mm high. The discharge was strongest in the gap between the inner and outer conductors. No tuning of the cavity appears to have been provided, which absorbed power of about 115 W, as estimated from the heat absorbed by the cooling gas. The dissociation fraction was about 85%, which increased from about 33% if water vapour was not added.

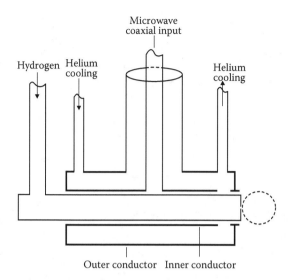

FIGURE 6.27 Miller's microwave source of atomic hydrogen. (Adapted with permission from Miller TM, 1974. Atomic beam velocity distributions with a cooled discharge source. *J. Appl. Phys.* **45** 1713–20. Copyright 1974, American Institute of Physics.)

A microwave source for hydrogen atoms that was within the vacuum envelope 'for experimental reasons' is described by Ding et al. (1977). Cooling of both the discharge tube, which was 300 mm long and 8 mm in diameter, and the microwave connecting cable was provided by circulating low absorption silicon oil. The discharge tube was treated with orthophosphoric acid and was replaceable since its efficiency for the production of atoms was found to decrease after about 20 hours of operation. The microwave power supply operated at 2.45 GHz. The cavity containing the discharge tube could be tuned with two tuning rods, only one of which is shown. It was necessary to initiate the discharge with a Tesla coil. A diagram of the apparatus is shown in Figure 6.28. The measured dissociation was about 80% when the tube was operated at a pressure of about 70 Pa.

Murphy and Brophy (1979) describe a variant of the microwave source described by Fehsenfeld et al. (1965) achieved by using 31.75-mm-outside-diameter brass tubing to increase the cavity length by 244 mm, namely, two wavelengths at the 2.45-GHz operating frequency. By so doing, it was possible for the beam exit aperture to be close to the discharge so reducing recombination in the discharge-free exit tube. The cavity was cooled by ice-cold water and compressed air to reduce recombination. The discharge tube was constructed from vicor, a 96% silica glass usable continuously up to temperatures over 1100 K. It was prepared for use by successive washing with hydrofluoric acid and distilled water. Their design is shown in Figure 6.29. The ignition electrode was used to initiate the discharge with a Tesla coil.

The microwave cavity described by Walraven and Silvera (1982) and operating at 2.45 GHz with a power of 25 W is shown in Figure 6.30a. The outer conductor was 153 mm long, being 1¼ wavelengths, and 28 mm in internal diameter. The slightly

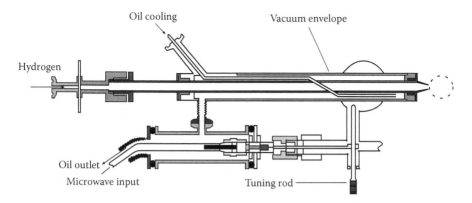

FIGURE 6.28 Ding et al.'s microwave source. (Adapted with permission from Ding A, Karlau J and Weise J, 1977. Production of H-atom and O-atom beams by a cooled microwave discharge source. *Rev. Sci. Instrum.* **48** 1002–4. Copyright 1977, American Institute of Physics.)

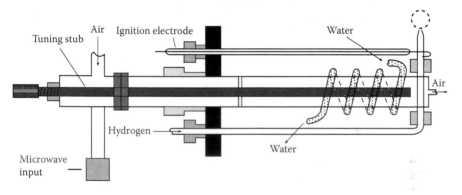

FIGURE 6.29 Murphy and Brophy's microwave source. (Adapted with permission from Murphy EJ and Brophy JH, 1979. Atomic hydrogen beam source: A convenient, extended cavity, microwave discharge design. *Rev. Sci. Instrum.* **50** 635–6. Copyright 1979, American Institute of Physics.)

shorter inner conductor had an external diameter of 14.8 mm. A fine adjustment screw enabled the outer conductor to be moved longitudinally relative to the inner conductor. Microwave power was fed into a quarter wave finger joint around the inner conductor which minimised electrical losses in the sliding contact. The discharge tube was of fused silica with an outside diameter of 12 mm. It passed through the inner conductor and the end cap of the outer conductor. When initiated with a Tesla coil, the discharge occurred in the 3.5-mm gap between the end of the inner conductor and the end cap of the outer. Figure 6.30b shows the cavity incorporated into the apparatus. A cooling jacket surrounded the dissociator, but unfortunately the design restricted efficient cooling of the discharge tube, which had to be replaced after a few hours use, since cleaning proved very difficult. It is reported that an alternative design with the diameter of the discharge tube reduced to 6 mm enabled air

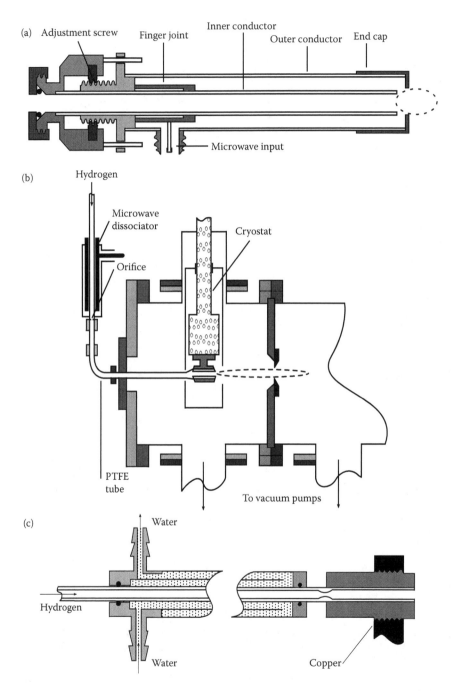

FIGURE 6.30 (a) Walraven and Silvera's microwave cavity. (b) The apparatus used by Walraven and Silvera. (c) Walraven and Silvera's beam temperature controller. (Adapted with permission from Walraven JTM and Silvera IF, 1982. Helium-temperature beam source of atomic hydrogen. *Rev. Sci. Instrum.* **53** 1167–81. Copyright 1982, American Institute of Physics.)

cooling in addition to the water cooling but made tuning of the cavity more difficult. The beam left the discharge region through an orifice that was 350 μm in diameter and 1 mm thick. Further beam transport and temperature control arrangements, which enabled beam temperatures to be obtained between 7.7 K and 600 K, are illustrated in Figure 6.30c. The authors give a detailed discussion of various designs for obtaining a cooled beam without requiring excessive coolant. The hydrogen pressure was operated at typically 67 Pa and the measured dissociation of the beam was 90%.

Hildebrandt et al. (1984) mention that they dissociated atomic hydrogen flowing through a fused silica tube with a 2.45-GHz, 100-W microwave generator. They found that it was necessary to coat the discharge tube with phosphoric acid [H_3PO_4/ P_2O_5], but since this only lasted a few days they also treated the tube with sodium hydroxide and nitric acid.

The factors that affect the efficiency of the cavity described by Fehsenfeld et al. (1965) for producing atomic beams are discussed by Spence and Steingraber (1988). In their case the beam emerged through a slit which was 5 mm long and 380 μm wide. It was found that the dissociation depended quite strongly on the internal diameter of the discharge tube with 8 mm found to be optimum. They only achieved about 10% dissociation of hydrogen without added water vapour and found that 0.5% of water vapour was optimum, leading to 95% dissociation, independent of whether the tube was made of Pyrex glass or fused silica. However, the microwave power level had some effect on the dissociation with 60 W of the 120 W available being optimum. Their tube preparation was simply to swill with acetone to remove dust, since running the discharge for 15 min to 20 min cleaned the tube adequately. A small distance between the discharge and the beam exit slit was found advantageous.

Donnelly et al. (1992) studied the performance of their 2.45-GHz microwave discharge source based on the designs of Fehsenfeld et al. (1965) and Murphy and Brophy (1979) discussed earlier in this section. They investigated various cleaning procedures that had been described by Murphy and Brophy, Ding et al. (1977), Slevin and Stirling (1981), Hood et al. (1978) and their own, namely, acetone followed by distilled water, chromic acid and finally distilled water. They found the hydrogen dissociation fraction was independent of these methods and that the measured dissociation was independent of the microwave power output in the range from 35 W to 100 W. They also compared fused silica tubes with those of pyrex glass of two differing thicknesses and preferred the thicker when it was coated with PTFE. Since Donnelly et al. were concerned with uncontaminated beams, none of their experiments involved addition of other gases or water vapour to the gases to be dissociated. Kikuchi et al. (1993) investigated the effect of both oxygen and water vapour on the dissociation of hydrogen produced by a microwave source operating at 2.45 GHz with a power output of 20 W. They concluded that the continuous introduction of water vapour passivates the surface of their fused silica discharge tube, increasing the hydrogen atom concentration by eighty times. They showed that this was neither due to hydroxyl [OH] radicals nor to oxygen atoms.

McCullough et al. (1993a, 1993b), Higgins et al. (1995a) and McCullough (1997) describe the use of copper slotted line radiators operating at 2.45 GHz which were based on the design of Lisitano et al. (1968) described earlier in this section. Each 1.5-mm-thick half cylinder had four slots 54 mm long and 2 mm wide with short interconnecting slots. They were clamped around a pyrex glass discharge tube, which was 150 mm long, 26 mm in external diameter, and 1.7 mm in wall thickness. A diagram of their arrangement is shown in Figure 6.31. The radiators were fed from the same 200-W power supply but with differing cable lengths so that a phase difference of π was expected. The beam impedance was 20 mm long and 1 mm in diameter, so $\Gamma = 20$. Measurements of the dissociation fraction on both source pressure and microwave input power showed a larger variation than Donnelly et al. (1992) found with their source.

A general introduction to the production of surface wave plasmas, but not their use in extracting an atomic beam, is given by Moisan and Zakrzewski (1991) and by Moisan et al. (1995). Although they did not extract an atomic beam, Rousseau et al. (1994a) made a detailed study of surface wave microwave plasmas in flowing hydrogen at a frequency of 2.45 GHz. Powers of up to 1.1 kW were available. They found that the atomic hydrogen concentration decreased from about 75% as the power input was increased and was constant at about 10% above powers of about 300 W. Following on from this work, Rousseau et al. (1994b) used a pulsed microwave power of 1 kW superimposed upon a continuous power of 250 W to obtain almost complete dissociation of hydrogen in their surface wave discharge. However, it was a poor 30% without the addition of the pulsed source. They interpret their results as being due to the increase of the recombination coefficient of hydrogen on the fused silica discharge tube walls with temperature. Since the discharge is

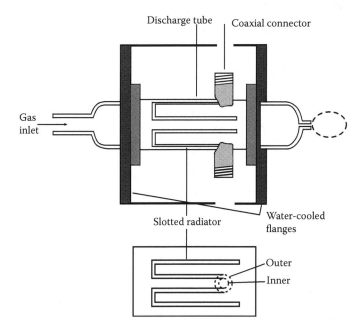

FIGURE 6.31 The slotted line radiators described in several papers. (Adapted from McCullough RW, Geddes J, Donnelly A, Liehr M, Hughes MP and Gildbody HB 1993. A new microwave source discharge source for reactive atom beams *Meas. Sci. Technol.* **4** 79–82. Copyright IoP Publishing 1993.)

produced in a flowing gas system, their approach is clearly suitable for producing a collimated beam of atomic hydrogen.

Paolini and Khakoo (1998) adapted the microwave source described by Murphy and Brophy (1979), but with an extension by five wavelengths rather than two, to provide a movable hydrogen beam within their vacuum system. The gas inlet to the discharge tube was through 3.2-mm-outer-diameter PTFE tubing and the exit beam passed through a capillary tube to 4-mm-external-diameter PTFE tubing. The authors detail the construction of the atom transport system. The glass and PTFE were soaked in orthophosphoric acid and then washed with deionized water in order to remove dust. Glass tubing that the discharge was not in contact with was coated with diluted PTFE solution. It was reported that the discharge could take up to an hour to stabilise and that a mains stabiliser assisted the discharge stability. Air cooling was applied and 99.999% purity hydrogen was admitted without palladium 'filtration' to a pressure between 38 Pa and 45 Pa. The microwave power input was 50 W. The dissociation of 80% was found to be largely independent of microwave power and gas pressure.

Koch and Steffens (1999) describe in detail a surface wave waveguide surfactron dissociator in which microwaves at a frequency of 2.45 GHz and power of up to 1.2 kW are produced by a magnetron. They are fed through a coaxial cable via a rectangular waveguide into a coaxial line at the end of which a surface wave is launched by means of a coupling slit. This transfers the transversal electromagnetic mode to the azimuthal-symmetric surface wave mode. Several tuning elements enabled the reflected power to be reduced to zero under all operating conditions. The authors give dimensional details of the waveguide. The fused silica discharge tube was 1.1 m long with an outer diameter of 14 mm and a wall thickness of 1.5 mm. It was cleaned successively with acetone, methanol and distilled water followed by a five-minute treatment with a mixture of 2 parts of 40% hydrofluoric acid to one part of 32% hydrochloric acid. Following extensive rinsing with distilled water, the tube was baked at a temperature of over 400 K for 24 hours.

In use the discharge tube was cooled by suction which caused ambient air to flow over the whole region of the discharge. The distance between the end of the plasma column and the beam exit nozzle was variable between 10 mm and 80 mm and was usually 45 mm. However, no difference in dissociation was normally measured as this distance was varied, which indicates that wall recombination was not taking place in that part of the discharge tube where no dissociation occurred. The hydrogen and deuterium admitted were 99.999% pure. Oxygen of 99.7% purity was added to improve the dissociation of hydrogen, which was about 80% at the lowest flow rates. Over 6 days of continuous operation, the degree of dissociation decreased slightly to 75%. The beam-forming nozzle could be maintained at any temperature between 40 K and 300 K, the lowest temperatures being obtained with a helium closed-cycle refrigerator. Several nozzle materials were investigated, including copper and anodised aluminium with only slight measured differences in dissociation, provided the nozzles were cooled so that ice formed upon them.

Kubo et al. (2000) describe a microwave discharge source for atomic hydrogen operating at 2.45 GHz. Hydrogen passed through the cavity in a 6-mm-internal-diameter fused silica tube which was constricted at both the entrance and exit to the discharge region to 1 mm in internal diameter. The discharge tube was surrounded by a cooling jacket through which dimethyl poly siloxane (poly (hexamethyl cyclo-trisiloxane)) [poly $(C_6H_{18}O_3Si_3)$] was circulated to maintain the temperature of the discharge tube at 283 K. The discharge tube was treated with orthophosphoric acid [H_3PO_4] before use, but no water needed to be added to the hydrogen. The beam was transported 300 mm in similarly treated PTFE tubing and emerged through an aperture that was 1 mm in diameter. Dissociation of 73% was measured at a gas pressure at the inlet of 20 Pa and microwave power of 45 W. The fraction dissociated decreased with increasing hydrogen pressure and peaked at powers in the range 50 W to 90 W.

Lisitano et al.'s (1968) slotted radiator design discussed earlier in this section was also used by Perry et al. (2002), whose design is shown in Figure 6.32. Their Pyrex glass discharge tube was 260 mm long with an outer diameter of 25.5 mm and inner diameter of 21.5 mm. The discharge tube was soaked in orthophosphoric acid [H_3PO_4] for 12 hours and rinsed with distilled water before use. It was surrounded by the radiator of 5-mm-thick copper. It was found necessary to run the source continuously to obtain dissociation beyond the initial 20%. Hydrogen of 99.9995% purity entered the discharge tube through a liquid nitrogen cold trap. The partially dissociated gas left the discharge tube through a 1-mm-diameter tube 20 mm long. It then passed through a curved 5.4-mm-internal-diameter PTFE tube, which was connected to a 5-mm-diameter and 100-mm-long aluminium tube which are not shown in the diagram. This tube was cooled by a 2-W closed cycle helium compressor to

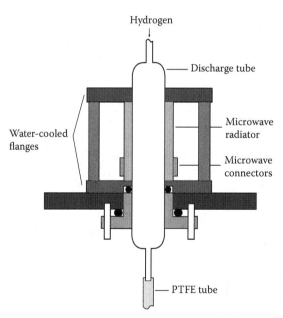

FIGURE 6.32 The hydrogen atom source used by Perry et al. (Adapted from Perry JSA, Gingell JM, Newson KA, To J, Watanabe N and Price SD, 2002. An apparatus to determine the rovibrational distribution of molecular hydrogen formed by the heterogeneous recombination of H atoms on cosmic dust analogues. *Meas. Sci. Technol.* **13** 1414–24. Copyright IoP Publishing 2002.)

a temperature of about 110 K and from its exit further transport was by means of PTFE tubing.

6.4 ANOTHER METHOD FOR PRODUCING A BEAM OF ATOMIC HYDROGEN

An unusual method of obtaining atomic hydrogen is described by Voronov and Martakova (1968). They first saturated a titanium disk 20 mm in diameter and 2 mm thick with hydrogen by cooling it from a temperature of 1500 K to laboratory temperature over a period of 5 hours in an atmosphere of hydrogen at about 27 kPa. The disk was found to contain 0.6 hydrogen atoms for each titanium atom. A neodymium laser with an energy of about 1 J and a pulse duration of about 30 ns was fired at the target after passing through an optical system designed to produce a uniform beam on the titanium disk. The emitted beam reached maximum intensity 10 μs after the laser pulse was fired and was found to contain atomic and molecular hydrogen and titanium atoms and ions. The dissociation percentage increased with the laser radiation density.

6.5 DISCUSSION OF ATOMIC HYDROGEN PRODUCTION

As introduced in Section 1.1, capillary arrays are used to collimate an atomic beam and so improve its quality. Collimated beams of atomic hydrogen have successfully been produced both by dissociation on tungsten metal foil arrays and, following dissociation, by passage through glass capillary arrays. The construction of the former is described in Section 10.2 and their use by Kleinpoppen et al. (1962) and Hils et al. (1966) is described in Section 6.2.3. As mentioned in Section 10.5, Roberts and Roberts (1970) state that collimated holes structures could be manufactured in tungsten. Section 6.3.2 includes an example of a glass capillary array being used in conjunction with already dissociated hydrogen by Ad'yasevich et al. (1965) using their array described in 10.4.6. Section 6.3.4 contains further examples, namely, by Rudin et al. (1961), Vályi (1968) and Slabospitskii et al. (1970).

As we mentioned in Chapter 1 and describe further in Chapter 9, beam intensity is not the only parameter of importance in comparing atomic beams. Since a measure of beam quality as defined in Chapter 9 has not been given by authors in this section, it is not possible to draw conclusions as to the preferred methods of obtaining an atomic hydrogen beam. Decisions will depend on the purpose for which the beam is required and the design of the apparatus may also limit the choice. Degrees of dissociation achieved appear to be comparable for the best executions of each type.

Before the beam reaches the experimental region with a discharge source it may be necessary to remove many other products of the discharge, such as electrons, metastable atoms and molecules and atomic and molecular ions and also photons if any of these would interfere with measurements. The thermally heated furnaces operating at temperatures up to 3000 K have the obvious disadvantages of requiring the handling of high powers and temperatures in vacuum and they usually have a limited operating life. They will be an intense source of continuous radiation as well as evaporated tungsten and if electron beam heating is used, stray electrons may interfere with experiments, either directly or because they interact with background gas. However, this method can be used to produce the purest beams of ground state atoms with very few hydrogen molecules and very little contamination present. Nevertheless some does occur, since Hucks et al. (1980) report that their beam contained about 0.1% tungsten when their tubular furnace with a central aperture was operated at a temperature of 2600 K and a pressure of 50 Pa. Discharge sources have the advantage that most things that are likely to go wrong are usually outside the vacuum envelope, so they can be repaired without the necessity of destroying the vacuum in the apparatus. Radiofrequencies at high powers can be difficult to screen from sensitive parts of the apparatus. However, the large currents used to heat furnaces may cause problems with the magnetic fields they create.

The tungsten source does not require the addition of another gas or water vapour to the admitted hydrogen, but this is not always required with a discharge. Discharge sources tend to have problems associated with the cleanliness of the discharge tube in which troublesome hot spots of recombination can occur. There is a risk that these can destroy the tube. Since experience with discharge tubes is so varied, it is postulated that microscopic metallic particles can sometimes be incorporated in glass tubing during manufacture.

Since extraction of an atomic beam from microwave discharges appears to be no more difficult than with any other type, they appear to be the preferred discharge source. Hydrogen beams cooled to below laboratory temperature have mainly been dissociated with radiofrequency discharges but beams from microwave sources have also been cooled.

6.6 ATOMIC OXYGEN BEAMS

The dissociation energy of oxygen is 5.17 eV. It can be dissociated in an iridium furnace, as is described briefly by Geddes et al. (1974), but they obtained only 6% to 8% dissociation when the furnace was operated at a temperature of 2100 K. We have already mentioned van Zyl and Gealy's (1986) furnace tube used to produce a beam of atomic hydrogen in Section 6.2.3. In order to dissociate oxygen they used an iridium tube that was 3.23 mm in internal and 4.24 mm in external diameter, but otherwise details of their apparatus are the same. They noted that although iridium melts at a temperature of 2650 K its useful long-term operating temperature is limited to 2150 K because of dislocation creep. However, it could be operated at 2250 K for short periods. At a temperature of 2100 K and pressure of 6.7 Pa 75% dissociation was measured. An iridium tube was also suggested by Horn et al. (1999) for producing atomic oxygen, but no experimental investigation appears to have been made.

Since no other furnace sources have been reported, the following discussion is concerned with discharge sources. In view of the relatively few papers to be discussed they are considered as a single group, rather than being separated into discharge types. As with atomic hydrogen, various cleaning and coating techniques have been employed with discharge tubes as well as additives to improve the fraction of molecules dissociated.

We have already briefly mentioned the work of Kurt and Phipps (1929) in Section 6.3.4, since they also produced a beam of atomic hydrogen with their apparatus, but its main purpose was to produce atomic oxygen. They mention that their first experiments with a Wood's tube were unsatisfactory because of the reaction between oxygen and the aluminium electrodes. Hence they employed an electrodeless ring discharge as shown in Figure 6.33. A 1-kW mains driven transformer had a secondary winding which delivered 25 kV to a 15-pF capacitor in series with a six-turn copper coil which was about 3 mm in diameter and which surrounded a Pyrex glass bulb of 2×10^5 mm³ volume. A spark gap was placed in parallel with the secondary of the transformer. It consisted of two zinc electrodes each of area 200 mm² and separated by 15 mm. The discharge operated at a pressure of 17 Pa, and it was found that the oxygen required saturating with water vapour in order to produce the highest atom concentration. The atomic beam emerged through two glass slits which were 1.2 mm long and 50 μm wide and separated by 66 mm with a pumped region between them.

Fite and Brackmann (1959) state briefly that they obtained 20% to 30% dissociated oxygen in an electrodeless radiofrequency discharge.

Kaufman and Kelso (1960) investigated the addition of nitrogen, nitric oxide [NO], hydrogen, argon, helium, carbon dioxide and nitrous oxide [N₂O] to oxygen in a microwave discharge operating at 2.45 GHz at powers up to 800 W. Although they did not form an atomic beam, their work does indicate the usefulness of some additives to improve the dissociation. Whereas helium, argon, and carbon dioxide had no effect, hydrogen and nitrogen and its oxides improved the dissociation.

Neynaber et al. (1961) used a radiofrequency discharge operating at 15 MHz and a power of 30 W to dissociate oxygen. They obtained 25% to 35% dissociation at a

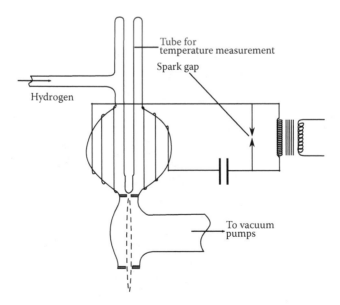

FIGURE 6.33 Kurt and Phipps' ring discharge source of atomic oxygen. (Adapted from Kurt OE and Phipps TE, 1929. The magnetic moment of the oxygen atom. *Phys. Rev.* **34** 1357–66. Copyright 1929 American Physical Society.)

pressure of 200 Pa when 2% to 3% of hydrogen was added to the oxygen. Their Pyrex glass discharge tube was 25 mm in diameter and 150 mm long and the beam was formed by a 1.4-mm-diameter aperture.

Brown (1967) studied the effect of additives on the dissociation of oxygen produced in the cavity number 5 described by Fehsenfeld et al. (1965). The addition of hydrogen or water vapour was shown to be more effective than any other molecules they tried.

Only brief details are given by Radlein et al. (1975) of their microwave source of atomic oxygen in the ^3P state. They obtained 10% dissociation when about 200 W of power were fed into oxygen at a pressure of about 130 Pa that was contained in a fused silica discharge tube. The atoms were estimated to be at a temperature of about 350 K. They travelled 600 mm in a pyrex glass tube before reaching a defining slit. The authors claim that metastable O and $O_2\left(^1\Sigma_g^+\right)$ were quenched before reaching it but that 5% $O_2\left(^1\Delta_g\right)$ was present in the atomic beam.

We have already mentioned Ding et al.'s (1977) microwave source of atomic hydrogen in Section 6.3.5. The same source was also used to produce atomic oxygen, which yielded a maximum of 33% measured dissociation at a pressure of 80 Pa. The discharge tube remained treated with orthophosphoric acid and only pure oxygen was admitted. As we shall see from later papers, higher values of dissociation should be achievable with a microwave discharge.

We include Gorry and Grice's (1979) microwave source of producing atomic oxygen, although the dissociation was carried out in the presence of excess inert buffer gas and the object was to obtain a supersonic beam, since they give extensive details of their design of cavity, which was based on the design of Fehsenfeld et al. (1965). The source is shown in Figure 6.34a. A mixture of about 15% oxygen in helium or neon at a pressure of 12 kPa was admitted to a fused silica tube which passed through the water-cooled microwave cavity and emerged through a 300-μm-diameter orifice. The tuning stub was adjustable from outside the vacuum system. About 100 W of power from a microwave power source was led by a coaxial cable and terminated in a coupling stub whose length was adjustable with a grub screw. The design of the stainless steel coupling stub is shown in Figure 6.34b. It is claimed to be a compromise between limiting impedance mismatches and providing heat-resisting insulation. Initial tuning was carried out outside the vacuum system using a sealed discharge tube in place of the gas inlet, then final tuning was accomplished when the source was mounted in the vacuum system. Some reflected power was found to be necessary to give the maximum discharge tube life of 100 h. Initially 35% dissociation of molecular oxygen was reported, but Gorry and Grice (private communication) obtained 70% dissociation with a shorter fused silica tube.

Sibener et al. (1980) give comprehensive details of their radiofrequency atomic oxygen beam source. It was, in fact, used to produce a supersonic beam so we concentrate on those aspects of the source which would be relevant if a thermal beam were produced, for example, in conjunction with a collimator. The authors needed to produce a plasma at a pressure in the region of 27 kPa and they found that it was necessary to mix oxygen with an inert gas such as helium or argon. Because it was intended to operate the discharge over a wide pressure range, the authors rejected the arrangement of the coil of a radiofrequency oscillator surrounding the discharge

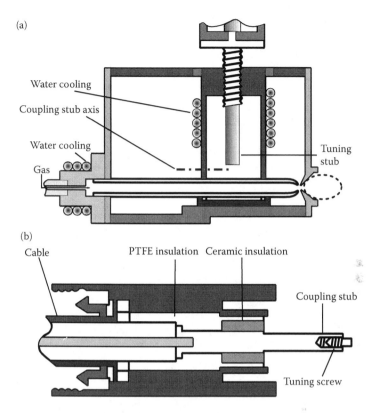

FIGURE 6.34   (a) Gorry and Grice's microwave source. (b) The coupling used by Gorry and Grice. (Adapted from Gorry PA and Grice R, 1979. Microwave discharge source for the production of supersonic atom and free radical beams. *J. Phys. E: Sci. Instrum.* **12** 857–60. Copyright IoP Publishing 1979.)

tube, because of the re-tuning required as the gas pressure was changed. Hence a variable frequency oscillator was decoupled from the plasma by a power amplifier. The oscillator was tuneable over a range of several hundred kHz around 14 MHz and could deliver 140 W into the power amplifier which could operate at up to 750 W. The discharge tube and coil are shown in Figure 6.35. That part of the coil nearest the beam exit consisted of 6.5 turns of 13.7-mm-internal-diameter copper tubing 3.2 mm in external diameter in series with a further five turns which were 50.8 mm in internal diameter. It was cooled with low conductivity water. The authors describe in detail the tuning arrangements, which involved a variable earthed tap on the coil surrounding the plasma tube and four variable capacitors. Measurements of the dissociation of the beam were made with both helium and argon used as seed gases. The 10% argon mixture gave dissociation in the range of 80% to 90% which was neither strongly dependent upon gas pressure nor radiofrequency power.

Samson and Pareek (1985) used a microwave discharge in a mixture of one part oxygen to four parts of helium at a pressure of about 27 Pa to form a beam of atomic oxygen. The apparatus is shown in Figure 6.36, from which it can be seen that the

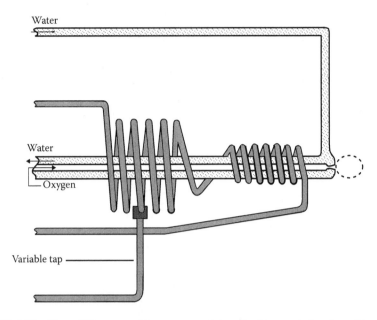

FIGURE 6.35 Part of Sibener et al.'s apparatus. (Adapted with permission from Sibener SJ, Buss RJ, Ng CY and Lee YT, 1980. Development of a supersonic O(^3P_J), O($^1D_2$) atomic oxygen nozzle beam source. *Rev. Sci. Instrum.* **51** 167–82. Copyright 1980, American Institute of Physics.)

microwave cavity was some distance from the beam-forming aperture. The flow tube was made of Pyrex glass, which was cleaned with a detergent, then rinsed and dried. A boiling saturated solution of boric acid was then poured in to coat all the internal surfaces and subsequently poured out. The tube was then heated to a temperature of about 500 K for at least four hours. Gas was pumped through this tube by a rotary pump with a speed of 60 l s^{-1}. The degree of dissociation of the oxygen molecules was estimated to be between 20% and 30%.

A brief description of a Penning-like discharge source of atomic oxygen is given by Clampitt and Hanley (1988). A potential difference of 5 kV was initially applied between a coaxial mesh anode and hemi-cylindrical cathodes. These were in a magnetic field produced by permanent magnets. When running, the discharge current was 50 mA at 600 V.

We have already mentioned Spence and Steingraber's (1988) investigation of the optimum conditions for obtaining a highly dissociated hydrogen beam in Section 6.3.5. In addition they also produced atomic oxygen beams and found 60% to 70% dissociation when 0.5% water vapour was added as before. Missert et al. (1989) used the microwave cavity described by Fehsenfeld et al. (1965) to dissociate oxygen. The input power was in the range 100 W to 300 W. The discharge tube was 1 m long and made of fused silica. It had an outside diameter of 10 mm and a bore of 8 mm. It was treated with a concentrated solution of boric acid before use. The oxygen pressure in the discharge was from 130 Pa to 270 Pa. Matijasevic et al. (1990) also used the microwave cavity described by Fehsenfeld et al. (1965) to produce atomic oxygen. The gas flowed through a 10-mm-outside-diameter fused silica tube which was

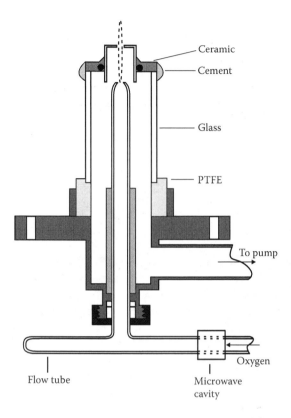

Ceramic

Cement

Glass

PTFE

To pump

Oxygen

Flow tube

Microwave
cavity

FIGURE 6.36 Samson and Pareek's microwave source. (Adapted from Samson JAR and Pareek PN, 1985. Absolute photoionization cross sections of atomic oxygen. *Phys. Rev. A* **31** 1470–6. Copyright 1985 American Physical Society.)

treated to reduce recombination of the atoms with a boiling concentrated solution of boric acid followed by heating to a temperature of about 900 K for several hours. Microwave power inputs were in the range 100 W to 300 W.

Donnelly et al. (1992) measured the dissociation of oxygen obtained with the radiofrequency source used by Slevin and Stirling (1981) and described in Section 6.3.4 and found it to be less than 15%. Unlike hydrogen, the oxygen dissociation fraction in their 2.45-GHz microwave discharge source, which was based on the designs of Fehsenfeld et al. (1965) and Murphy and Brophy (1979) discussed in Section 6.3.5, was strongly pressure dependent. The highest dissociation was 60% at the lowest pressure and hence at the lowest intensity. It is interesting that they found no improvement in cleaning with boric acid, but it is not clear whether they used the extensive baking procedure described in the next paragraph that was used by Yu-Jahnes et al. (1992).

The microwave cavity described by Fehsenfeld et al. (1965) was also used by Yu-Jahnes et al. (1992) to dissociate oxygen. They found it superior to an electron cyclotron resonance source. The cavity was operated at a microwave power of 250 W. Their fused silica discharge tube was 460 mm long and 12.7 mm in outside diameter. It was cleaned by the method described by Kern and Puotinen (1970) (which Yu-Jahnes et al. call the RCA method). Initially Kern and Puotinen removed gross organic residues with hot trichlorethylene (trichloroethene) [C_2HCl_3]. Next a solution consisting of approximately one part 27% ammonia [NH_4OH] to between one or two parts of 30% unstabilised hydrogen peroxide [H_2O_2] and respectively up to five or seven parts of water distilled in fused silica was heated in the tube for 10 min to 20 min at a temperature of about 350 K in order to remove organic contaminants and some metals. Following rinsing with more water distilled in fused silica to remove heavy metals, the tube was treated in the same way with a second solution consisting of one part of 37% hydrochloric acid and between one or two parts of hydrogen peroxide with respectively between six to eight parts of silica-distilled water. After a final rinse, Yu-Jahnes et al. immersed their tube in a saturated solution of boric acid at a temperature of 363 K for 30 minutes and dried it overnight. The tube was next heated with an oxy-hydrogen torch when oxygen was flowing through it. The authors describe in detail the subsequent annealing of the tube in an atmosphere of oxygen by heating to temperatures up to 800 K for four days. They stress that the clear and colourless coating should only be exposed to air for short periods; otherwise water vapour would turn the coating milky. It was found that the addition of about 4% of nitrogen to the oxygen considerably improved the amount of atomic oxygen produced. The dissociation was estimated to be between 25% and 30%.

McCullough et al.'s (1993a, 1993b), Higgins et al.'s (1995a) and McCullough's (1997) copper slotted line radiator source which we have described in Section 6.3.5 with hydrogen was also used for oxygen. It was found that the highest dissociation of 54% was obtained at the lowest source pressure and therefore at the lowest beam intensity, as also stated by Donnelly et al. (1992).

A discharge tube of high purity alumina was used by Imai et al. (1995) in conjunction with a radiofrequency generator operating at 13.56 MHz and capable of delivering power of 500 W. The tube was 150 mm long and 25 mm in internal diameter and sleeved with pyrolytic boron nitride. Power was fed through a cable to radiofrequency vacuum feedthroughs since the water-cooled coil surrounding the discharge tube was within the vacuum system. An automatic matching device was used to tune the system. The oxygen pressure in the discharge tube was less than 67 Pa. The beam emerged through up to 25 200-μm-diameter holes in the 300-μm-thick pyrolytic boron nitride end plate of the system. The degree of dissociation was not measured.

Kanik et al. (2001) added 5% of nitrogen to their oxygen which was dissociated in an extended air-cooled microwave cavity through which the mixture flowed in a fused silica tube. The source was similar to that described by Murphy and Brophy (1979) to produce atomic hydrogen. They obtained between 20% and 35% dissociation of oxygen. Stable dissociation was obtained by cleaning the tube with ortho-phosphoric acid. Recombination at the tube exit was reduced by coating it with PTFE. The beam emerged through a 2-mm-diameter aperture in a PTFE plug. Further experiments were conducted with the same atomic oxygen source by Noren et al. (2001) with 10% nitrogen added. In neither experiment was any atomic nitrogen observed. Noren et al. found that the fraction of oxygen atoms dissociated was very sensitive to the tuning of the microwave discharge, which was dependent upon both the size and shape of the resonant cavity. Over periods in excess of 24 hours needed for their measurements, temperature variations caused slow variations in the dissociation fraction.

In conclusion, it is more difficult to obtain high degrees of dissociation of oxygen than of hydrogen and the cleaning of the discharge tube generally needs to be more thorough. Coating of the discharge tube has proved advantageous as well as additives to the gas, which are not restricted to water vapour.

6.7   BEAMS OF ATOMIC NITROGEN

Nitrogen has the extremely high dissociation energy of 9.80 eV, which makes obtaining a highly dissociated beam difficult. Little discussion of the methods of cleaning discharge tubes is found for nitrogen dissociation, but there are some reports of additives improving the degree of dissociation. As in the case of oxygen, in view of the relatively few papers to be discussed, they are considered as a single group, rather than being separated into discharge types. We also include papers discussing dissociation without forming a beam, since they may assist in choosing a method for producing one.

Jackson and Broadway (1930) were able to obtain some dissociation using a water- and air-cooled Wood's tube with the beam slit close to the discharge. The method of slit construction is described in Section 3.4.3. It was 50 μm wide and both 1.5 mm long and high.

We have already described Hendrie's (1954) tungsten furnace in Section 6.2.3 and mentioned that he also unsuccessfully used it to obtain atomic nitrogen, thus establishing a lower limit for the dissociation energy of 8.80 eV and so establishing the value given above.

After unsuccessfully attempting to dissociate nitrogen in the radiofrequency discharge as used by Neynaber et al. (1961) and discussed in the previous section, Smith et al. (1962) obtained about 20% dissociation using a pulsed direct current discharge. Their discharge tube is shown in Figure 6.37. A direct current power supply operated at 1.6 kV and delivered a current of 50 mA through a series resistance of 18 kΩ to a capacitor of 3 μF to which the aluminium electrodes within the discharge tube were connected. The current pulse is described as similar in shape to a one-cycle damped

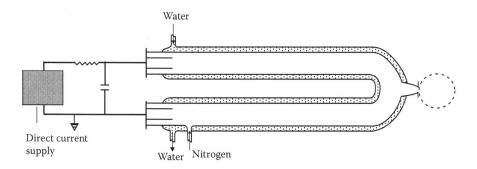

FIGURE 6.37   The nitrogen discharge tube used by Smith et al. (Adapted from Smith ACH, Caplinger E, Neynaber RH, Rothe EW and Trujillo SM, 1962. Electron impact ionization of atomic nitrogen. *Phys. Rev.* **121** 1647–9. Copyright 1962 American Physical Society.)

sine wave with a peak positive current of 750 A, a peak negative current of 350 A, a duration of about 26 µs and a repetition interval of about 66 ms. The U-shaped discharge tube was made of Pyrex glass and was surrounded by a water cooling jacket. Each leg of the tube was about 300 mm long. The discharge operated at a pressure of about 27 Pa. The 20-mm-long beam exit tube was in the middle of the U tube, so the beam emerging through a slit in its end, which was either 35 µm or 70 µm wide and 2.5 mm long, was from the discharge afterglow and not from the discharge column. The authors considered that their beam was unlikely to contain any significant concentrations of metastable nitrogen atoms or molecules.

McCarroll (1970) modified Fehsenfeld et al.'s (1965) one quarter wavelength microwave cavity described in Section 6.3.5 by extending it by one half wavelength to improve the stability of the discharge. It was tested in nitrogen with input powers up to 97 W, and whereas no beam was extracted, it was thought that the design was more efficient for the dissociation of nitrogen. Miller's (1974) microwave discharge source already described in Section 6.3.5 for hydrogen was also used by him to obtain atomic nitrogen, which was 10% dissociated, with 11% possible for short periods with an overrun microwave generator.

Bickes et al. (1976) produced a supersonic nitrogen beam by adopting the arc source used by Way et al. (1976), described in Section 6.3.3, which is itself a development of that described by Young et al. (1969). Their source is shown in Figure 6.38. An adjustable thoriated tungsten rod forms the cathode and the anode is of copper which contains an aperture 400 µm in diameter through which the beam emerges. Both electrodes are water cooled. An air-cooled electromagnet produces an axial magnetic field in the arc region. The arc operates at currents of up to 100 A at up to 15 V at a pressure of 100 kPa. The arc was initially struck in argon at a pressure of 8 kPa, using a Tesla coil, before the magnetic field was applied. When nitrogen was added to the argon about 60% dissociation was obtained. The authors do not discuss whether the arc would operate without the addition of argon.

In Sections 6.3.5 and 6.6, we have already mentioned Spence and Steingraber's (1988) studies of the cavity described by Fehsenfeld et al. (1965). However, for nitrogen, only 4% dissociation was measured, when yet again 0.5% water vapour was added to the admitted gas. Meikle and Hatanaka (1989) briefly describe the use of a 2.45-GHz microwave cavity operating in the TE_{10} mode in conjunction with a magnetic field applied perpendicular to their 33-mm-internal-diameter Pyrex glass discharge tube. This combination meant that the nitrogen plasma was generated by electron cyclotron resonance. With 15 W of power the ratio of atomic to molecular nitrogen was at least 12% at a pressure of about 670 mPa. Although the object of their experiments was not to obtain a beam of atomic nitrogen, they indicate that the efficiency of dissociation is at least 70 times greater when the magnetic field is present.

The radiofrequency dissociator for nitrogen described by Hoke et al. (1991) for MBE employed a discharge tube constructed of pyrolytic boron nitride. Up to 500 W of power at 13.56 MHz was supplied to a water-cooled coil within the vacuum system via a matching unit, which reduced the reflected power to less than 1 W. An effective radiofrequency shield within the vacuum system surrounded the high frequency components. Two alternative beam exit plates, for which no constructional details are given, were also made from pyrolytic boron nitride. Measurements were

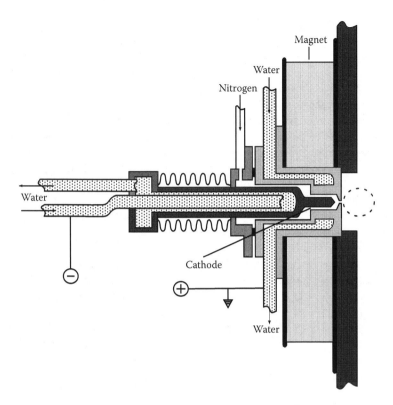

FIGURE 6.38 The arc source developed by Bickes et al. (Adapted with permission from Bickes RW, Newton KR, Herrmann JM and Bernstein RB, 1976. Utilization of an arc-heated jet for production of supersonic seeded beams of atomic nitrogen. *J. Chem. Phys.* **64** 3648–57. Copyright 1976, American Institute of Physics.)

made which showed that the relative concentration of atomic nitrogen increased both with the gas flow rate and the radiofrequency power.

Donnelly et al. (1992) state briefly that they found no more that 4% nitrogen dissociation using the microwave discharge described in Section 6.3.5 and developed by Fehsenfeld et al. (1965) and the radiofrequency discharge of Slevin and Stirling (1981) described in Section 6.3.4.

McCullough et al.'s (1993b) copper slotted line radiator source which we have described in Section 6.3.5 with hydrogen and Section 6.6 for oxygen was also used with nitrogen. As is explained in the five papers cited below, following on from the work of Meikle and Hatanaka (1989), a magnetic field in conjunction with a microwave cavity enables useful degrees of dissociation of atomic nitrogen to be obtained. Geddes et al. (1994) and McCullough et al. (1996) placed a coil carrying a current of 140 A of internal diameter of 140 mm, external diameter of 280 mm and length of 72 mm around the discharge assembly to produce the magnetic field. They obtained a maximum dissociation of 67% with 200 W of microwave power at their lowest pressure of 400 mPa. Higgins et al. (1995a, 1995b) and McCullough (1997) describe the addition of ring magnets separated by aluminium alloy rings. A diagram of the modifications to the slotted line radiator source shown in Figure 6.31 to include the ring magnets is shown in Figure 6.39. It was found that a maximum dissociation of 62% was obtained when an axial magnetic field of 65 mT was applied, but that this decreased with increasing source pressure. Without the field, only 4% dissociation was measured. The enhancement was not obtained with the other gases mentioned in preceding sections.

An arc discharge for producing an atomic nitrogen beam is described by Xu et al. (1997). Their cathode was a water-cooled sharp tip of thoriated tungsten which could be moved about 10 mm axially relative to a water-cooled molybdenum anode, by an adjustment that is not described. The atomic beam was formed by an aperture 500 µm in diameter in the conical end-plate of the anode. An axial magnetic field in the discharge region of 20 mT was generated by a water-cooled electromagnet that was external to the vacuum system. The cathode was surrounded by a fused silica cylinder which provided radiation shielding. This cylinder was in turn surrounded by a Pyrex glass cylinder forming the vacuum envelope. Their source is shown in Figure 6.40.

To produce the arc, a stable glow discharge was first obtained without the magnetic field present at a nitrogen pressure of about 2.7 kPa when a current of less than 200 mA flowed at 240 V. When the pressure was increased to 14 kPa, an arc was obtained with a current of 9.5 A to 10 A at 50 V to 55 V. Stable operation was obtained with the magnetic field present at pressures in the range 4 kPa to 47 kPa and arc currents greater than 7 A. Xu et al. were only able to estimate the dissociation degree to be between 50% and 60%. A similar arc discharge source to obtain an atomic nitrogen beam is briefly described by Grunthaner et al. (1998), who used a replaceable rhenium beam-forming tube, which was 8 mm long and 500 µm in diameter, so $\Gamma = 16$. Their arc current of 200 mA at a pressure of 22.7 kPa is much lower

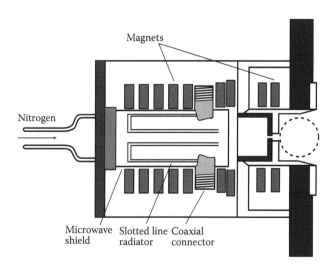

FIGURE 6.39 McCullough et al.'s design including ring magnets. (Adapted from McCullough RW, Geddes J, Donnelly A, Liehr M, Hughes MP and Gilbody HB, 1993. A new microwave discharge source for reactive atom beams. *Meas. Sci. Technol.* **4** 79–82. Copyright IoP Publishing 1993.)

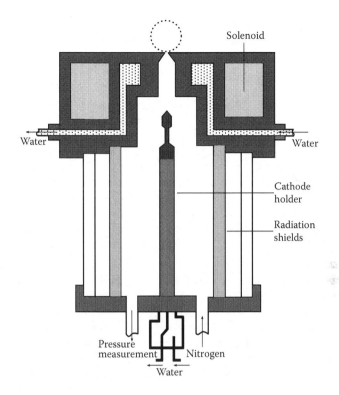

FIGURE 6.40   The arc discharge source used by Xu et al. (Adapted with permission from Xu N, Du Y, Ying Z, Ren Z and Li F, 1997. An arc discharge nitrogen atom source. *Rev. Sci. Instrum.* **68** 2994–3000. Copyright 1997, American Institute of Physics.)

than that in Xu et al.'s experiment and the estimate of the degree of dissociation is only in the few percent range.

Voulot et al. (1998, 1999) measured the nitrogen dissociation in a radiofrequency atomic nitrogen source. The discharge was generated at 13.56 MHz with radiofrequency power in the range 200 W to 500 W in a cylindrical pyrolytic boron nitride discharge tube which was 133.6 mm long and 16.8 mm in diameter. The nitrogen of 99.998% purity was admitted at one end of the tube and the partially dissociated nitrogen emerged through a 2.8-mm-diameter aperture. A measured dissociation of 40% was obtained at the highest input power and lowest discharge pressure of 9 Pa.

Godfroid et al. (2003) studied the dissociation of nitrogen in a pulsed surface wave plasma of a type that was mentioned in Section 6.3.5. Since the atomic nitrogen source was required for surface studies, a gas mixture of argon and nitrogen was used. A pulsed mode at 2.45 GHz was employed and detailed studies showed that the dissociation of nitrogen increased with increasing gas flow and decreasing amount of nitrogen in the mixture. The highest degree of dissociation recorded was 43%. Further details of the apparatus and their measurements are contained in their paper.

We conclude that producing a highly dissociated intense beam of nitrogen is still challenging, as is to be expected from its high dissociation energy. Highest dissociation appears to be associated with the lowest discharge operating pressure and hence the lowest beam intensity.

6.8 SOURCES OF HALOGEN BEAMS

The dissociation energies of the halogens are quite low, being 1.65 eV for fluorine, 2.51 eV for chlorine, 2.32 eV for bromine, and 2.21 eV for iodine. Since the low dissociation energies mean that they should, in principle, be easy to dissociate and also because many authors treat them as a group, we do the same and also do not separate the discussion by dissociation methods. Both dissociation of halogens in furnaces and in discharges have been successful. In the latter case, some authors find coating the tube walls and the use of additives advantageous. Dissociation of halogen-containing compounds has also been employed.

Rodebush and Klingelhoefer (1933) used a coil of ten turns of 3-mm-diameter copper tubing wound on a 2×10^5 mm^3 spherical pyrex glass bulb to obtain atomic chlorine. A high frequency discharge was obtained with a spark gap in series with the coil and the secondary of a 25-kV transformer, across which a capacitor was connected. The transformer primary was fed via a resistance from the mains supply. Schwab and Friess (1933) report briefly that they obtained atomic chlorine by using a Wood's tube made of fused silica with water-cooled iron electrodes that were supplied with alternating current. The microwave source of atomic hydrogen used by Davis et al. (1949b) and described in Section 6.3.5 was also used to dissociate chlorine. The discharge occurred in the chlorine 'for a centimetre or so directly behind the slit.' The authors report 'a good ratio' of atoms to molecules in the emerging beam.

Garvin et al. (1958a) report briefly that the vapour pressure of iodine at laboratory temperature is sufficient for a radiofrequency discharge at 450 kHz to be maintained. Their 6-mm-diameter discharge tube was made of fused silica and coated with aquadag to obtain good capacitative coupling to the radiofrequency leads. The authors also describe the electronic circuitry which enabled a stable discharge to be obtained. The iodine beam emerged through a slit which was about 1 mm long and 120 µm wide. The beam was considered to be over 80% dissociated. A similar dissociation degree was reported for bromine, but no further details are given. Astatine has no stable isotopes, but Garvin et al. (1958b) succeeded in producing a beam which was believed to be 70% to 80% dissociated. It was formed from an astatine iodine complex which was placed in a glass tube heated to a temperature of about 400 K. From there the vapour passed through a 'slow' leak into a platinum tube heated to about 1000 K by electron bombardment, where it was dissociated.

Sherwood and Ovenshine (1959) found almost 100% dissociation of iodine in an oven at a temperature of about 950 K, but operated it at 100 K to 150 K hotter than this. Since iodine is quite volatile at laboratory temperatures, when not in use the iodine was maintained at ice temperatures outside the vacuum system. The vapour entered it through a tube. They found that the only material which did not react with the iodine vapour was solid gold.

Although he did not form the halogen atoms into beams, Ogryzlo (1961) studied the production of halogen atoms produced by a 2.45-GHz 100-W microwave generator. After the discharge tube was cleaned, it was filled with hot concentrated potassium hydroxide solution and allowed to stand for 30 min. It was then washed with distilled water and filled with a 10% solution of potassium hydroxide for one hour. This was followed by evacuation for two hours at a temperature of about 520 K. In order to obtain atomic chlorine and bromine Ogryzlo describes trying a number of surface coatings which were applied in an unspecified manner. Experiments with fluorine and iodine were unsuccessful.

Using a graphite oven heated to a temperature of 1450 K, Beck et al. (1968) report briefly that they obtained 40% dissociation of chlorine. Lee et al. (1969) briefly mention that their graphite oven, which was 6.3 mm in diameter and 1 mm in wall thickness, was heated by a current of 160 A at 2 V to temperatures between 1600 K and 1700 K. They mention at least 75% dissociation of chlorine and 95% for bromine at pressures in the range 130 Pa to 270 Pa.

A double oven constructed of glass and monel metal for producing a beam of iodine atoms was briefly described by Fluendy et al. (1970). The iodine was maintained at a pressure of about 130 Pa in the first oven operating at a temperature of about 300 K and which was connected to the second with a glass tube. At a temperature of about 800 K, dissociation was more than 90%. The beam emerged through a gold 'micro mesh' aperture. Grover and Lilenfeld (1972) briefly describe the production of a beam of a radioactive isotope of astatine from hydrogen astatide which was dissociated in a platinum tube heated to a temperature of 1300 K. Preparation of this is described by Grover et al. (1971), who also obtained a beam of hydrogen astatide using a capillary array.

A microwave discharge cavity of the type described by Fehsenfeld et al. (1965) and discussed in Section 6.3.5 was employed by Kolb and Kaufman (1972) to dissociate carbon tetrafluoride (tetrafluoromethane) $[CF_4]$ to produce fluorine atoms. The cavity operated at a power of up to 100 W at 2.45 GHz. They also used a mixture of fluorine with argon with an argon to fluorine ratio in the range from 5 to 20. The gases passed through the cavity in a 12.7-mm-diameter cast alumina tube at a pressure in the range 27 Pa to 80 Pa for carbon tetrafluoride and total pressures in the range 67 Pa to 200 Pa for the gas mixture. Other possibilities for the discharge tube material discussed by the authors are Pyrex glass, with which fluorine reacts rapidly, and fused silica, which has about a ten-hour life within the cavity, but which produces a higher dissociation degree than alumina. Outside the discharge region, PTFE liners or coatings were used. The latter were suitable for 50 h of use. When carbon tetrafluoride was used, the only other species present in addition to the undissociated molecule and atomic and molecular fluorine was hexafluoroethane (perfluoroethene) $[C_2F_6]$.

Thermal dissociation of fluorine was employed by Parson and Lee (1972) to obtain a beam of atomic fluorine, using a nickel oven heated to a temperature of about 1000 K in which the pressure of fluorine was up to 130 Pa. No other details of the operation are given. Valentini et al. (1977) used a graphite oven to dissociate chlorine and bromine. It was used to produce a seeded supersonic beam, but their oven could also be used to produce a thermal beam. The construction of the oven, which has novel features, is shown in Figure 6.41. An outer tube which was 22.2 mm in outer diameter with a 3.2-mm wall thickness was constructed from high-density graphite. This was screwed into a water-cooled copper mounting block that formed

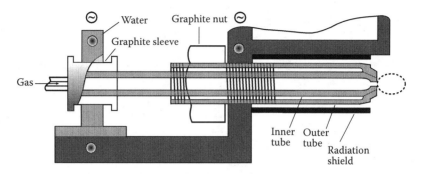

FIGURE 6.41 Valentini et al.'s graphite oven used for chlorine and bromine. (Adapted with permission from Valentini JJ, Coggiola MJ and Lee YT, 1977. Supersonic atomic and molecular halogen nozzle beam source. *Rev. Sci. Instrum.* **48** 58–63. Copyright 1977, American Institute of Physics.)

one of the current leads and held in place with a graphite nut. The inner graphite tube had the same wall thickness and was 12.7 mm in outside diameter. It contained the 79-μm beam aperture and had a conical cone to fit into the conical bore at the end of the outer tube. The rear support for this tube was a sliding fit into a water-cooled copper support which provided the other current lead. It was fitted with springs (not shown) to maintain the two tubes in contact near the orifice. Electrical insulation was with mica for hotter components and poly (methyl methacrylate) [poly ($C_5H_8O_2$)] for cooler ones. With one tantalum radiation shield, an alternating current of 450 A at 7 V rms enabled temperatures of up to 1700 K to be achieved.

De Leeuw et al. (1978) dissociated fluorine, chlorine and bromine in a 100-W microwave discharge. Fluorine was dissociated in an uncoated alumina tube by running a discharge in carbon tetrafluoride (tetrafluoromethane) [CF_4]. However, the tube became porous after a few days' use. A chlorine or bromine discharge was obtained in fused silica or alumina discharge tubes. The authors found that it was necessary to coat their discharge tubes with orthophosphoric acid [H_3PO_4]. The coatings were produced by applying a phosphorous oxide [P_2O_5] layer to the tube and then passing a small amount of water vapour through it. Attempts to produce iodine atoms with the discharge were unsuccessful. We have already described Gorry and Grice's (1979) source in Section 6.6, since, in addition for use with oxygen, it was used to obtain about 80% dissociation of chlorine in helium buffer gas.

Schwalm (1983) dissociated fluorine in a sapphire tube that was 150 mm long and 8 mm in internal and 12.7 mm in external diameter and surrounded at the end closest to the beam exit by a cylindrical microwave cavity. The beam was emitted through a 1-mm-diameter aperture in a 1-mm-thick sapphire disk. Schwalm reported problems with reactions of atomic fluorine with metallic components in his vacuum system. Almost complete dissociation was obtained with 25 W of microwave power. Neumark et al. (1985) report briefly that they obtained a beam of fluorine atoms from a nickel oven that was resistively heated to a temperature of 920 K. The gas pressure within the oven was 270 Pa. The beam was estimated to be about 50% dissociated. It emerged through a rectangular aperture with dimensions 1.5 mm by 1.8 mm.

Geis et al. (1986) calculated the degree of dissociation of fluorine, chlorine and bromine molecules on tungsten. Their curves show that over 90% dissociation could be obtained at temperatures respectively of about 1300 K, 2100 K, and 1700 K. Although this and later work (Geis et al. 1987, 1988) is concerned with producing atomic halogens for etching surfaces, their experience of the temperature dependence of the compatibility of atom beams with rhenium, a tungsten–rhenium alloy and iridium is valuable for the design of apparatus for halogen beam production.

Stinespring et al. (1986) dissociated fluorine in a 10-mole-percent mixture with helium using the microwave cavity described by Fehsenfeld et al. (1965) and discussed in Section 6.3.5. The gas mixture flowed through an alumina tube at a pressure in the range 6.7 Pa to 1.3 kPa. The power required was in the range of 30 W to 60 W and the beam emerged through a 25-µm-diameter orifice. Below a pressure of 270 Pa the fluorine was found to be completely dissociated, but the degree of dissociation decreased at higher pressures. Despite the small dissociation energies, Spence and Steingraber (1988) were unable to detect any halogens dissociated from their molecular forms using the microwave cavity described by Fehsenfeld et al. (1965). Beams of fluorine and chlorine were obtained by Freedman and Stinespring (1992), using essentially the same apparatus design described by Stinespring et al. (1986). They used a gas mixture of 5% fluorine in argon at a pressure of 270 Pa in an alumina discharge tube. At 70-W power, 'nearly' 100% dissociation of fluorine was obtained. A gas mixture of 5% chlorine in argon, however, produced dissociation in a fused silica discharge tube only in the range 50% to 70% after the tube had been coated with a halocarbon wax.

McCullough et al.'s (1993b), Higgins et al.'s (1995a) and McCullough's (1997) copper slotted line radiator source which we have described in Section 6.3.5 for hydrogen, Section 6.6 for oxygen and Section 6.7 for nitrogen and McCullough et al.'s (1993a) source described in Section 6.3.5 for hydrogen and Section 6.6 for oxygen was also used for chlorine. It was found that dissociation of over 96% was obtained over a wide source pressure. However, the dissociation decreased with increasing pressure and hence with increasing beam intensity.

Ericson et al.'s (1994) production of a supersonic beam of fluorine is of interest for the materials they investigated that would be compatible with atomic fluorine. They noted that nickel was only usable up to a temperature of about 980 K. Calcium fluoride [CaF_2] was usable up to, but not above this temperature and barium fluoride [BaF_2] proved too fragile to use. However, magnesium fluoride [MgF_2] was successful up to 1300 K.

Wang et al. (1994) briefly mention that they produced a beam of iodine atoms by dissociating a mixture of methyl iodide (idomethane) [CH_3I] and argon in a microwave discharge operating at a power of up to 100 W. The gas flowed through an 8.5-mm-internal-diameter fused silica discharge tube which had been treated with orthophosphoric acid [H_3PO_4].

Faubel et al. (1996) consider possible materials for the construction of a long-lifetime fluorine oven by referring to detailed published information on its corrosive properties. They present calculated curves of the degree of dissociation of fluorine with temperature. At a pressure of 170 Pa dissociation should be almost complete at a temperature of 1300 K, but higher temperatures are needed at higher pressures. After reviewing previous work producing supersonic beams with oven materials such as pure nickel, they investigated artificial sapphire [Al$_2$O$_3$] and magnesium fluoride [MgF$_2$] in detail. Ovens of these materials were constructed from single crystals. That of magnesium fluoride was 100 mm long, 10 mm in external diameter and 4 mm in internal diameter. The sapphire crystal was similar in size. It was heated with thermocoax and surrounded by three nickel radiation shields and mounted in a water-cooled copper block. The use of a commercially available magnesium oxide paste coating on the heating coils and hottest parts of the radiation shields enabled temperatures up to 1300 K to be produced. The measured degree of dissociation of their seeded supersonic fluorine beam was 50%. The authors conclude that both nickel and alumina are suitable for temperatures up to 920 K and magnesium fluoride up to 1300 K, with increased lifetimes at lower temperatures.

Dharmasena et al. (1997) refer to previous work on substances for use with atomic fluorine. Fluorine is only up to 15% dissociated at the maximum operating temperature for an oven made of nickel. Iridium had also been found to be impracticable for an oven material. They describe a tube that was 140 mm long and made from a single boule of magnesium fluoride. It was heated by a 100-μm-thick tantalum ribbon which was supplied with 150 W of alternating current power. A careful arrangement of radiation shields enabled the 1-mm aperture for the beam exit to be the hottest part of their oven. The source had proved to be stable for 10 h of continuous running and at temperatures of over 1300 K. The measured dissociation of the fluorine was 86% at an oven temperature of 1300 K. A supersonic beam was produced, but the design is suitable for producing a thermal beam. Keil et al. (1997) give further constructional details of this source.

Wang et al. (1998) tested several means of dissociating bromine in a 100-W microwave source, using an 8.5-mm-internal-diameter fused silica discharge tube which had been pre-treated with phosphoric acid. They found that one part of silicon tetrabromide (tetrabromosilane) [SiBr$_4$] to three parts of either argon or helium produced the best results.

An apparatus for the thermal dissociation of chlorine is described by Levinson et al. (2000) and shown in Figure 6.42. Chlorine was fed into a non-porous graphite tube which was 48 mm long and nearly 5 mm in diameter. It was heated by electron bombardment from a 250-μm-diameter tungsten filament surrounding the tip of the graphite tube and insulated from it with boron nitride. The internal diameter of the tube's tip was 500 μm for a length of 11 mm. Temperatures of about 2300 K were reached with one tantalum radiation shield, a power input of 35 W to heat the filament and 30 W to accelerate the electrons to the last 5 mm of the tube. However, almost complete dissociation of the chlorine molecules was reached at a temperature of 1900 K. Their measurements agree with calculations involving the Gibbs free energy.

Rusin and Toennies (2006) improved Faubel et al.'s (1996) fluorine seeded atomic beam source described earlier by using an oven constructed of a single crystal of calcium fluoride [CaF$_2$] which was so mounted to reduce mechanical strains during the heating cycle. The apparatus is shown in Figure 6.43. Operating lifetimes of 800 h were achieved at temperatures of about 1300 K, but these reduced to about 500 h at temperatures 200 K hotter. They give many further details of the oven construction, which included calcium fluoride granules between the thermocoax heaters and the crystal in order to provide good thermal contact with reduced thermal strains.

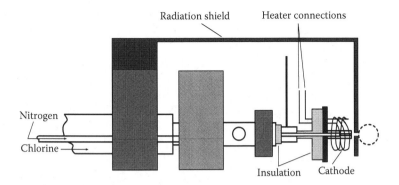

FIGURE 6.42 Levinson et al.'s method of producing a beam of atomic chlorine. (Adapted from Levinson JA, Shaqfeh ESG, Balooch M and Hamza AV, 2000. Ion-assisted etching and profile development of silicon in molecular and atomic chlorine. *J. Vac. Sci. Technol. B* **18** 172–90. With permission from the American Vacuum Society, Copyright 2000.)

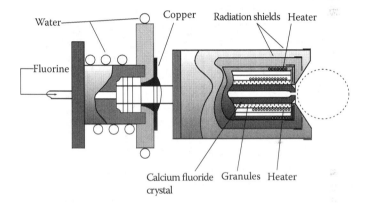

FIGURE 6.43 Rusin and Toennies' calcium fluoride oven for producing fluorine. (Adapted from Rusin LY and Toennies JP, 2006. An improved source of intense beams of fluorine atoms. *J. Phys. D: Appl. Phys.* **39** 4186–93. Copyright IoP Publishing 2006.)

In conclusion, high degrees of dissociation of the halogens have been obtained both in ovens and in discharge tubes. In the former, careful choice of oven materials is essential and in the latter, surface treatment is important in addition.

6.9 BEAMS OF OTHER RADICALS

The beams of free radicals described in this section have nearly all been produced by thermal dissociation.

The production of beams of the methyl [CH_3] and ethyl [C_2H_5] radicals is described by Fraser and Jewitt (1937). The starting point was the corresponding lead compounds, tetramethyl lead [$C_4H_{12}Pb$] and tetraethyl lead [$C_8H_{20}Pb$]. These are liquid at laboratory temperature, so their vapour pressures were controlled by cooling in a freezing mixture at temperatures between 263 K and 273 K, which rather stretches the definition of an oven. The vapour passed into a fused silica tube which was heated externally by a nichrome ribbon to produce the free radicals, which then passed through a 500-μm-diameter aperture to form the beam. Brief details of the production of a beam of the methyl radical are given by Kalos and Grosser (1969). They dissociated dimethyl mercury [C_2H_6Hg] and trimethyl bismuth [C_3H_9Bi] in three tantalum tubes that were made from 25-μm-thick foil. They were heated both by radiation and electron bombardment from tungsten ribbon filaments to temperatures of up to 1800 K. A beam of the methyl radical was produced by McFadden et al. (1972) by dissociation of either dimethyl cadmium [C_2H_6Cd] or dimethyl zinc [C_2H_6Zn] at a temperature of about 1500 K and pressure of about 100 Pa in a tantalum tube. This was 18 mm long, 5 mm in internal diameter and 25 μm in wall thickness. Beam collimation was obtained with a 'crinkly foil' spiral which was inserted into the end of the tube. It was estimated that the beam contained from about 7% to 20% of the methyl radical CH_3, with other components being the constituent metals and hydrocarbons.

Brief details of the production of a molecular beam of the methylene [CH_2] radical are given by Porter and Grosser (1980) who passed a methylene halide beam from a hypodermic needle through an oven containing an alkali metal vapour. Other components in the resulting beam were much slower and could therefore be removed with a velocity selector. Sodium and caesium were reacted both with methylbromide (bromomethane) [CH_3Br] and methyliodide (iodomethane) [CH_3I] and in addition, rubidium was reacted with methyl bromide.

Garvey and Kuppermann (1984) report that an arc source similar to that used by Way et al. (1976) described in Section 6.3.3 also produced metastable H_3 molecules when operated at pressures in the range 7.3 kPa to 11 kPa.

We have already mentioned Geis et al.'s (1986, 1987, 1988) production of halogen atomic beams in the previous section. They also dissociated bromomethane (methylbromide) [CH_3Br], carbontetrafluoride (tetrafluoromethane) [CF_4], bromotrifluoromethane [$CBrF_3$], hexafluoroethane (perfluoroethane) [C_2F_6] and azomethane (dimethyldiazene) [$C_2N_2H_6$] in a tungsten tube which was directly heated up to a temperature of 2300 K to form radicals. They reported that sulphur hexafluoride [SF_6] could not reliably be produced this way because of its reaction with hot tungsten. Since they were concerned with etching surfaces, they did not produce collimated radical beams, but their method could be adapted to produce them. Geis et al. (1987) also dissociated butane [C_4H_{10}], dimethyl ether [C_2H_6O] and acetone [C_3H_6O] on hot tungsten to produce mixtures of radicals.

Another approach to the production of a beam of methyl radicals has been used by Peng et al. (1992) who used azomethane (dimethyldiazene) [$C_2N_2H_6$] which flowed either through a fused silica tube which was 1 mm in internal and 3 mm in external diameter or an alumina tube which was 800 μm in internal and 1.6 mm in external diameter. Typically a section near the exit of the 227-mm-long tube was heated to up to a temperature of 1300 K for a length of 20 mm by passing less than 10 A of current through a 250-μm-diameter tantalum wire heater which was closely wound on the tube and enclosed in an alumina-based cement. The authors considered that the final 12 mm of the tube provided cooling for radicals with high thermal velocities. Some tubes of other dimensions were investigated.

Mass spectrometric studies showed that azomethane is completely decomposed at a temperature of 1173 K.

We have already mentioned Horn et al.'s (1999) furnace for the production of a beam of atomic hydrogen in Section 6.2.3. They suggest it is also useful for producing radicals such as the methyl radical from azomethane (dimethyldiazene) $[C_2N_2H_6]$ in ultrahigh vacuum. Schwarz-Selinger et al. (2001) formed the methyl radical both by the decomposition of methane $[CH_4]$ at 1650 K and also dimethyldiazene at 1150 K in a tungsten tube. It was 50 mm long, 1 mm in internal and 1.6 mm in external diameter, so $\Gamma = 50$. It was heated by the method described by Horn et al. The methyl radical intensity was found to be about an order of magnitude higher from azomethane than from methane. Jacob et al. (2003) describe how dimethyldiazene can be admitted to their tube of the same design by heating an azomethane–copper chloride $[CuCl_2]$ complex at a temperature of about 320 K. Details of synthesizing this stable complex are given. Analysis by a mass spectrometer showed that several molecules were also present in their methyl radical beam.

Tu et al. (2009) obtained a beam of the SrF radical by evaporating a mixture of the powders of strontium fluoride $[SrF_2]$ and a 10% mole fraction of boron in a graphite oven that was 20 mm long and 10 mm in outside diameter. A current of 310 mA at 700 V enabled it to be heated by electron bombardment to a temperature of 1550 K. It was shielded by a single stainless steel radiation shield. A running time of 4 h was obtained with 800 mg of mixture. The radical beam emerged through a 4-mm nozzle in the oven, of which no further details are given and was collimated by two circular apertures.

6.10 BEAMS OF MOLECULES OF GASEOUS SPECIES

In general, production both of molecular beams from gaseous molecules and beams of inert gas atoms at laboratory temperature do not cause any particular problems. However, we mention Gordon et al.'s (1955) production of a beam of ammonia molecules using a crinkly foil array of the type discussed in Section 10.2 to which anhydrous ammonia was admitted from a cylinder at a pressure of several hundred Pascal. They corrugated a 6.35-mm-wide strip of either stainless steel or nickel foil that was 25 μm thick and wound it on a mandrel with alternately plane strips. This produced an array that was about 10 mm in diameter that contained channels about 50 μm by 300 μm. Pumping of the beam was assisted by cooling internal surfaces of their vacuum system with liquid nitrogen. However, this has the disadvantage for long term operation that solid ammonia builds up on the surfaces.

The development of a technique for producing a cooled beam of molecular oxygen by flowing it through a small cryogenic cell containing helium has been briefly described by Patterson and Doyle (2007). The helium is pumped by activated charcoal cryopumps and differential pumping is used to produce the oxygen beam.

7 Gases

7.1 INTRODUCTION

Two topics that logically follow a chapter on gases are included here, namely, the recirculation of gases used to form beams and the levitation of solids. This was originally undertaken with gas beams, but we take liberties with the chapter title by including other methods that may prove useful for atomic beam production. Details of the vacuum systems in which the beam is produced, such as pumping and differential pumping, have not, in general, been included when discussing experiments, since these tend to be closely associated with the experiment for which the beam is required. Neither have general vacuum techniques been discussed, since these are covered in the many textbooks devoted to them. We nevertheless include details of methods that have been or could be used to recycle gases used in atomic beam experiments, since these tend not to be included elsewhere. It is also not our practice to discuss atomic beam applications such as those that are covered by the review articles referred to in Section 1.2, but we make an exception for levitation by atomic beams, since we are not aware of any review that includes these. We then add other levitation techniques that might prove useful for producing beams of refractory metals.

7.2 RECIRCULATION OF GASES

Recirculation of gases is necessary when the material is expensive, but it is best avoided otherwise, since the complexity of purifying the gases for reuse is not justified. It is also essential if the gas is radioactive and so cannot be discharged to the atmosphere. We describe methods that have been used for hydrogen and its isotopes, particularly tritium, as well as helium-3, neon-21 and cyclopropane.

A hot palladium thimble was used by Nelson and Nafe (1949) to purify tritium in their gas recirculation and recovery system. Only brief details of their pumping system are given, but they mention that they had problems both with loss of tritium and evolution of hydrogen when their discharge was operated to form the atomic beam.

Weinreich et al. (1953) describe briefly how 3×10^3 mm^3 of He^3 could be recirculated using a mercury diffusion pump. Liquid air traps removed all condensable material, and copper oxide heated to a temperature of about 800 K removed hydrogen. The aluminium electrodes of the discharge tube removed air leaking into the vacuum system. Weinreich and Hughes (1954) chose a zirconium filament heated to a temperature of about 1600 K to purify their He^3, since zirconium adsorbs hydrogen and combines with oxygen and nitrogen. The aluminium cathode of the discharge also provided an efficient means of removing impurities. They also found that copper oxide heated to a temperature of about 800 K was essential to oxidise hydrogen to water and methane to water and carbon dioxide, which were frozen out in liquid nitrogen traps. The authors give a schematic diagram of their gas handling system as well as many other details of their experience with circulating He^3.

In a short note, Grosof and Hubbs (1956) indicate that Ne^{21} disappeared into the electrode system of a Wood's tube. They therefore used an electrodeless discharge tube. Since they found radiofrequency sources difficult to shield, they employed a frequency of 500 Hz at an operating voltage of 5 kV when a current of 20 mA passed through the discharge tube. This was 750 mm long and 7 mm in diameter, with 70-mm pyrex glass spheres at both ends, to which aquadag provided electrical contact. Hubbs and Grosof (1956) used titanium heated to a temperature of about 1200 K to remove hydrogen, oxygen and nitrogen from their 10^4 mm^3 sample of Ne^{21} in their recirculating system using the discharge system described above.

We have already mentioned Prodell and Kusch's (1957) discharge source in Section 6.3.4. They briefly state that their recirculating system for tritium involved passing it through palladium. They reported that running their discharge caused a serious reduction in the available volume of gas.

Prada-Silva et al. (1977) describe a molecular beam system in which gas is recycled using a mercury diffusion pump that was backed by a mercury booster pump. Stainless steel bellows reciprocating pumps completed the oil-free circulating system. Only an ice-cooled trap was necessary in this system to condense mercury vapour. However, the beam of cyclohexane $[C_6H_{12}]$ was not subject to any means of dissociation.

The next two papers do not describe gas recycling in conjunction with atomic beams, but they describe recirculation systems of a scale that would make them suitable for beams. Schlindwein et al. (1990) briefly describe a turbomolecular pump modified for tritium applications in conjunction with an oil-free backing pump and an adsorber. Bertl et al. (1995) discuss metals that reversibly absorb hydrogen and its isotopes to form hydrides, thus forming a storage medium. They selected a proprietary alloy that was supplied in steel bottles equipped with valves. The gas absorption is effective at laboratory temperature and desorption occurs at temperatures up to over 500 K. The hydrogen in their circulating system was also purified by a palladium purifier, with water vapour being removed by a liquid nitrogen trap.

In conclusion, the circulation of gases requires clean vacuum pumping techniques in conjunction with appropriate gas purification.

7.3   ATOM BEAM LEVITATION

A general introduction to the various physical means of producing levitation is given by Brandt (1989). In particular he discusses aerodynamic levitation of which the most familiar example is the suspension of a table tennis ball in the exhaust stream of a domestic vacuum cleaner. The review by Santesson and Nilsson (2004) updates Brandt's article and then concentrates on the acoustic method. Levitation by acoustic, optical, electric, magnetic, radiofrequency and superconducting means is considered together with applications of each of these techniques.

The possible use of levitation for containerless production of atomic beams from refractory materials (Section 5.17) obviously requires less power than direct electron beam heating from molten metal in a water-cooled container. This is first discussed in Section 3.2.5 and in connection with the work of Ohba and Shibata (1998) in Section 5.9. Some subsequent sections consider other applications. Hence we next discuss experiments to achieve levitation both by electromagnetic and electrostatic means.

Heating an oven by electromagnetic induction is discussed in Section 3.2.3. However, by arranging radiofrequency coils so that there are more turns of smaller diameter at the base of the heating area and one turn at the top in which current flows in the reverse direction, van Audenhove (1965) produced a potential well in which levitation of the metallic sample to be evaporated was achieved. The water-cooled copper coil that was 1.5 mm in external diameter was supplied with up to 4 kW of radiofrequency power at a frequency of 1 MHz. The coil, which was 12.5 mm in external diameter and 9 mm long, was found to be preferable to a larger design because a glow discharge was likely to be formed. Successful evaporation of samples of aluminium, cobalt, copper, iron, silver, titanium and uranium, each weighing up to 1 g, was achieved.

Details of apparatus for electrostatic levitation of samples that were up to 4 mm in diameter at temperatures of over 2000 K in high vacuum have been given by Rhim et al. (1993). They describe the electrode system and servo feedback systems. The position of the sample is detected by illumination with two lasers in conjunction with position detectors. It is heated by radiation from a 1-kW high-pressure xenon arc lamp. Details of the sample preparation and its electrostatic charging are given. Evaporation was not attempted, but melting of aluminium, antimony, bismuth, copper, germanium, indium, nickel, lead, tin and zirconium was achieved, with most experiments being performed on the latter. Ishikawa et al. (2005) review subsequent work. They employed lasers to melt electrostatically several metallic and refractory elements at temperatures up to 2800 K.

We now discuss levitation of microspheres by arrays of capillaries, which were introduced in Section 1.1 and which are discussed in more detail in Chapter 10.

Cahn (1964) discusses the conferences and publications concerned with the production of micropellets as fuel for inertial confined fusion reactors, in which extremely powerful lasers are fired at tiny spherical shells containing a mixture of deuterium and tritium at a pressure of 5 MPa. He postulated that because of possible military applications, much information was then no longer being published. However, the proposed European High Power Laser Energy Research Facility (HiPER) will be a civilian facility. Levitation of microballoons is used to coat them by, for example, sputtering.

The following papers discuss the levitation of microspheres using the gas flow from capillary arrays in order that they may be coated with metals and dielectrics. Lowe and Hosford (1979) used a collimated holes structure (Section 10.5) that was both 6.35 mm in diameter and thickness, which contained channels that were 50 μm in diameter. The structure was dimpled by electrical discharge machining to a depth of 3.175 mm and a claimed diameter of 9.525 mm. In one experiment, argon gas at a pressure of 12 Pa passing through this structure levitated glass microballoons within the dimple with good lateral stability when the structure was carefully levelled. Their diameter was 200 μm, the wall thickness was 2 μm, and they weighed about 2 μg. Following on from these experiments, Crane (1980) showed how deflecting the periphery of the beam toward the centre of the plane array produced radial stability of the levitated sphere.

A focussing glass capillary array (Section 10.7) was successfully used for levitation of glass microspheres by Varon and Goldstein (1981). Their plate was 9.5 mm in diameter, and the 50-µm-diameter pores pointed to a point 25 mm above the surface of the plate, which was 3 mm thick. It was mounted in a funnel-like structure, which had a 30° lip that deflected the beam inward in the same way as in Crane's (1980) design. An argon pressure of 130 mPa to 1.3 Pa levitated the hollow spheres that were from 150 µm to 400 µm in diameter. The authors indicate that levitation over a wide pressure range and sphere size was found possible. A similar technique is briefly reported by Glocker et al. (1982), who added a bevelled washer to the array to improve the lateral stability.

Crane et al. (1982) made more detailed experiments on the levitation of microspheres of several different diameters as a function both of the input pressure to the array in the range of 13 Pa to 1.3 kPa and of the diameters of the capillaries, which were 10 µm, 25 µm and 50 µm. Since the atom density produced by a capillary array is at a maximum in its centre, this is clearly not inherently a stable configuration for levitation. The authors suggest that a minimum in the density on the axis of the array may be achieved either by having a lower gas feed to its centre as shown in Figure 7.1a or by having longer capillaries in the centre as shown in Figure 7.1b. However, the use of a centering ring to deflect the periphery of the beam, as shown in Figure 7.1c, was found to be preferable. However, Rocke (1982) found that a centering ring was not necessary with his glass capillary array, which had 25-µm-diameter capillaries. Levitation of his hollow glass microspheres, which weighed less than 1 µg and which were from 100 µm to 200 µm in diameter, was in the height range of 50 µm to 200 µm. Finally, Katayama et al. (1990) used a plane capillary array with capillary diameters of 5 µm in conjunction with a concave centering ring to levitate microspheres that were 900 µm in diameter. They used acetylene (ethyne) $[C_2H_2]$ at a pressure of 100 Pa as the levitating gas. Measurements were made of the stable pressure region for levitation, which was in the range of 10 Pa to 2 kPa.

In conclusion, radiofrequency and electrostatic levitation may have applications in atomic beam production, and an unusual use of either metal or glass capillary arrays has been the gas flow through them in order to successfully levitate small glass spheres for coating.

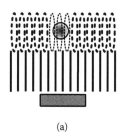

(a)

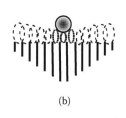

(b)

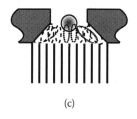

(c)

FIGURE 7.1   Some levitation configurations used by Crane et al. (a) Lower gas feed in the centre, (b) longer capillaries in the centre, (c) centering ring. (Adapted from Crane JK, Smith RD, Johnson WL, Jordan CW, Letts SA, Korbel GR, and Krenik RM, 1982. The use of molecular beams to support microspheres during plasma coating. *J. Vac. Sci. Technol.* **20** 129–33. With permission from the American Vacuum Society. Copyright 1982.)

8 Theory of Collimated Atomic Beam Formation

8.1 INTRODUCTION

In Section 1.1 we introduced the subject of atomic and molecular beam formation by stating that a beam with a cosine distribution is obtained when the beam-forming impedance is a circular orifice and that a narrow slit has also been employed to produce an improved beam. Most relevant theoretical papers use the convention of referring to 'molecules' for both atoms and molecules, but we continue to retain the consistent use of 'atoms' to mean both atoms and molecules unless a distinction is necessary. Following the introduction of the basic equations of the kinetic theory of gases in Section 2.2, the relatively simple expressions for the beam formed by an orifice were introduced in Section 2.3. It was also mentioned in Section 1.1 that when the impedance is a single tube a much narrower collimated beam is obtained and that by employing an array of tubes a greatly improved beam intensity is obtained. In this chapter we present the theory of beam formation to include the more difficult case of flow through a tube. We only briefly mention calculations that only apply when $\Gamma \le 10$ where Γ is the ratio of length l to diameter d (Section 2.5). These are of little interest for atomic beam formation, but are needed for calculations of Knudsen cell operation, which are outside the scope of this work. Following the discussion of the basic theory, we present in the next chapter simplifications that make the design of an atomic beam and the predictions of its performance easier. However, we warn that an extensive comparison of theoretical results with experiment in Chapter 11 points to an unsatisfactory disagreement between theory and experiment for the beam half-width, especially in the case of gaseous rather than vapour beams and also particularly when the beam-forming impedance is a focussing capillary array.

The historical development of the application of the kinetic theory of gases to flow through tubes is covered in detail by the books and review articles mentioned in Section 1.2. Knudsen's (1909a,b) formulae for the throughput both through an orifice and through cylindrical tubes of circular cross-section and large Γ form the basis of continuing developments. As well as Knudsen, early contributions were made by Dushman (1962) but originally published in 1920 and in a series of papers by Clausing that were published between 1926 and 1932 and which are listed by Clausing (1932). His derivation of the flow through tubes of any Γ provides the basis for most subsequent treatments of the subject. Clausing (1930) derived the angular distribution of the beam produced by a tube, provided that no collisions occurred within it.

Before discussing the development of the theory of atomic beam formation and detailing the results achieved and its limitations, we extend the basic assumptions of the kinetic theory of gases already used in Section 2.4. These are that atoms can be treated as perfectly elastic spheres that occupy a small volume compared with the volume of the container they occupy. When considering the flow of gas through an impedance in the form of a collimating tube, the additional assumptions that are made are as follows. First, the pressure of the gas is assumed to be so low that collisions between the individual atoms of the gas in the tube can be neglected. The atoms move independently of one another and their motion is determined by the interactions they have with the walls of the tube. They travel in straight lines between collisions. This type of flow is unfortunately referred to by many alternative expressions, namely, free-molecule flow, molecular flow, collision-free flow, diffusive flow, Knudsen flow, transparent mode and in this chapter also as free-atomic flow. Diffusive flow does not imply that surface diffusion is taking place. The condition for this flow is that the mean free path λ of the atoms must be at least as large as the characteristic dimension of the system. Thus as already mentioned for an orifice, the mean free path must be at least as large as the diameter d, whereas for a tube of length l the condition is that $\lambda \geq l$. However, the optimum operating conditions for a collimator are at an input pressure such that $\lambda < l$, but $\lambda \geq d$ so that collisions between the individual atoms need to be taken into account. These are considered in Sections 8.3.5, 8.4.5 and 8.5.4.

In this chapter to save repetition, it will be assumed, unless stated otherwise, that the mean free path is always greater than or equal to the tube diameter. When this condition is not met, namely, at slightly higher gas pressures, transitional flow is said to take place. The transition is to viscous flow when quite different equations governing mass flow at high pressures apply. This is also referred to as continuum flow and drift flow. At even higher pressures, as introduced in Section 1.1 and discussed in more detail in Section 11.7, supersonic flow occurs.

Another assumption concerns the mode of interaction of the atoms with the tube walls. Knudsen assumed that the walls were rough on a microscopic scale so that the atoms colliding with the wall lose all trace of their original motion and are reflected in all possible directions, the probability of a given angle of emergence being given by the cosine law. It is referred to as the Knudsen law of diffuse scattering and is considered further in Section 12.2. An alternative way of viewing the surface interaction is that atoms are adsorbed, losing all memory of the previous motion and are then re-emitted with a cosine distribution. These atoms are also sometimes described as accommodated.

Some of these assumptions may not necessarily be valid and may cause significant discrepancies between theory and experiment. Some effects such as specular reflection from smooth surfaces, backscatter from those that are rough on a macroscopic scale, and surface diffusion may be important. In this latter case atoms are adsorbed and migrate along the surface before being re-emitted. A detailed review of surface diffusion has been made by Levdansky et al. (2008), who also refer to previous reviews. They consider the effect of surface diffusion on the flow of gases through tubes under free-atomic flow conditions, but the effect on the angular distribution of the beam does not appear to have been calculated. Hence Knudsen's assumptions concerning collisions with surfaces are usually assumed in calculations and are the basis we use initially. It is apparent from Chapters 4 and 5 that for the production of beams from vapours the impedance is almost invariably held above laboratory temperature. For the production of atoms by thermal dissociation discussed in Chapter 6, considerably higher temperatures are usually involved. As discussed in Section 8.3.5 an atom makes many collisions with the walls of the impedance before it is emitted. It is therefore assumed that the temperature of the atoms is the same as that of the impedance. The discussion is also restricted to steady flow through the impedance. The gas in the first vessel V_P is assumed to be in equilibrium so that the atom velocities are distributed according to the Maxwellian velocity distribution and results from the kinetic theory of gases that were stated in Chapter 2.2 apply.

Since the same cosine law is valid for the diffuse reflection of radiation, where it is known as Lambert's law, it follows that the methods developed for radiation may be applied to atoms. However, the results presented by Buckley (1927, 1928) and Hottel and Keller (1933) only apply to tubes with small values of Γ, and they have also been superseded by more accurate numerical calculations. However, Walsh's (1920a,b) treatment of radiation was used as the basis of Clausing's (1932) approach and Usiskin and Siegel (1960) provide simple expressions for which evaluations can be made that can be tested for their suitability when Γ is large.

For ease of reference we will refer to the first vessel at a pressure p to be V_P and that at the output of the impedance where the beam is formed to be V_0. Many other terms have been employed for V_p including 'source', 'reservoir', 'source chamber', 'entrance' and 'input'. We consider the impedance between these two vessels V_P and V_0 to be a right circular cylinder of length l and diameter d. We also assume that the pressure in the second vessel V_0 is zero and hence may be referred to just as vacuum. Because of the normally large difference in pressure between the two vessels, assuming zero pressure in V_0 is a good approximation in practice.

8.2 CLAUSING'S TREATMENT OF FLOW THROUGH A TUBE

Clausing's (1932) derivation is of fundamental importance, since in order to calculate the properties of the atomic beam produced by a tube, a knowledge of the rate of incidence of atoms on its walls and the number of atoms transmitted by it are required. Subsequent derivations or discussion of the Clausing equation has been given by Beijerinck et al. (1976), Bröhl and Hartmann (1981), Patterson (in Gadamer et al. [1961]), German (1963), Ivanov and Troitskii (1963), Yamamoto et al. (1980), Davies and Lucas (1983), Iczkowski et al. (1963), Cercignani (1988) and Mohan et al. (2007). Several numerical solutions of Clausing equations for axi-symmetric tubes other than right circular cylinders have also been made. It is evident that these have of necessity small values of Γ to be of any practical use and hence fall outside the scope of this work. The methods of construction of capillary arrays described in Chapter 10 indicate that although the tube cross-section is usually circular, there are cases, for example, for metal collimators, where this is no longer true. However, in these cases the cross-section is also ill defined. Hence we limit our theoretical treatment to the right circular cylinder.

We can now introduce what is now usually known as the Clausing function $\eta(X)$, although Clausing (1932) called it the escape probability. It is also known as the normalised incident rate or normalised collision density. The normalisation refers to the reference to the collision rate at the opening of the tube being related to Equation 2.9 for the flow through an orifice and hence the rate of impingement upon it. We illustrate the terminology we are using by reference to Figure 8.1.

However, before we consider the Clausing equation, we note that Clausing (1932) related distances from the high pressure end of the tube to be at $x = l$, whereas subsequent treatment has assumed this to be at $x = 0$ as in Figure 8.1. This leads to his simple equation for $\eta(x)$, namely, that

$$\eta(x) = \alpha + (1 - 2\alpha)\, x/l \tag{8.1}$$

where α is a constant dependent only upon Γ. When $x = l/2$ it follows from Equation 8.1 that $\eta(x) = 0.5$ and this is independent of the value of α. More generally the symmetry of the Clausing function is defined by $\eta(x) + \eta\,(l - x) = 1$. Or in words the Clausing function at the tube exit is the same as that amount at the tube entrance that is less than unity. Also the Clausing function at a distance x from the input of the tube added to the function at $l - x$ is unity. This is illustrated in Figure 8.2.

Clausing only trusted the validity of Equation 8.1 for tubes with small values of Γ. It assumes that the Clausing function is linear, and we discuss the linearity later, but it is sufficient to state at present not only that the linearity is a good approximation but also that the expression applies to tubes of any length.

Bröhl and Hartmann (1981) made two deductions that concern this symmetry about the mid point of a tube. They deduce that equal numbers of atoms colliding with the same surface area at both x and $l - x$ in the same time will move directly or indirectly into V_0, without returning at all to V_p. Secondly equal numbers of atoms colliding with the tube from the V_0 side of $l/2$ and from the V_p side of $l/2$ will move to V_p directly or indirectly.

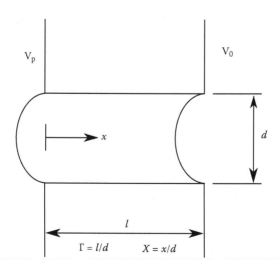

FIGURE 8.1 Atomic beam terminology.

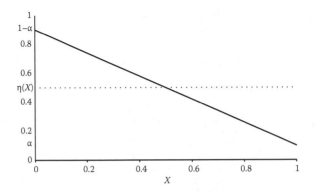

FIGURE 8.2 Symmetry of the Clausing function.

Equation 8.1 can be used as a check on the accuracy of numerical evaluations of η or it can be assumed initially, so that η need only be calculated for half the channel length. It is necessary to keep this symmetry behaviour in any approximate analytic solution that is obtained. However, Equation 8.1 is sometimes overlooked, for example, by Usiskin and Siegel (1960) in some solutions, Ivanov and Troitskii (1963), Olander and Kruger (1970) and Flory and Cutler (1993). Olander and Kruger retain the possibility of an asymmetry in Equation 8.1 in their discussion, which leads to expressions for the angular distribution of a beam in the collisionless case. However, it is a trivial matter to modify their equations to restore the symmetry. Since it has been established from numerical calculations, including those that show that $\eta(x)$ is non-linear with x, we disregard other contributions that ignore it.

In order to be consistent with subsequent work, we retain Equation 8.1 but define the value of the Clausing function at the entrance of the tube to be $\eta(0)$ and at the exit $\eta(l)$ so that when $x = 0$, $\eta(0) = 1 - \alpha$ and when $x = l$, $\eta(l) = \alpha$. We have used actual dimensions in the above discussion since they are more intuitive. It is now, however, usual and convenient to measure lengths in terms of the tube diameter so the distance along the tube is measured in units of d, so that we introduce a dimensionless X where $X = x/d$, and $\Gamma = l/d$. as before. Hence we rewrite Equation 8.1:

$$\eta(X) = 1 - \alpha - (1 - 2\alpha) X/\Gamma \qquad (8.2)$$

so that $\eta(l) = \alpha$ or in words α is the value of the Clausing function in the plane of the exit to V_0 and from symmetry $\eta(0) = 1 - \alpha$ so $1 - \alpha$ is its value in the plane of the entrance V_P. We retain α when discussing Clausing's applications of Equation 8.1 but use $\eta(X)$ more generally.

Since we redefine the input of the tube at V_P to be at $x = 0$ and the output at V_0 to be at $x = l$ it follows that X takes values within the tube from 0 to Γ and the mid point of the tube is given by $x = l/2$ so $X = \Gamma/2$. However, when we wish to compare values of $\eta(X)$ for tubes with different Γs we follow the convention that X takes values from 0 at the tube input to 1 at its exit so that results can be superimposed.

We write Clausing's (1932) integral equation for $\eta(X)$ in the form

$$\eta(X) = \eta_0(X) + \int_0^\Gamma \eta(X')K\left(|X - X'|\right)dX' \qquad (8.3)$$

where $\eta_0(X)\, dX$ is the probability of the first atomic collision with the walls occurring between X and $X + dX$. $K(|X - X'|)\, dX$ is the probability that an atom that has made its first collision with the wall at X' will make its next collision between X and $X + dX$. We have

$$\eta_0(X) = \frac{X^2 + 1/2}{(X^2 + 1)^{1/2}} - X, \qquad (8.4)$$

and

$$K(X) = 1 - \frac{X^3 + 3X/2}{(X^2 + 1)^{3/2}}. \qquad (8.5)$$

$K(X)$ is referred to as the kernel and Equation 8.3 as the Clausing equation and its solution, $\eta(X)$, as the Clausing function. The functions $\eta_0(X)$ and $K(|X - X'|)$, as given by Equations 8.4 and 8.5, are shown in Figure 8.3 in terms of X/Γ for two representative values of Γ, namely, $\Gamma = 10$ and $\Gamma = 100$. It can be seen that as Γ increases, the effective range of the kernel $K(|X - X'|)$ of the integral equation narrows. The function $\eta_0(X)$ also decreases more rapidly away from the tube entrance as Γ increases. The figure also shows the discontinuous slope of the kernel, which means that it is

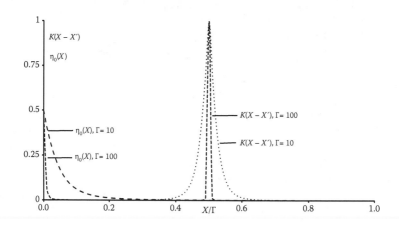

FIGURE 8.3 Functions contained in the Clausing equation.

sometimes described as a split kernel. These factors all combine to illustrate why numerical integrations prove to be quite difficult, especially when Γ is large.

8.3 SOLUTIONS OF CLAUSING EQUATION

8.3.1 INTRODUCTION

The Clausing equation, Equations 8.3 to 8.5, has no exact solution but various means have been used to determine $\eta(X)$. We are interested in Section 8.3 in methods that lead to results specifically for $\eta(X)$. There is some overlap with similar methods that are applied to obtaining solutions for the transmission probability. These are deferred to Section 8.4.

Before discussing solutions of the Clausing equation in chronological order we present our results using the Chebyshev polynomial method for $\eta(l) = \alpha$ from $\Gamma = 10$ in increments of 10 up to $\Gamma = 100$ and then in increments of 100 up to $\Gamma = 1000$ which are given in the second column of Table 8.1 with the corresponding values of Γ in the first. We use these to form a reliable basis for comparing both other numerical calculations and simple approximations to the Clausing function, which were mainly made without knowledge of the true solution.

TABLE 8.1

Calculated Values of $\eta(l)$ and the Slopes, Intercepts and Standard Deviations of Straight Lines Fitted to Calculations of $\eta(X)$

Γ	$\eta(l)$	Slope	Intercept	Standard Deviation
10	0.048094	0.888775	0.944387	2.62×10^{-3}
20	0.025963	0.933504	0.966752	2.14×10^{-3}
30	0.017831	0.951269	0.975634	1.76×10^{-3}
40	0.013591	0.961065	0.980533	1.50×10^{-3}
50	0.010984	0.967353	0.983677	1.30×10^{-3}
60	0.009219	0.971766	0.985883	1.16×10^{-3}
70	0.007943	0.975049	0.987525	1.04×10^{-3}
80	0.006979	0.977597	0.988799	9.52×10^{-4}
90	0.006223	0.979637	0.989819	8.76×10^{-4}
100	0.005615	0.981310	0.990655	8.13×10^{-4}
200	0.002843	0.989452	0.994728	4.84×10^{-4}
300	0.001904	0.992495	0.996251	3.53×10^{-4}
400	0.001431	0.994119	0.997066	2.80×10^{-4}
500	0.001147	0.995138	0.997569	2.34×10^{-4}
600	0.000957	0.995842	0.997921	2.02×10^{-4}
700	0.000821	0.996358	0.998180	1.78×10^{-4}
800	0.000718	0.996755	0.998379	1.60×10^{-4}
900	0.000639	0.997069	0.998538	1.45×10^{-4}
1000	0.001333	0.997325	0.998667	1.33×10^{-4}

Our results in Table 8.1 for $\eta(l)$ overlap with the two largest Γs calculated by Davies and Lucas (1983). We also use these data as a basis of comparison with standard deviations of the approximate solutions discussed in Sections 8.3.2 to 8.3.4 from the Chebyshev solution. In addition we have calculated the straight lines fitted by the least squares method to the 101 values of $\eta(X)$ for each value of Γ and the slope and intercept of each line is shown respectively in the next two columns of Table 8.1. As already discussed the symmetry condition implied by Equation 8.2 means that this line must pass through $\eta(X) = 1/2$, when $\Gamma = 1/2$ irrespective of the linearity of the Clausing function, so the fitted line is automatically forced through this point. The standard deviations of the lines from the Chebyshev calculations for each value of Γ are shown in the last column of Table 8.1. They are obtained from tabulating the difference between the fitted line and the Chebyshev solution at each of the 101 points. Examples of the differences Δ_{LIN} between the Chebyshev solutions and the fitted lines for $\Gamma = 10, 50, 100$ and 1000 are shown in Figure 8.4. It illustrates the symmetry as well as the non-linearity of the Clausing function and also shows that the linearity of the Clausing equation increases with Γ. We return in Section 8.3.3 to the question of where the maximum deviation from linearity occurs.

For atomic beam calculations analytic approximations to the Clausing function have been employed rather than numerical evaluations, because of their simplicity. It is therefore important to establish how reliable these are as the first stage in the calculation of the beam properties. However, since the methods of calculation have been superseded, we do not consider them in detail. We discuss the approximations to the Clausing function that have been obtained in a variety of ways. In Section 8.3.3 we compare them with our accurate numerical solutions from $\Gamma = 10$ to $\Gamma = 1000$. These extend those given by Davies and Lucas (1983) for $\Gamma \leq 20$ which were then compared with analytical methods given by Clausing (1930) using his short tube approximation, by Helmer (1967a), and by the two approximations of Neudachin et al. (1972). We retain our usual chronological order of presentation, independent of the type of approach used.

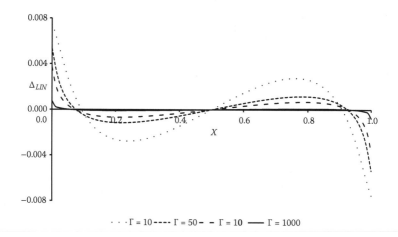

FIGURE 8.4 The Clausing function compared with the linear fit.

8.3.2 EARLY SOLUTIONS

Early solutions of the Clausing equation using exponential approximations to the kernel were only tested at the time to obtain solutions for small values of Γ, so we do not discuss them in detail unless they lead to an algebraic expression that is worth testing at the larger values of Γ that are used for beam formation. If accurate values should be required for small values of Γ, the Chebyshev polynomial method discussed in Section 8.3.4 converges very rapidly and has superseded all other methods. As mentioned in Section 8.1 the approaches of Buckley (1927, 1928) and Hottel and Keller (1933) are now therefore only of historic interest.

Clausing (1926) showed that when Γ is small, the value of α in Equation 8.1 is given by

$$\alpha = \frac{\Gamma^2 + 1 - \Gamma\sqrt{\Gamma^2 + 1}}{1 + \sqrt{\Gamma^2 + 1}} \tag{8.6}$$

We note that as $\Gamma \to 0$, $\alpha \to 0.5$ which is to be expected from the symmetry condition of Equation 8.1 and when $\Gamma \to \infty$ $\alpha \to 0$. Hence for these two extremes of Γ, $\eta(X)$ is symmetrical about the mid point of the tube and it is therefore a reasonable assumption that this is also true for any intermediate value of Γ. Equation 8.6, with α used in Equation 8.1 is what Clausing refers to as the short tube approximation for $\eta(X)$, which of course means that it is applicable when Γ is small. We discuss this table in Section 8.3.3 when we have discussed how the other columns have been derived. The mean standard deviation σ of results obtained by this approximation from the Chebyshev solutions is given in Table 8.2. They are averaged over the values of Γ in Table 8.1. We again defer discussion until the discussion of the method of derivation of all columns is complete.

Clausing (1932) obtains the following more detailed expression for α, namely,

$$\alpha = \frac{(1-X)\sqrt{\Gamma^2(1-X)^2 + 1} + \Gamma(2X-1) - X\sqrt{\Gamma^2 X^2 + 1}}{\Gamma^2\left[X^2(\Gamma^2 X^2 + 1)^{-1/2} - (1-X)^2\{\Gamma^2(1-X)^2 + 1\}^{-1/2}\right] + (1-2X)\left[\sqrt{\Gamma^2(1-X)^2 + 1} + \sqrt{\Gamma^2 X^2 + 1}\right]} \tag{8.7}$$

TABLE 8.2
Standard Deviations of Approximations to the Clausing Function

Author	Year	Equations	σ
Clausing short	1926	8.6 and 8.1	2.624×10^{-3}
Helmer	1960	8.9	2.621×10^{-3}
NPS1	1974	8.11	1.722×10^{-3}
Clausing long	1929	8.7 and 8.8	1.540×10^{-4}
Fitted line			8.599×10^{-4}
NPS2	1972	8.12	5.164×10^{-4}

Clausing tabulated α as given by Equation 8.7 for some small values of Γ against which the calculation of $\eta(X)$ can be checked with only the first value for $\Gamma = 0.05$ in disagreement with our calculations up to the fifth significant figure. He noted that for $\Gamma \ll 1$ α is practically independent of X, so when $\Gamma \ll 1$, $X = 0$ could be used, for example, in Equation 8.7, which reduces to Equation 8.6.

However, instead of substituting $X = 0$ in Equation 8.7, Clausing suggests

$$X = \sqrt{7}/(3\Gamma + \sqrt{7}) \tag{8.8}$$

which he calls his long tube approximation but he considered it to be a good approximation for any value of Γ. Except when referring to these two Clausing's approximations we avoid the use of the expressions 'short tube' and 'long tube', since it is of course Γ, the ratio of the length l to the diameter d of the tube that is important in determining atomic flow not the tube length. The mean standard deviation for $\eta(X)$ is included in Table 8.2.

There are two other approaches that might offer useful means of calculating the Clausing function that we do not consider further. In his extensive treatment of Knudsen flow problems, DeMarcus (1956) considers in detail two other methods of solving linear integral equations. Note that DeMarcus and Jenkins (1957) list the errata to DeMarcus (1956, 1957), and for brevity we do not refer to their paper again. DeMarcus called the first 'squeezing', which is an iterative method of constructing two functions with the property of one being always greater than and the other always less than the true solution. Thus the true solution is squeezed between the two functions. He called the second 'scissoring', which is a variation of and improvement upon squeezing. Two functions are constructed such that the solution of the integral equation lies in the two triangular regions between them. However, neither method has been extended by DeMarcus or by others to present results for the Clausing function for large values of Γ. However, we refer to them again in Section 8.4.2 since they have been used successfully for calculations of the transmission probability.

The first of Usiskin and Siegel's (1960) methods we consider is the derivation of a relatively simple expression for $\eta(X)$ which is worth comparing with other treatments. Figure 8.3 shows that both the kernel $K(|X - X'|)$ and the function $\eta_0(X)$ have approximate shapes of exponential functions. Usiskin and Siegel made use of this similarity in a paper dealing with thermal radiation in which they had obtained an integral equation of exactly the same form as Equations 8.3 to 8.5. They approximated $K(X)$ by the function $\exp(-2X)$ and $\eta_0(X)$ by $(1/2) \exp(-2X)$ and obtained

$$\eta(X) = [0.5 + \Gamma(1 - X)]/(1 + \Gamma). \tag{8.9}$$

We note that Patterson (in Gadamer et al. [1961]) applied the 'scissor' method to obtain a linear solution for $\eta(X)$ but it was only successful for values of $\Gamma \leq 2$.

Without apparently being aware of the work of Usiskin and Siegel (1960), Helmer (1967a) also derived the expression for $\eta(X)$ given by Equation 8.9 as an approximate solution to the Clausing equation for small values of Γ. The standard deviations of Helmer's and Usiskin and Siegel's (1960) solution for $\eta(X)$ from the Chebyshev solutions are given in Table 8.2, which, for brevity, we attribute just to Helmer. We again defer discussion until the discussion of the method of derivation of all columns is complete.

We consider in a group variational methods for solving the Clausing equation. The basic technique of the method is to obtain accurate values of a particular quantity by means of a fairly simple guess at the function. The success then rests upon the ability to find a suitable functional whose stationary value is related to the quantity of interest and for which variations in the function give rise to only second order variations in the functional. DeMarcus (1956, 1957) and Usiskin and Siegel (1960) applied the method for tubes with values of Γ of 10 or less and Neudachin et al. (1972) for Γ up to 80. It was mentioned in Section 8.3.1 that the Clausing function is nearly linear, particularly when Γ is large. DeMarcus, Usiskin and Siegel and Neudachin et al. all used a trial function of the form

$$\eta(X) = 1/2 + C(\Gamma/2 - X) \tag{8.10}$$

and find the value of C from the extremium in the functional they use. DeMarcus does not give explicitly the value of C. It is, however, given by Usiskin and Siegel and Neudachin et al., namely,

$$C = \frac{4}{3} \frac{\Gamma^3 - 2 + (2 - \Gamma^2)(1 + \Gamma^2)^{1/2}}{\Gamma(1 + \Gamma^2)^{1/2} - \sinh^{-1}\Gamma}. \tag{8.11}$$

As mentioned in Section 8.2 the symmetry relation, Equation 8.1, is not satisfied by another value of Usiskin and Siegel's constant. We do not include it since the symmetry relation is well established. The standard deviations of these values of $\eta(X)$ from the Chebyshev solution are included in Table 8.2 and are designated as NPS1. Neudachin et al. (1972) carried the variational method further and obtained

$$\eta(X) = 0.5 + \frac{\Gamma}{4} \times \frac{2a_1 a_{22} - a_2 a_{12}}{a_{12}^2 - 4a_{11}a_{22}} \{0.5 - X\} + \frac{\Gamma^2}{4} \times \frac{2a_2 a_{11} - a_1 a_{12}}{a_{12}^2 - 4a_{11}a_{22}} \{0.5 - X\}^3$$

where

$$a_1 = \frac{2}{3}\left\{\Gamma^3 - 2 + (2 - \Gamma^2)\sqrt{1 + \Gamma^2}\right\}$$

and

$$a_2 = \frac{1}{10}\left\{\Gamma^5 - 10\Gamma^2 + 16 + 5\Gamma^2(3\Gamma^2 + 7)\sqrt{1 + \Gamma^2} - 16(1 + \Gamma^2)^{\frac{5}{2}} + 15\Gamma\sinh^{-1}\Gamma\right\}$$

with

$$a_{11} = \frac{1}{8}\left\{\sinh^{-1}\Gamma - \Gamma\sqrt{1 + \Gamma^2}\right\}$$

and

$$a_{12} = \frac{1}{48}\left\{ 24\Gamma(1+\Gamma^2)^{5/2} +18\Gamma(2-\Gamma^2)(1+\Gamma^2)^{3/2} \right.$$
$$-3\Gamma(18+23\Gamma^2+2\Gamma^4)\sqrt{1+\Gamma^2}$$
$$\left. -3(2+3\Gamma^2)\sinh^{-1}\Gamma+8\Gamma^3 \right\}$$

and

$$a_{22} = \frac{1}{3840}\left\{ 960\Gamma(1+\Gamma^2)^{7/2} +240\Gamma(7\Gamma^2-2)(1+\Gamma^2)^{5/2} \right.$$
$$-10\Gamma(258\Gamma^4-140\Gamma^2+153)(1+\Gamma^2)^{3/2}$$
$$+5\Gamma(201-688\Gamma^2-922\Gamma^4-12\Gamma^6)\sqrt{1+\Gamma^2}$$
$$\left. +45(1-2\Gamma^4)\sinh^{-1}\Gamma+96\Gamma^5 \right\} \tag{8.12}$$

and determining the coefficients that make the functional a maximum. The symmetry relation, Equation 8.2, implies that there can be no terms quadratic in $(\Gamma/2 - X)$. The standard deviations of the differences between Neudachin et al.'s values of $\eta(X)$ and the Chebyshev solution are designated NPS2 in Table 8.2. Their solution is the only one in this section that allows for the non-linearity of the Clausing function.

Figure 8.5 shows for $\Gamma = 50$, 100, 500 and 1000 examples of how the differences Δ_{NPS2} between values calculated from Equation 8.12 and our calculations by the Chebyshev method vary with distance from the entrance to the tube. As with Figure 8.4

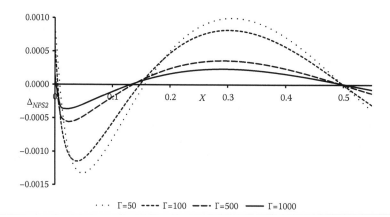

FIGURE 8.5　The difference between Neudachin et al.'s and the Chebyshev solution.

the difference decreases as Γ increases for all values of X. Davies and Lucas (1983) show a curve for smaller values of Γ which show that the difference also decreases as Γ decreases. We have already shown how the linearity of the Clausing equation increases with increasing Γ and this figure shows how NPS2 gives a better fit as Γ increases. It is also apparent that terms of higher order than the cubic would be needed to obtain an improved fit.

8.3.3 Discussion of above Solutions

We can now discuss the evaluation of approximations to the Clausing equation contained in Table 8.2 now that it is complete. For convenience of comparison it is arranged in order of increasing accuracy. As mentioned in the previous section, since Usiskin and Siegel's (1960) and Helmer's (1967a) formulae are identical we refer to them as Helmer's. We note that Clausing's (1932) short tube approximation gives almost identical results to these for the standard deviations from the Chebyshev solution. Helmer's function is, however, marginally superior. Neudachin et al.'s (1972) linear approximation, NPS1, is better than both Clausing's short tube and Helmer's approximations. The Clausing (1932) long tube approximation is a slight improvement over these three approximations. However, its use is hardly justified because of the complexity of the algebraic function used to derive it.

Table 8.2 shows that Neudachin et al.'s (1972) non-linear approximation NPS2 is the best approximation to the Clausing function, being superior to the linear fit when averaged over the values of Γ used to calculate it. Examination of the actual standard deviations for each value of Γ rather than the means shows that NPS2 is an improvement over the linear fitted line to the Chebyshev solution for $\Gamma \leq 100$. Otherwise our linear fit is superior. Figure 8.4 shows that the standard deviations of our fitted straight lines to the Chebyshev solution decrease with increasing Γ. Hence the Clausing function becomes more linear as Γ increases. Therefore if estimates of $\Gamma > 1000$ are required, an analytical approximation should be satisfactory. The non-linearity of the Clausing function was mentioned in Section 8.3.1. As indicated by Figure 8.4 its curvature is such that it dips below the fitted line close to the tube entrance and therefore, from the symmetry given by Equation 8.1, falls above it near the exit. The physical basis of the curvature appears to be that when Γ is small, collisions with the tube wall are dominated by atoms entering the tube that have not yet made a wall collision. However, when Γ is large, atoms already in the tube are little affected by the small number entering it. Both of these contributions are very nearly linear as a function of position within the tube, but with different slopes. When the two effects are comparable the non-linearity becomes apparent.

As a tube tends to an orifice, namely, when $\Gamma \to 0$ $\eta(X) = 1/2$ for all X, so the Clausing function is also linear. It can be assumed from our calculations that as $\Gamma \to \infty$, $\eta(0) = 1.0$ and the function is linear. Hence at some value of Γ, $\eta(X)$ will have a maximum deviation from linearity. A measure of the curvature is obtained from the standard deviation of the fitted straight lines from the Chebyshev solutions for each calculated value of Γ. The value of Γ for maximum deviation does not appear to be a particularly important quantity to establish but it probably lies between $\Gamma = 10$ and $\Gamma = 20$. Since NPS2 is expected to be an advantage when the Clausing function itself is more non-linear, the superiority of our linear fit when Γ is large is not particularly surprising. As shown by Davies and Lucas (1983) NPS2 is an improvement over linear approximations when Γ is small. Our linear fit is of course the only one discussed that takes advantage of knowing in advance the Chebyshev solution to which it is fitted, so its improvement over other linear fits is perhaps to be expected. Hence Table 8.1 gives a quick and fairly accurate means of obtaining respective values of $\eta(l)$ and $\eta(X)$ for the stated values of Γ without performing the Chebyshev calculation with its heavy demand on computing power when Γ is large.

It follows that if approximations to the Clausing functions are required at the values of Γ used in our calculations, the linear fitted parameters offer the simplest and usually the most accurate approach. However, the small range of σ in Table 8.2 of only just over a factor of 5 should be noted. High accuracy is more of relevance to those interested in theoretical approaches than to those requiring data to compare with experiment. In view of the success of the linear least square fits, it is of interest to examine whether cubic fits would be an even greater improvement, since they would allow for the curvature of the Clausing function. Although the standard deviation is a slight improvement over the linear fit, it was not judged sufficient to include values of the coefficients in Table 8.1 nor the values of σ in Table 8.2. The lack of significant improvement could either be due to the accuracy of the Chebyshev coefficients used for the fits or the accuracy of the spreadsheet calculation of the cubic least squares fit. Since rounding the input data produced very little degradation of the standard deviations, it is apparent that these calculations of fits would require higher numerical precision. If these could be made, the possibility exists of producing the equation of the Clausing function surface showing the dependence of $\eta(X)$ on both Γ and X, thus enabling interpolation between the calculated values.

8.3.4　Numerical Solutions

Some of the early numerical treatments were only for small values of Γ. These include those of Clausing (1932), Hottel and Keller (1933), Usiskin and Siegel (1960), Richley and Reynolds (1964), de Leeuw and Gadamer (1967), Helmer (1967b) and Moore (1972). However, Sparrow et al. (1963) calculated $\eta(X)$ up to $\Gamma = 24$ and Garelis and Wainwright (1973) for Γ up to 25 using iterative techniques. The increased speed and storage capacity of computers mean that the Chebyshev polynomial method now enables numerical solutions to the Clausing equation to be obtained up to any value of Γ and to any required accuracy, though the resources required increase rapidly as Γ is increased. The method is even preferable, for small values of Γ where computation is rapid. Numerical solutions are important because they are the only means by which the accuracy of the analytical approximations can be established.

Davies and Lucas (1983) used the Chebyshev polynomial method of solving the Clausing equation, as described by El-Gendi (1969) and implemented by the National Algorithms Group (NAG) program D05AAF, which is recommended for the split kernel. This was found to be more efficient than their improvements to the program given by de Leeuw and Gadamer (1967). However, difficulties arose for large values of Γ, since a large number of polynomial coefficients were required so that both storage space and computing time became excessive. A reasonable practical limit was then reached with $\Gamma = 20$ which was comparable with the largest values considered by other authors at that time. However, it unfortunately meant that there was some uncertainty in the value of the Clausing function in the range most likely to be needed in practice where capillary arrays can be constructed with Γs in the range of hundreds to thousands. We have already mentioned our update that was possible with the availability of increased computing power in Section 8.3.1, since we needed it there as the basis of comparison of other results. For a given accuracy we found that the number of Chebyshev coefficients required was proportional to Γ and the computing time proportional to Γ^3. The accuracy achieved for each number of coefficients employed was established by making use of the symmetry of $\eta(X)$ given by Equation 8.1 and calculating the difference from 1 of $\eta(0) + \eta(l)$. The number of coefficients was increased within the program until the desired six-figure accuracy was achieved.

Lobo et al. (2004) used the same method for 10 values of Γ up to 20. They used 2000, 3000 and 5000 Chebyshev polynomials for each value of Γ in order to determine the uncertainties in $\eta(0)$ of 1 part in 10^{-10}, from the differences. They quoted results to ten significant figures without discussing why this extreme accuracy for $\eta(0)$ is needed. As is to be expected, their results agree with those obtained by Davies and Lucas (1983) and with our more recent calculations to the accuracies quoted in each case.

A detailed numerical comparison of the methods of solving integral equations is given by Delves and Mohamed (1985). We conclude from their study that for solving the Clausing equation for large Γ the Chebyshev polynomial method is the most suitable and we have not found any more recent methods that would be more appropriate.

Before we leave this discussion of numerical solutions of the Clausing equation, we make brief mention of the direct-simulation Monte Carlo method. Applications of this method to atomic flow problems have usually dealt with determinations of the transmission probability (Section 8.4.3). However, Beijerinck et al. (1976) used this approach to determine the Clausing function. Along the axis the tube was divided into a large number of slices (typically, of the order of 100Γ) and the number of collisions with each slice, A_i, was recorded. The results show considerable scatter and a statistical error $A_i^{1/2}$ in the number of collisions per slice is assumed. Beijerinck et al. produced a 'smoothed' version of their results by normalising with respect to the Clausing function in the tube entrance cross-section and then substituting the Monte Carlo results into the right-hand side of the Clausing equation and evaluating the integral numerically. This only eliminates statistical noise with a period smaller than the effective range of the kernel $K(X)$ and therefore works best for short channels. Beijerinck et al. were only able to obtain accuracies in the percent range for small values of Γ. However, increases in computing power enabled Lobo et al. (2004) to quote an error of 0.03% for their Monte Carlo calculations. They used up to 3×10^{11} atoms at the same values of Γ for their Chebyshev polynomial method referred to above.

8.3.5 Solutions at Higher Pressures

We have not yet considered the effect on the Clausing function of interatomic collisions within the tube. Pollard and Present (1948) and Beijerinck and Verster (1975) indicate that each atom makes an average of d/λ collisions with other atoms between successive wall collisions. Hence at the highest pressures used for beam formation,

TABLE 8.3
Calculations of W

| | Chebyshev | | Monte Carlo | | | | | | | | |
| | Gómez-Góñi and Lobo (2003) | Mohan et al. (2007) | Szwemin and Niewiński (2002) | | | Gómez-Góñi and Lobo (2003) | | | Kersevan and Pons (2009) | | |
Γ	W	W	W	$2\sigma_{MC}$	N_{MC}	W	$2\sigma_{MC}$	N_{MC}	W	$2\sigma_{MC}$	N_{MC}
10	0.1093207144	0.109320714420	0.1093193	5.9×10^{-6}	1.1×10^{10}	0.10932	2.0×10^{-6}	1.0×10^{11}	0.1092860	6.12×10^{-5}	1×10^{8}
15	0.0769377994	0.0769377993847	0.0769408	5.1×10^{-6}	1.1×10^{10}	0.07694	4.9×10^{-6}	1.2×10^{10}	0.0769090	5.22×10^{-5}	1×10^{8}
20	0.0594504191	0.0594504191367				0.05945	3.1×10^{-6}	2.4×10^{10}	0.0594415	4.63×10^{-5}	1×10^{8}
25	0.0484764500	0.0484764500443	0.0484807	4.2×10^{-6}	1.0×10^{10}	0.04848	6.8×10^{-6}	4.0×10^{9}	0.0484607	4.21×10^{-5}	1×10^{8}
30	0.0409393067	0.0409393066694	0.0409399	3.9×10^{-6}	1.0×10^{10}	0.04094	6.5×10^{-6}	3.7×10^{9}	0.0409377	3.88×10^{-5}	1×10^{8}
35	0.0354393800	0.0354393799725				0.03544	6.4×10^{-6}	3.3×10^{9}	0.0354403	3.62×10^{-5}	1×10^{8}
40	0.0312472930	0.0312472930382				0.03125	6.4×10^{-6}	3.0×10^{9}	0.0312529	3.41×10^{-5}	1×10^{8}
45	0.0279452038	0.0279452037190				0.02794	6.2×10^{-6}	2.8×10^{9}	0.0279465	3.23×10^{-5}	1×10^{8}
50	0.0252763636	0.0252763635354010	0.0252781	3.1×10^{-6}	1.0×10^{10}	0.02528	6.2×10^{-6}	2.6×10^{9}	0.0252784	3.08×10^{-5}	1×10^{8}
100		0.0129447752360							0.0129592	2.22×10^{-5}	1×10^{8}
250		0.0052630252523092							0.0052708	1.42×10^{-5}	1×10^{8}
500	0.002647618	0.00264761594328				0.00265	3.2×10^{-6}	1.06×10^{9}	0.0026505	1.01×10^{-5}	1×10^{8}
5000		0.000266427697695									

when $\lambda = d$, the number of collisions of atoms with other atoms at the entrance to the tube is comparable to the number of atom collisions with the wall. From the measurements of Kurepa and Lucas (1981) discussed in Section 12.4, for the purposes of this discussion we can assume that the density within the tube decreases approximately linearly toward its exit. Hence on average along the tube length there will be more collisions of an atom with the tube wall than with other atoms even when the input pressure is such that $\lambda = d$. As the pressure decreases and so λ increases, atom collisions with the wall will predominate. Nevertheless since interatomic collisions do occur, the usual assumption for atomic beam calculations that the Clausing function also applies when collisions take place within the tube, is questionable. As discussed in Section 8.3, it has only been derived in the collision-free case.

An attempt has been made by Flory and Cutler (1993) to calculate the Clausing function when collisions between atoms in a tube with $\Gamma = 5$ are also taken into account. Unfortunately their curve for the absence of interatomic collisions labelled 'no scattering' does not obey the symmetry relation, Equation 8.2, since it is concave upward and also does not satisfy the condition that $\eta(\Gamma/2) = 0.5$ at the mid point of the tube. However, another curve for low pressure but labelled 'scattering' does exhibit the correct behaviour. The curves shown at higher pressures have $\lambda = d/2$ and $\lambda = d/20$, so like Γ, are outside the usual scope of this work. The presence of collisions has led to increasing concavity of the curves. It would be clearly of interest for atomic beam production for results to be obtained for larger values of Γ and at pressures such that $\lambda \geq d$. Some further discussion by comparison with the measurements of the density distribution by Kurepa and Lucas (1981) is in Section 12.4.

8.4   TRANSMISSION PROBABILITY

8.4.1   INTRODUCTION

In order to calculate the throughput through the beam-forming impedance, namely, the number of atoms flowing through the tube per second, it is necessary to know the transmission probability W, which was introduced in Section 1.1. It is a dimensionless factor that depends only on the geometry of the tube. Its value is 1 for an ideal orifice, that is, when the length of the tube tends to zero.

As shown by Cercignani (1988) W is given by

$$W = \left(\frac{1}{2\Gamma}\right) \int_0^{2\Gamma} \left(X^2 + 2 - X\sqrt{(X^2 + 4)} \right) \eta(X) dX \qquad (8.13)$$

where $\eta(X)$ is given by Equations 8.3 to 8.5.

The conductance C of the impedance, namely, the quotient of the volume of gas flowing through the tube and the input pressure is related to W by

$$C = C_0 W \qquad (8.14)$$

where C_0 is the conductance of an orifice of the same cross-sectional area as the tube. An expression for C_0 is given by Equation 2.12 in Section 2.3. The relationship between the transmission probability W and the throughput N that was introduced in Section 2.2 is discussed in Section 8.5.2. We note that early discussions of the flow through tubes did not introduce the concept of transmission probability.

In Section 8.3.1 we found it useful to introduce accurate values of the Clausing function as a basis for comparison with earlier calculations. In the next section we follow the same approach for the transmission probability by first presenting highly accurate values. We then discuss in chronological order numerical and empirical calculations of W with various degrees of sophistication for a cylindrical tube. Since we wish to compare both types in Table 8.4, we do not separate those obtained by approximating the Clausing expressions and those that achieve results by empirical means. However, Monte Carlo calculations are sufficiently numerous to be given their own Section, 8.4.3. We briefly mention other tube shapes in Section 8.4.4. All these methods assume that no interatomic collisions occur in the tube. Attempts to include these are discussed in Section 8.4.5.

As in the case of the Clausing function much particularly early work did not extend to large values of Γ. However, unlike previous Clausing function tabulations, data has frequently been included up to $\Gamma = 500$. Formulae and tables for W have been more frequently expressed in terms of the ratio of tube length to radius, namely, 2Γ, rather than diameter and $1/\Gamma$ has also been employed. For consistency with other chapters we retain the use of Γ. Some of the early expressions were of course developed for use without the benefits of electronic means of calculation. Where tabulated data and formulae both exist for the same approximation, we have therefore not normally drawn attention to small discrepancies in the least significant digit. We assess various expressions in order to compare their accuracy and complexity so that intermediate values may be calculated for values of Γ where no tabulated value exists. Although many expressions are superseded by improved values, we nevertheless include them since they are frequently quoted in publications. Many of the formulae are also valid from $\Gamma < 10$ right down to $\Gamma = 0$, but we have resisted the temptation both to include values outside the range needed for comparison with measurements using tubes employed for atomic beam formation and to examine accuracy much beyond that currently needed for comparison with experiment. An exception is made in Tables 8.3 and 8.5. In the former case this data is also needed to produce Table 8.4, and in the latter, the data would be meaningless with fewer significant figures. Much work on transmission probabilities has been a theoretical exercise without concern for current experimental requirements where accuracies less than 1% are likely to be better than experimental uncertainties. Our comparison calculations have been made using the Excel spreadsheet on a computer with a Pentium 4 processor, which provides 14 decimal digit accuracy.

8.4.2 NUMERICAL AND EMPIRICAL CALCULATIONS

As mentioned in Section 8.4.1 we commence with highly accurate values of the transmission probability W and these have been calculated by Gómez-Goñi and Lobo (2003) and Mohan et al. (2007). Both of these are included in Table 8.3. Gómez-Goñi and Lobo tabulate results obtained by the Chebyshev polynomial method for small values of Γ and also for Γ from 10 to 500. Note, however, that they did not use one of the NAG D05AA program series that allows for the split kernel. They repeated their calculations with 5000 in place of 3000 coefficients in order to determine the accuracy. They also made Monte Carlo calculations and these are included in the same table and discussed in the next section.

Mohan et al. (2007) compare several methods of calculating W. They solved the Clausing integral equation using a 32-point Gauss–Legendre quadrature method for each of 16 equal-length segments of the tube. They tabulate values for small Γ and also for $\Gamma = 25, 50, 500$ and 5000. However, as they point out, this method is unreliable for large values of Γ, since an impossible negative value is obtained when $\Gamma = 5000$. We also do not include calculation details and results of the two approximate methods they used for the same range of Γ, which involved integration over the tube length, since they were also rather poor for large Γ compared with their preferred method. For this they tabulate results for smaller values of Γ as well as 13 values of Γ from 10 to 5000 using a singularity subtraction method modified to smooth the kernel to remove the singularity and dividing the tube length into 2^n segments of equal length where n is incremented by 1 from 0 to 5. Their final results are presented to 12 significant figures, although for $\Gamma \geq 50$, convergence with increasing n is not obtained.

We note that their results for large Γ are consistent with

$$W \rightarrow 4/(3\Gamma) \tag{8.15}$$

as $\Gamma \rightarrow \infty$ with Mohan et al.'s (2007) calculated value for $\Gamma = 5000$ differing by less than 2.4×10^{-7} from that given by Equation 8.15 and we shall see that most expressions valid for large Γ are consistent with it. DeMarcus (1957) proved that this asymptotic form is correct.

Our method of comparison with all other analytic expressions for W is to calculate the difference between each and Mohan et al.'s (2007) tabulation at each value of $\Gamma \geq 10$ that they used, namely, from $\Gamma = 10$ in increments of 5 until $\Gamma = 50$ and then for $\Gamma = 100, 250, 500$ and 5000 as shown in Table 8.3. This of course is not a very even distribution of calculated values, but since they approach $4/(3\Gamma)$ for large values of Γ, the error in calculating the standard deviation σ from only these differences is unlikely to be significant. The results are collected in Table 8.4 and they will be discussed in chronological order for each approximation. In the table we have, however, ordered them in decreasing value of σ in order to make their comparison easier. We feel that this approach of tabulating standard deviations is more useful and compact than the usual method of tabulations of W as a function of Γ then listing percentage differences for each value of Γ.

Early estimates of W have received extensive review, both in the general references given at the beginning of this chapter and in the specific papers referred to for each formula. These are now mainly of historic importance, but we retain the expressions since many have been employed in atomic beam calculations. Originally published in 1920, Dushman (1962) developed a simple equation which has the correct behaviour of $W = 1$ for an orifice when $\Gamma = 0$ and Equation 8.15 as $\Gamma \rightarrow \infty$, namely,

$$W = \frac{4}{(3\Gamma + 4)} . \tag{8.16}$$

TABLE 8.4

Standard Deviations σ for Transmission Probabilities Compared with Mohan et al. (2007)

Author	Date	Equation	σ
DeMarcus lower bound	1957	8.21	9.85×10^{-3}
Knudsen	1909	8.15	6.76×10^{-3}
Kennard	1938	8.20	5.47×10^{-3}
DeMarcus mean	1957	8.21	3.65×10^{-3}
Zugenmaier	1966	8.24	3.21×10^{-3}
Zugenmaier approx. large Γ	1966	8.25	3.20×10^{-3}
DeMarcus upper bound	1957	8.21	2.54×10^{-3}
Dushman	1920	8.16	2.38×10^{-3}
Clausing early	1926	8.17	1.78×10^{-3}
Clausing short tube	1932	8.7 and 8.19	1.78×10^{-3}
Clausing long tube	1932	8.7, 8.8 and 8.19	1.23×10^{-3}
Kennard	1938	8.18	9.34×10^{-4}
Gottwald	1973	8.27	3.99×10^{-4}
Yamamoto et al.	1980	8.29	3.74×10^{-4}
Henning	1978	8.28	1.79×10^{-4}
Santeler	1986	8.30	1.08×10^{-4}
Berman approx. large Γ	1965	8.23	2.05×10^{-5}
Berman general	1965	8.22	2.01×10^{-5}
Neudachin et al. 2nd approx.	1972	8.26	2.36×10^{-6}

It indicates an approximate hyperbolic form for the dependence of W upon Γ. Comparison of Equation 8.16 with Mohan et al.'s (2007) tabulation is included in Table 8.4.

Clausing (1926) must be credited with the first detailed expression for W as follows:

$$W = \frac{1}{3\Gamma^3}\left\{8 + 8\Gamma + 12\Gamma^2 + 7\Gamma^3 - 2\Gamma^4 - 2\Gamma^5 - 4\Gamma^6 - 2\sqrt{1+\Gamma^2}\left[4(1+\Gamma+\Gamma^2) - \Gamma^4 - 2\Gamma^5\right]\right\}.$$

(8.17)

Clausing notes that $W \to 7/(3\Gamma)$ as $W \to \infty$ rather than the value obtained from Equation 8.15. The results are labelled 'Clausing early' in Table 8.4 and agree closely with his later short tube approximation, requiring more significant figures to distinguish them. Clausing (1932) shows that Equation 8.15 also follows from Knudsen's (1909a,b) and von Smoluchowski's (1910) derivation of the throughput through a tube. Values calculated from Equation 8.15 are labelled 'Knudsen' in Table 8.4 and appear near the top of it, so they are now only of historical importance except when Γ is very large.

Hottel and Keller (1933), in dealing with radiation, present algebraic expressions from which W can be derived. However, the results are poor for large values of Γ

leading to an incorrect asymptotic value, so we have included neither the equations nor the evaluation.

There are also some other simple formulae for the calculation of W. Kennard (1938) quotes Clausing's thesis dating from 1928 which offers

$$W = \frac{8\Gamma + 10}{6\Gamma^2 + 19\Gamma + 10} \qquad (8.18)$$

which is labelled with 'Kennard' and the above equation number. It is the first in Table 8.4 with a $\sigma < 10^{-3}$.

By substituting X from Equation 8.8 in Equation 8.7, Clausing (1932) obtained the value of α to be substituted in the following equation to obtain W for any value of Γ:

$$W = 1 + \frac{2(1 - 2\alpha)}{3\Gamma}[1 - (\Gamma^2 + 1)^{1/2}] + (2/3)(1 + \alpha)\Gamma[\Gamma - (\Gamma^2 + 1)^{1/2}]. \qquad (8.19)$$

By comparison with the corresponding derivation of $\eta(X)$ that is given in Section 8.3.2 we refer to this as the long tube approximation for W, although Clausing considered it to be the best approximation to Equation 8.13 for any value of Γ. Clausing also substituted $X = 0$ in Equation 8.7 to obtain a value of α to substitute in Equation 8.19 which he refers to as his short tube approximation. It is slightly worse over this range of Γ, as its name suggests.

As can be seen from Table 8.4 all the Clausing values are an improvement on those of Knudsen and Dushman, which is to be expected since they contain more terms.

The variational methods described in Section 8.3.2 were used initially by Kennard (1938) to obtain values of the transmission probability. Any polynomial approximation for $\eta(X)$ enables Equation 8.13 to be evaluated analytically. For example, the simple expression for $\eta(X)$ given by Equation 8.9, gives

$$W = 1/(\Gamma + 1). \qquad (8.20)$$

As can be seen from Table 8.4 Kennard's values using the above equation, which were intended as an approximation only for small values of Γ, are only a slight improvement over those of Knudsen (1909a,b) and are inferior to those of Clausing (1932).

In Section 8.3.2 we mention DeMarcus' (1956) 'scissoring' method. This was used to produce accurate upper and lower bounds for the transmission probability by DeMarcus and Hopper (1955) although only for $\Gamma \le 1$. Their quoting W to at least six significant figures, however, seems to have started a trend for calculating results to high accuracy, which is of use for comparing the reliability of calculations but is normally unimportant for practical applications.

DeMarcus (1957) also used the variational approach described in Section 8.3.2 to obtain values for W to five significant figures but only for $\Gamma \le 10$. DeMarcus did not, unfortunately, give any explicit formula by which W for larger values of Γ could be evaluated. However, he also proved that for large Γ

$$\frac{4}{3\Gamma} - \frac{\ln 2\Gamma}{2\Gamma^2} + \dots \geq W \geq \frac{4}{3\Gamma} - \frac{1.96788982416 \ln 2\Gamma}{\Gamma^2}. \tag{8.21}$$

Although it was not intended as a means of calculating W for large values of Γ, it is of interest to determine how DeMarcus' bounds compare with those of others. His lower bound has the largest value of standard deviation from Mohan et al.'s values shown in Table 8.4, and his upper bound is even inferior to Dushman's (1962) value.

Berman (1965) evaluated the integrals given in the DeMarcus variational approach to obtain analytical expressions for W. He found that

$$W = 1 + \Gamma^2 - \Gamma(\Gamma^2 + 1)^{1/2} - \frac{2\left[(2 - \Gamma^2)(\Gamma^2 + 1)^{1/2} + \Gamma^3 - 2\right]^2}{9\left[\Gamma(\Gamma^2 + 1)^{1/2}\right] - \sinh^{-1}\Gamma}. \tag{8.22}$$

This leads to an impressive $\sigma = 2.01 \times 10^{-5}$. Lund and Berman's (1966a,b) and Szwemin and Niewiński's (2002) tables of Berman's values have small discrepancies from our calculated values that are presumably caused by the limited precision of their calculations. Since our standard deviation from Mohan et al.'s values is less than from their data, we assume our calculation is correct.

Berman also gave asymptotic forms of this equation. For large Γ

$$W = \frac{4}{3\Gamma} - \frac{\ln 2\Gamma}{2\Gamma^2} - \frac{91}{72\Gamma^2} + \frac{4\ln 2\Gamma}{3\Gamma^3} + \frac{1}{3\Gamma^3} - \frac{(\ln 2\Gamma)^2}{2\Gamma^4} \tag{8.23}$$

which leads to a very slightly inferior $\sigma = 2.05 \times 10^{-5}$, but Equation 8.23 is hardly easier to employ. As can be seen from Table 8.4, Equation 8.22 gives highly accurate values for σ, and W is calculated relatively easily, so we include them in Table 8.5. Its purpose is to provide the best values of W at the same values of Γ for consistency that were used in Table 8.1 for $\eta(l)$. Table 8.5 shows values of W to 7 decimal places in order that differences become apparent for comparison purposes. Seven of the values of Γ overlap those shown in Table 8.3 which are of course superior.

Zugenmaier (1966) derived the following expression for the transmission probability:

$$W = \frac{4\Gamma^3 + 6\Gamma + 4 - 4(\Gamma^2 + 1)^{3/2}}{2\Gamma^3 + 6\Gamma + 2 - 2(\Gamma^2 + 1)^{3/2}}. \tag{8.24}$$

As Table 8.4 shows this is not a particularly good approximation, but it is important since Zugenmaier used it to derive the atomic beam properties discussed in Section 8.5.4 and evaluated by Lucas (1973a) and used in Chapter 9 as the basis of expressions useful for estimating atomic beam properties. These are then applied

TABLE 8.5

Calculations of W Using Expressions Given by Neudachin et al., Berman and Santeler Using, Respectively, Equations 8.26, 8.22 and 8.30

Γ	NPS2 8.26	Berman 8.22	Santeler 8.30
10	0.1093241	0.1093838	0.1093117
20	0.0594567	0.0594907	0.0597205
30	0.0409453	0.0409652	0.0411801
40	0.0312524	0.0312652	0.0314420
50	0.0252807	0.0252895	0.0254338
60	0.0212297	0.0212361	0.0213553
70	0.0183001	0.0183049	0.0184049
80	0.0160824	0.0160861	0.0161712
90	0.0143450	0.0143479	0.0144212
100	0.0129469	0.0129492	0.0130131
200	0.0065607	0.0065612	0.0065850
300	0.0043950	0.0043952	0.0044079
400	0.0033046	0.0033047	0.0033127
500	0.0026478	0.0026479	0.0026534
600	0.0022089	0.0022089	0.0022130
700	0.0018948	0.0018948	0.0018980
800	0.0016589	0.0016589	0.0016615
900	0.0014753	0.0014753	0.0014774
1000	0.0013283	0.0013283	0.0013300

in Chapter 11 to compare with experimental measurements. For these purposes the accuracy of Equation 8.24 is sufficient. It is an improvement on Giordmaine and Wang's (1960) use of the large Γ approximation given by Equation 8.15. We mention here for comparison purposes which approximations have been used by others for atomic beam calculations. For example, rather than using Clausing's (1932) long tube Equations 8.7, 8.8 and 8.19, Jones et al. (1969) interpolated between values that are traceable to Clausing's accurate four-significant figure tabulation using them, whereas Becker (1961b) and Beijerinck and Verster (1975) used Dushman's Equation 8.16.

Zugenmaier also gave approximate expressions for W both valid for small and large Γ. The latter is shown in Equation 8.25, which in our range of comparison is marginally superior to the above expression as Table 8.4 shows.

$$W = \frac{16\Gamma - 6}{12\Gamma^2 + 8\Gamma - 3}. \tag{8.25}$$

In a research note Pao and Tchao (1970) give asymptotic expressions for W for large Γ but they neither tabulate values nor offer algebraic expressions from which W

can be calculated. Values of W have also been tabulated by Moore (1972) but only for small values of Γ.

Neudachin et al.'s (1972) calculations of the Clausing function are discussed in Section 8.3.3. They also give expressions for W using their cubic variational approximation $\eta(X)$. If their method is used to calculate $\eta(X)$ it is relatively straightforward to use in addition to the following equation to obtain W:

$$W = 1 + \Gamma^2 - \Gamma\sqrt{1+\Gamma^2} - \frac{a_1^2 a_{22} + a_2^2 a_{11} - a_1 a_2 a_{12}}{4\left(a_{12}^2 - 4a_{11}a_{22}\right)} \tag{8.26}$$

The coefficients a_n and a_{nn} where $n = 1, 2$ are given as a function of Γ as part of Equation 8.12. Their results are the most accurate included in Table 8.4, so we have also included them in Table 8.5.

Values of W have also been calculated by Garelis and Wainwright (1973) using diffusion theory. They tabulate W for small values of Γ and for a few values up to $\Gamma = 50$ and also for $\Gamma = 500$. Up to the four significant figures they give, there is general agreement with Mohan et al. (2007). Gottwald (1973) calculated W for tubes with $\Gamma \le 10$ and also obtained

$$W = 4/(3\Gamma + 7) \tag{8.27}$$

for larger values of Γ. The standard deviations are included in Table 8.4. Unpublished results of Nawyn and Meyer have been tabulated by van Essen and Heerens (1976) for small values of Γ and also for $\Gamma = 10, 15$ and 20. To their quoted five decimal place accuracy they also agree with those of Mohan et al. (2007).

The variational method mentioned in Section 8.3.2 was developed still further by Cole and Pack (1975) and by Cole (1977a,b). They used two different functionals to provide upper and lower bounds for the transmission probability and in each case both a linear and a cubic trial function for $\eta(X)$, of the correct symmetry behaviour. Cole (1977a) obtained values of W for small values of Γ and from $\Gamma = 10$ in increments of 5 up to $\Gamma = 50$ and then for $\Gamma = 80$ and $\Gamma = 500$. He quotes the accuracy of each but Cole (1977b) gives the upper and lower bounds for W. The lower bounds are all less than Mohan et al. (2007), and the upper bounds are greater. These values have been compared with their own by Mohan et al., who show that Cole's data agree within his quoted uncertainties in this range of Γ. As in the case of DeMarcus' (1957) results the upper bounds are closer to Mohan et al. than the lower bounds. Since Cole did not use the full range of values of Γ used by Mohan et al., we cannot include a meaningful value for his standard deviation in Table 8.4.

Henning (1978) obtained various empirical formulae for Γ by fitting to the numerical data of Nawyn and Meyer as quoted by van Essen and Heerens (1976). For tubes with $\Gamma \le 100$ Henning proposed a modification of Dushman's formula given in Equation 8.16 to give

$$W = 25/(19\Gamma + 40) \tag{8.28}$$

which is an improvement in σ by more than a factor 13 over Equation 8.16 in our range of Γ as Table 8.4 shows despite it not giving quite the correct value when $\Gamma \to \infty$. Equation 8.28 is similar to Equation 8.27 obtained by Gottwald (1973), but Henning appears not to have been aware of his work.

Yamamoto et al. (1980) also solved an integral equation numerically but it is not given explicitly. Their solution for the transmission probability is given by

$$W = \frac{4}{3\Gamma} + \frac{1}{\Gamma^2}\left(\frac{8}{3}\mu - \frac{1}{2}\ln\Gamma\right) + \frac{1}{\Gamma^3}$$
$$\left\{\frac{3}{16}(\ln\Gamma)^2 - \left(2\mu + \frac{3}{16}\right)\ln\Gamma + \mu + \frac{3}{16}\mu^2 + \frac{3}{32}\right\} + 0\left(\frac{(\ln\Gamma)^3}{\Gamma^4}\right)$$

(8.29)

where $\mu = -0.6323$.

As can be seen from Table 8.4 the value of σ is more than twenty times larger than is obtained from Berman's (1965) results obtained some years previously. The first two terms of their expressions are identical with those of Equation 8.23, due to Berman, but there are differences in the coefficients of higher order terms. This is not unexpected since the Berman formula assumes a linear variation for the Clausing function whereas Yamamoto et al. did not. However, it should be noted that there is a large discrepancy between their graph of W for smaller values of Γ than those considered here and our evaluation, so the possibility of a misprint in their printed equation cannot be ruled out.

Santeler (1986) developed a relatively simple approximation, which amazingly in view of its simplicity is only surpassed in accuracy by those of Berman (1965) and Neudachin et al. (1972). Noting that the simple Dushman expression for large Γ given by Equation 8.16 and Kennard's even simpler one for small Γ in Equation 8.20 gave very good approximations to W as $\Gamma \to \infty$ and as $\Gamma \to 0$, respectively, he chose a mathematical function which has the asymptotic form of these equations in the appropriate limit. He determined the values of the parameters by computer fit to Cole's (1977a) data to obtain an equation which simplifies to

$$W = 2 (2\Gamma + 7)/(3\Gamma^2 + 18\Gamma + 14).$$ (8.30)

This equation is the simplest of three that Santeler derived, since the other two involve the substitution of six-figure constants to give an expression for W involving two constants, one of which is an exponent, which could be chosen to minimize either the absolute or the percentage errors. However, one gives a larger value of σ than Equation 8.30 and the other about 20% less. Since the advantage of Equation 8.30 is its accuracy combined with its simplicity, we do not include the others. We note that it is also the one he tabulates to compare with Cole (1977a) and Dushman (1962) which he quotes as 1949.

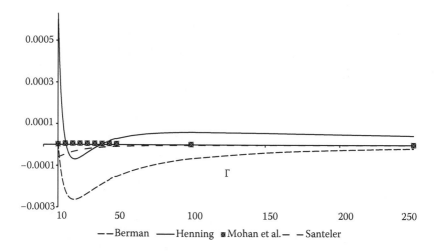

FIGURE 8.6 The difference between Neudachin et al.'s values of the transmission probability and those of Berman, Henning, Santeler and Mohan et al.

We conclude from Table 8.4 that Neudachin et al.'s expressions are the best available for calculating W over the range from $\Gamma \geq 10$, and if they are being used in any way to calculate $\eta(X)$, it is a simple next step to calculate W using Equation 8.26. Otherwise Berman's general Equation 8.22 is much easier to employ but with over a factor of 10 loss in accuracy. Santeler's Equation 8.30 at slightly above 0.01% accuracy is unlikely to be confirmed or disproved by experiment in the foreseeable future. All other expressions are less accurate than Santeler's, and since his is so easy to calculate, their use is now superseded, but they have been included since they are often quoted and employed, so this detailed comparison should be useful.

It is not surprising that where comparison of Table 8.2 for the Clausing function standard deviations with Table 8.4 is possible, the same order of increasing accuracy for each approach is apparent. However, the much larger range of σ compared with Table 8.2 should be noted. It is partly due to the inclusion in Table 8.4 of some very simple early formulae and also the availability of accurate numerical data that enabled some empirical solutions to be developed.

Figure 8.6 gives examples of how the difference between Neudachin et al.'s (1972) values of the transmission probability and some other authors vary with Γ over the range of Γ from $\Gamma = 10$ to $\Gamma = 250$. Since Table 8.4 shows that their values are by far the closest fit to Mohan et al. (2007), they are a useful means of obtaining a continuous curve, and it can be seen how Mohan et al.'s calculated points are in close agreement. All values are converging upon the asymptotic value given by Equation 8.15. All other curves than those shown behave in a similar way, in that the difference either decreases as Γ decreases, as Berman's curve does, or behave like Henning's and Santeler's curves.

8.4.3 MONTE CARLO CALCULATIONS

Direct-simulation Monte Carlo methods of calculating the Clausing function are mentioned briefly in Section 8.3.4. The transmission probability may be obtained directly by calculating the paths of a large number of atoms entering a tube and determining how many leave its exit. Davis (1960) and Beijerinck et al. (1976) give detailed descriptions of the method. They show that it is conceptually easy to formulate, especially for more complex geometries. The latter authors also include the modification of the wall collision process to include adsorbing walls, though only for small values of Γ. The increasing accuracy of Monte Carlo calculations in later publications, especially for larger values of Γ, reflects the increasing power of computers. The accuracy of W especially for large values of Γ is, however, usually exceeded by Chebyshev methods. This is because the standard deviation of the Monte Carlo results is given by

$$\sigma_{MC} = [W(1 - W)/N_{MC}]^{1/2}, \tag{8.31}$$

where N_{MC} is the number of paths calculated. For example, Gómez-Goñi and Lobo (2003) used up to 1×10^{11} paths for $\Gamma = 10$ to achieve $\sigma_{MC} = 1 \times 10^{-6}$. As Γ increases, the computing time for a given accuracy increases, not only because of the large number of atoms that must be considered but also because of the average greater number of wall collisions per atom.

The Monte Carlo method was first used for the calculation of transmission probabilities by Davis (1960) who was asked to calculate flow rates from a large vacuum tank through a cylindrical elbow to a pump. Davis presents results for a variety of geometries, including the cylindrical tube of circular cross-section of interest here, but only up to $\Gamma = 2.5$. For this tube he obtained an accuracy of about 1.6% with 10,001 paths. Davis et al. (1964) extended the calculations of Davis (1960) by calculating the effect of what they describe as 'rougher-than-rough' surfaces on the transmission probabilities of small Γ tubes.

Talley and Whittaker (1969) present graphical results for the transmission probability for steady and transient flow in a tube for $\Gamma \leq 50$, with and without absorption on the walls, and for steady flow in a converging-diverging channel.

Füstöss (1970) suggests that the computing time could be reduced by considering a tube to be made up of two shorter ones and combining their transmission probabilities as is mentioned in Section 8.4.4. His results are also limited to small values of Γ. An alternative approach is suggested by Choong (1971), using a modified technique which makes use of the probability functions appearing in the Clausing equation. His results are accurate to better than 0.1%, so they are an improvement on those of Davis, but they are also given only for the range $\Gamma \leq 2.5$. However, these two approaches have not been used subsequently.

We have already referred to Beijerinck et al.'s (1976) Monte Carlo approach to determine the Clausing function $\eta(X)$ in Section 8.3.4 and hence the angular distribution $f(\theta)$ which is treated in Section 8.5. They also obtained values of W for Γ up to 25, taking 1.2×10^5 trajectories to yield an accuracy of just over 1%. Carette et al. (1983) use 15 values of Γ up to 50 but unfortunately made an error in dealing with the emission angle at the entrance to the tube so that their results are invalid. Their error is acknowledged in an erratum. Adamson et al. (1988b) calculated W for Γ up to 50 using typically 10^6 trajectories with an accuracy better than 1%. To within this their values agree with those of Mohan et al. (2007). We refer again to Adamson et al.'s

Monte Carlo method in Section 8.5.2 since they used it to determine the angular distribution of an atomic beam. Eibl et al. (1998) also calculated W for four values of Γ between 37.5 and 75. Their results, which are quoted to three decimal places, have errors of a few percent. Szwemin and Niewiński (2002) employed the method with 10^{10} trajectories for Γ up to 50. The standard deviations varied with Γ but when $\Gamma \geq 10$, σ_{MC} did not exceed 3×10^{-6}. Their results are shown in Table 8.3.

In addition to their numerical calculations of the Clausing function which are discussed in Section 8.3.4, Gómez-Goñi and Lobo (2003) also used the Monte Carlo method to obtain values of W. They used variable numbers of trajectories in excess of 10^9 and with σ_{MC} less than 3.4×10^{-6}. Their results for $\Gamma \geq 10$ are also shown in Table 8.3. Lobo et al. (2004) also undertook Monte Carlo calculations in addition to their numerical methods over the same range of Γ, which were described in Section 8.3.4. They used up to 3×10^{11} trajectories to achieve accuracies stated to be 0.03% so they only presented results graphically. Casella et al. (2009) also made calculations for several different tube geometries, including right circular cylinders, but only for small values of Γ. Shi et al. (2012) also used the Monte Carlo method to calculate W for $\Gamma \leq 100$ and present graphical results. They do not discuss the accuracy of their calculations, which were also extended to other tube shapes than cylindrical.

As discussed in Section 8.4.2, for the simple geometry of a cylindrical tube, highly accurate calculations of the transmission probability using the Chebyshev method are available at selected values of Γ, and for other values, accurate formulae are available. Hence as suggested by Talley and Whitaker (1969) and Szwemin and Niewiński (2002), the results of more accurate calculations can be used to ascertain the correct functioning of the complete Monte Carlo calculations. They can then be applied to more complex geometries, which cannot easily be dealt with otherwise, and to investigations of different wall collision processes. This approach was also used by Kersevan and Pons (2009) who discuss their two types of Monte Carlo calculations. They tabulate their results for Γ up to 500, which agree within their errors to those of Gómez-Goñi and Lobo (2003) and Mohan et al. (2007). They show how their method is applicable to calculating pumping speeds in vacuum systems. Hence it could be useful for determining the ambient gas pressure surrounding an atomic beam using Equation 2.8.

Interatomic collisions within the beam-forming impedance are discussed in Section 8.4.5, but Monte Carlo calculations of the transmission probability that include them, which would clearly be of interest, do not appear to have been made.

8.4.4 OTHER ASPECTS OF FLOW

We only briefly discuss the flow through tubes in series since it is largely outside the scope of this review because beam-forming impedances aim to be of uniform bore. They are, however, part of a vacuum system that is used either for gas inlet or for production of a vapour in an oven. An important problem is the determination of the pressure at the input to the impedance when the pressure measurement is made some distance from it. Rather than treat the subject in detail we refer to those reviews that include a discussion of the transmission probability of connected tubes.

Steckelmacher (1966, 1986) has discussed flow resistances in his reviews and Santeler (1986) has also contributed calculations. The physical basis of the problem is that in contrast, for example, to electrical resistances, in atomic flow when two tubes are connected in series the presence of the second tube affects the flow through the first and vice versa. For example, atoms emerging from the exit of the first tube may be reflected back into that tube because of the presence of the second tube. If atoms emerge into a vacuum this will not occur. In addition the transmission probability of each tube is calculated on the assumption of a cosine distribution of the directions of the gas atoms at the input. However, the beaming effect produced by the first tube (Section 8.5) means that this assumption cannot be valid for the second.

Unlike other theoretical work discussed in this chapter, which can be directly or indirectly compared with experiment, there are no experimental data of which we are aware that can be used to test the validity of the above range of theoretical treatments.

Another aspect of flow to be considered briefly is the effect of the cross-sectional shape of impedances on the transmission probability. As mentioned in Section 8.2 this is not treated in detail in this chapter. However, we mention that Steckelmacher (1978a,b) has made a detailed critical review of the effect of cross-sectional shape on the transmission probability of tubes with regular geometries. He showed that W is within a few percent of that calculated for a cylindrical tube of the same area. This is confirmed by the Monte Carlo calculations of Li et al. (2013), who considered tubes with regular polygon cross-sections. They used up to 10^8 trajectories and tested their method on a polygon with 10,000 sides where their results agreed with the calculations of Gómez-Goñi and Lobo (2003), that are discussed in Section 8.4.2, to within their own estimated errors. Li et al. considered tube lengths up to 50 times the polygon diameters. Their calculations for triangular, square and hexagonal cross-sections are of particular interest for atomic beam collimation.

8.4.5 ALLOWING FOR INTERATOMIC COLLISIONS

So far in all the work on transmission probability we have considered, we have assumed that the pressure in the tube is so low that no collisions between atoms occur in the tube. Since for optimum atomic beam formation, as we show in Chapter 9, this is not the case, it is important to consider the effect on W of higher pressures. We remain only interested in the cases where Γ is suitable for good atomic beam collimation. There are two cases we wish to consider. When the pressure is such that collisions occur in a tube, but that $\lambda \geq d$ we need to consider if W is modified from the collisionless case so that Equation 8.14 and the results from Section 8.5.2 are no longer an accurate prediction of the throughput. In the second case, if a tube with a value of Γ that is one or more orders of magnitude greater than that considered so far is used at input pressures such that $\lambda < d$ we would like to ascertain whether the quality of the beam is an improvement on those obtained with shorter tubes at lower pressures. Unfortunately although extensive theoretical studies of the flow through tubes at higher pressures have been made, they do not consider the angular distribution of the beam formed under these conditions. Hence a detailed review of the throughput in the transition region is outside the scope of this work.

We have already mentioned Flory and Cutler's (1993) calculations of the Clausing function in Section 8.3.5. They essentially used the same approach to calculate the transmission probability up to high pressures for the same tube with $\Gamma = 5$. As the pressure increases from zero to that where $\lambda = d$, they found that W decreases from the accepted value by about 25%. There has been a considerable amount of theoretical work on the throughput, from which W at higher pressures can be deduced. A detailed review, with 178 references, of the theory of flow through tubes of all lengths and shapes and at all pressures has been given by Sharipov and Seleznev (1998). Beskok and Karniadakis (1999) extend previous work with their own calculations with results presented in graphical form. The general picture is that as the pressure is increased from that where only diffusive flow occurs so that collisions occur in the tube, the drift component becomes important and the conductance decreases slightly before increasing rapidly when $\lambda < d$. Hence it follows from Equation 8.14 that W decreases slightly when collisions occur. A typical decrease in C and hence W is about 30% as λ approaches d when Γ is large, but the effect decreases with decreasing Γ. Hence, the subject is hardly worth a detailed discussion. In order to indicate its existence conductance on a linear scale, or a quantity proportional to it, is usually plotted against pressure. Hence the usual plot of throughput against pressure on a double logarithmic scale for atomic beam purposes, for example, as employed by Lucas (1973a), would hardly show any departure from linearity if the decrease were included. At pressures just above those of interest for atomic beam formation, there is a much discussed minimum in the conductance, known as the Knudsen minimum, which Lund and Berman's (1966a,b) model explains, but the minimum is outside the scope of this work.

8.5 ANGULAR DISTRIBUTION OF ATOMIC BEAM

8.5.1 INTRODUCTION

As mentioned in Section 1.1, in collision-free flow, the atoms emerging from an impedance in the form of an ideal orifice have a cosine distribution, as shown in Figure 1.1. When the impedance is a tube, rather than an orifice, the distribution of the emerging atoms becomes narrower as a result of the collisions that the atoms make with the walls of the tube on their passage through it. The distribution then takes on the shape shown in Figure 1.3. In practice atomic beam collimators are usually used in conditions where $\lambda < l$ so that the treatment of the angular distribution must consider interatomic collisions within the tube. The collisionless theory is of use where a high-quality beam is not required in practice. In addition the equations that neglect interatomic collisions are simpler to apply without the need for integration. Also there is no uncertainty about the end effects given by values of Z_P and Z_V that are discussed further in Section 8.5.4. The equations also form a useful test of the theory that takes interatomic collisions into account since they should be consistent with the collisionless theory when $\lambda \gg l$.

8.5.2 COLLISION-FREE THEORY

In Section 2.3 the theory of the flow of atoms through an orifice is developed. In particular, Equation 2.11 gives an expression for the axial intensity $I(0)$. We now introduce the relative angular distribution $f(\theta)$. Its use simplifies the expressions for the angular distribution, since normalisation to the tube input conditions can be employed just as for the Clausing function discussed in Section 8.2. In free-atomic flow the axial intensity is the same for a tube as for an orifice of the same diameter so that

$$f(\theta) = I(\theta)/I(0) \tag{8.32}$$

The formula for $f(\theta)$ can be obtained from simple geometrical considerations and was first given by Clausing (1930). Dayton (1957, 1958) also derived it but only presented results for $\Gamma \leq 5$. Beijerinck et al.'s (1976) and Steckelmacher et al.'s (1978) derivations were evaluated for $\Gamma \leq 25$. Tschersich and von Bonin (1998) also calculated the angular distribution for $\Gamma = 64$. Their beam profiles in atomic hydrogen are discussed in Section 11.2.4. Other forms of Clausing's equations have been given by Beijerinck and Verster (1975) and Olander and Kruger (1970). We later follow the latter approach, which is based upon Clausing's, but use α for the value of the Clausing function in the plane of the exit to V_0 and from symmetry $1 - \alpha$ as its value in the plane of the entrance V_P, rather than Olander and Kruger's expressions which do not assume this symmetry, as is discussed in Section 8.2.

However, we start with Clausing's (1930) derivation. For $\theta < \theta_0$, where $\tan \theta_0 = \Gamma^{-1}$, the beam will be made up of two contributions. The first arises from atoms that have passed straight through the tube without colliding with the walls. The second arises from atoms that have collided with the walls one or more times before emerging from the tube exit. For $\theta > \theta_0$ there can only be a contribution from atoms that have undergone wall collisions since no straight through paths are possible. Clausing derives the following expressions:

(i) for $0 \leq \theta \leq \theta_0$,

$$f(\theta) = (1-\alpha)\cos\theta + \frac{2}{\pi}\cos\theta\left\{(1-\alpha)R(p) + \frac{2}{3}(1-2\alpha)[1-(1-p^2)^{3/2}]p^{-1}\right\}$$

$$\tag{8.33}$$

where $R(p) = \cos^{-1} p - p(1 - p^2)^{1/2}$ and $p = \Gamma \tan \theta$, and

(ii) for $\theta_0 \leq \theta \leq \pi/2$,

$$f(\theta) = (1 - \alpha)\cos\theta + 4(1 - 2\alpha)\cos^2\theta/(3\pi \Gamma \sin\theta) \tag{8.34}$$

Hence the angular distribution of the beam when interatomic collisions do not take place may be calculated from Equations 8.33 and 8.34 for any chosen value of $\alpha = \eta(l)$ such as those from Table 8.1.

Clausing did not appreciate the use of the term halfwidth H to provide a single parameter for describing the shape of an atomic beam. Its use was introduced much later by Giordmaine and Wang (1960), who were apparently unaware of Clausing's derivation of the angular distribution. It can be easily obtained by putting $\theta = 0$ in

Equation 8.33 and then finding the value of θ from Equation 8.34 such that $f(\theta) = f(0)/2$, provided that H is both large enough for Equation 8.34 to be applicable and small enough to approximate $\sin\theta$ by θ. From our definition of H being the full width at half height, $H = 2\,f(\theta) = 16/(3\pi\Gamma) = 1.70/\Gamma$. Clausing's equations use his value of α that is given by Equation 8.6 of Section 8.3.2, since his general equations for $\eta(X)$ were not published until two years later. However, H is remarkable in being independent of α. The same value of H is also obtained from Giordmaine and Wang's (1960) equations and is consistent with their quoted numerical value of $1.68/\Gamma$, if we assume that their calculation is approximate. It will become clear by comparison with experiment as is discussed in Chapter 11 that theoretical halfwidths are about half those measured experimentally. Although we are still considering the case of no interatomic collisions, any acceptable theoretical treatment that considers these has to converge on the above results and for that reason they are particularly important. Hence to avoid any doubt, it is clear from Giordmaine and Wang's figure 7 that they use the mathematical cone semi-angle as the halfwidth, which equates to $H/2$, and we use H throughout. When converted to degrees, $H = 97.3/\Gamma$, which agrees closely with $H = 97.2/\Gamma$ given by the evaluation of Equation 9.9 using Zugenmaier's (1966) theory in Section 9.2. Other treatments of the angular distribution that include interatomic collisions usually start with Clausing's (1930) equations that neglect them and so lead to the same result.

A check on the accuracy of the angular distributions obtained using different approximations for $\eta(X)$ can be obtained by noting that integrating $I(\theta)\,d\omega$ over all forward solid angles ($0 \le \theta \le \pi/2$) will give the throughput N, which was introduced in Section 2.2. Since N is also the throughput of an orifice multiplied by the transmission probability W, it follows that

$$W = \frac{1}{\pi}\int_{\omega} f(\theta)\,d\omega$$

Writing $d\omega = \sin\theta\,d\theta\,d\phi$, $0 \le \theta \le \pi/2$, $0 \le \phi \le 2\pi$, we obtain

$$W = 2\int_{0}^{\pi/2} f(\theta)\sin\theta\,d\theta. \tag{8.35}$$

This equation has been frequently used for checking whether the derivation of the angular distribution gives the value of the transmission probability that is assumed in that derivation.

Clausing (1930) and Dayton (1957, 1958) only gave numerical results for $f(\theta)$ for small values of Γ. Dayton also derived the gas flow lobes in V_p due to those atoms which return to it after entering the tube. Dayton showed that the shapes are complementary since their sum gives the cosine distribution, but this is not needed for atomic beam formation.

Neudachin et al. (1972) give algebraic expressions for calculating the angular distribution using both their linear approximation for $\eta(X)$ given by Equation 8.10 and their cubic approximation, Equation 8.12. The former leads to $H = 16/(3\pi\Gamma)$ as noted previously for other authors, and the latter is consistent with it if the term in $(0.5 - X)^3$ in Equation 8.12 is small, which it normally is, since the function is nearly linear. They tabulate $f(\theta)$ for only four angles for $\Gamma = 20, 40$, and 80 and show that there is little difference between the results obtained using the two approximations. This is consistent with the independence of H on $\eta(X)$ that is noted above.

Other calculations for the collision-free case were carried out by Ivanov and Troitskii (1963), Zugenmaier (1966) and Beijerinck and Verster (1975). We return to Zugenmaier's treatment in Section 8.5.4 since he also allowed for collisions. The unsatisfactory nature of Ivanov and Troitskii's approach has already been discussed in Section 8.2. Beijerinck and Verster's derivation also leads to $H = 16/(3\pi\Gamma)$.

The angular distribution has also been treated by the direct simulation Monte Carlo method, which was discussed in Section 8.3.4. Ward et al. (1970) made Monte Carlo calculations for tubes with $\Gamma \leq 5$ to include specular reflection and surface diffusion and show that each slightly broadens the angular distribution.

Bellman and Raj (1997) show graphs of Monte Carlo calculations for their beam of lithium tantalate [$LiTaO_3$] that was produced by the array which is described in Section 10.4.8. Their calculations allowed for the finite size of the array by considering 100 trajectories through each of the 3020 capillary tubes and at three different distances from its surface. They unfortunately only give brief details of their calculations which allowed both for collisions within the array and absorption on its walls.

Eibl et al. (1998) describe their Monte Carlo calculations of the angular distribution produced by molecular hydrogen through tubes with $\Gamma = 50$ and $\Gamma = 100$ without intermolecular collisions. They continued by considering their atomic hydrogen source that is described in Section 6.2.3, in which dissociation occurred along the tube. Just under 3% of the molecules passed directly through the tube without hitting its surface and so were not dissociated. Hence at small angles the beam was largely molecular, but atoms predominated at all larger angles. Hence the dip in atomic hydrogen intensity that was found by Tschersich and von Bonin (1998) and Schwarz-Selinger et al. (2000) and discussed in Section 6.2.3 is expected theoretically.

We have made several references to Beijerinck et al. (1976) and in particular their Monte Carlo method is described in Section 8.3.4. They used it to calculate $f(\theta)$. For $\Gamma = 2$ and $\Gamma = 25$, in the latter case with 6×10^4 trajectories to yield a halfwidth that is consistent with $1.70/\Gamma$. Without apparently being aware of previous work that is covered in this Section, Nanbu (1985) used Monte Carlo methods to calculate $f(\theta)$ but only for small values of Γ.

We have already referred to Adamson et al.'s (1988b) Monte Carlo calculations of the transmission probability in Section 8.4.3. They (1988a,b) also used the method to obtain angular distributions of the beam for tubes having values of Γ up to 54 using about 10^6 trajectories. Unfortunately their treatment of the results, which in any case have relative intensities, to compare with their experimental conditions, means that comparison with the beam formed by a tube alone is not possible. They also made measurements of the angular distribution, and those with $\Gamma > 10$ are discussed in Section 11.2.4.

Murphy (1989) was concerned with the beam formed when reactions take place within the impedance. He presents diagrams of the angular distribution of the beam formed by a single tube using Equations 8.33 and 8.34 with α given by Equation 8.6. For several tubes including tubes with $\Gamma = 10$, 20, and 100, he compares the relative angular distribution of atoms which have and have not collided with the tube walls. Murphy also considers wall collisions in the case of capillary arrays followed by collimating apertures.

8.5.3   Density Distribution in a Tube

Just as the Clausing function is normalised to the rate of impingement at the tube entrance and is also referred to as the normalised wall collision density, so the density of atoms within the tube is normalised to the density in the first vessel V_P away from the tube entrance. We reiterate the statement made in Section 2.2 that we would drop the term 'number' from density so that the normalised density varies from 1 at V_P away from the tube entrance to 0 in the second vessel V_0 away from the tube exit. We emphasise away from the tube entrance and exit since, as we shall see later, the densities in the planes of the entrance and exit of the tube are treated separately.

In order to calculate the angular distribution of an atomic beam at operating pressures such that $\lambda < l$, it is necessary to take into account the interatomic collisions within the beam-forming impedance. As discussed in Section 8.5.1 the equations then involve not only the Clausing function but also the density within the impedance. Calculations of the density variation within a tube for the collision-free case have been made by Sparrow and Haji-Sheikh (1964) and Townsend (1965). Neither presents any numerical results for the density. Both papers allow the temperature of the atoms in V_P to be different from that of the tube wall and the density beyond the tube exit in V_0 to be non-zero, so that there is also a contribution from beyond the exit plane. In this situation, Sparrow and Haji-Sheikh considered the number of parameters too large for a concise presentation of results. However, de Leeuw and Gadamer (1967) presented detailed calculations and numerical results for the density both in and beyond the exit plane for a range of downstream-to-upstream density ratios for $\Gamma \leq 1$.

Davies and Lucas (1983) used Sparrow and Haji-Sheikh's (1964) formulation to calculate the density throughout the tube in the practical case. This is where flow is into vacuum so the pressure in V_0 is zero and there is no temperature difference between the atoms and the tube walls. Their calculation was for a range of Γ values using both Neudachin et al.'s (1972) cubic approximation to the Clausing function and their Chebyshev solution. Their results showed that the density is very nearly constant over the cross-section of the tube except near the entrance and exit. As Γ increases the variation in density across the tube becomes less and takes place closer to the tube wall. Moreover it was found that the variation of the density along the tube axis is practically identical to the variation of the Clausing function along the length of the tube. When Γ is large enough for useful beam collimation, the deviation is negligible. De Leeuw and Gadamer (1967) also calculated the density but only in the exit plane and just downstream of that plane for a range of downstream-to-upstream density ratios. Their tabulated results for $\Gamma \leq 1$ and a downstream-to-upstream ratio of zero agree with those of Davies and Lucas to the four significant figures given.

8.5.4   Effect of Interatomic Collisions

The discussion in the previous section refers to collisionless flow. Olander and Kruger (1970) state that when interatomic collisions occur within the beam-forming impedance, the density distribution has to be guessed. However, from the discussion in Section 8.3.5 of the Clausing function, we would expect little change as λ decreases. It is suggested therefore that in angular distribution calculations when interatomic collisions occur, a simple model for the density distribution would be to assume that it varies only with position along the tube and that it is constant over the cross-section of the tube. The calculations discussed in Section 8.5.3 show that this is a very good approximation for the large Γ values used in atomic beam collimators. The available measurements of these quantities are deferred to Section 12.4.

The use of beam-forming impedances at input pressures such that $\lambda < l$, so interatomic collisions within the tube occur is well established. Hence theories that do not neglect these are particularly important. We refer to Section 8.3.5 where the relative importance of atom–atom collisions compared with atom–wall collisions is discussed.

The effect of interatomic collisions on the shape of the angular distribution function can be understood, admittedly with some hindsight. Since those atoms that pass straight through the tube must have shorter paths before leaving the tube than those that have collided with the tube walls it follows that they have less probability of colliding with other atoms. Their intensity will, however, increase since the pressure at the input of the tube is greater since λ is decreasing from $\lambda \geq l$ up to $\lambda = d$. The atoms that have made collisions with the walls and with other atoms will be increased in number for the same reason, but when the mean free path is only slightly less than the tube length, atom collisions with the wall will predominate, as discussed in Section 8.3.5 and this means that the angular distribution from these atoms will be little affected. So as the pressure is increased beyond $\lambda = l$, the beam intensity increases linearly with pressure but initially the angular distribution of the beam is little changed. As the input pressure is increased, more interatomic collisions take place and these have the effect of the tube walls no longer dominating the beam formation, so the beam is broadened.

Theories of beam formation that employ the assumptions of the kinetic theory of gases and take collisions into account must satisfy a number of requirements. First satisfactory expressions for the Clausing function and the transmission probability in the region of Γ of interest must be employed. These quantities do not, however, necessarily appear explicitly in the final equations for the angular distribution. Until satisfactory inclusion of modifications to the functions when collisions take place is available, collisionless solutions that exhibit the correct symmetry for $\eta(X)$ and have the correct asymptotic value of W are the minimum requirements, provided that the highly accurate numerical solutions are modelled in any theoretical treatment with an accuracy at least as high as needed for comparison with experiment. Finally a satisfactory model is required for the density distribution along the axis of the beam-forming impedance. Other effects to be considered beyond the kinetic theory treatment include partial specular reflection and surface diffusion.

The theory taking interatomic collisions into account was developed by Giordmaine and Wang (1960) and independently of them by Troitskii (1962) and

Ivanov and Troitskii (1963). Troitskii only considered rectangular channels that did not produce narrow angular distributions. As discussed in Section 8.2, Ivanov and Troitskii did not have the correct symmetry of the Clausing function. Zugenmaier (1966) based his treatment on that of Giordmaine and Wang who extend the collisionless theory by first allowing for the attenuation of those atoms that have made a last collision with the tube walls before leaving the tube. They also consider the additional contribution from those atoms that have collided with other atoms and so are deflected into the beam by collisions.

The modified equations involve both the Clausing function $\eta(X)$ and the density $n(X)$. As already discussed, in collision-free flow, the Clausing function $\eta(X)$ is known to be very nearly linear, with the linear approximation being given by Equation 8.2, which assumes the symmetry of $\eta(X)$. This cannot be assumed for the analysis of the flow in tubes with interatomic collisions, so a linear function is assumed of the form

$$n(X) = Z_p - (Z_p - Z_V)X/\Gamma, \qquad (8.36)$$

where Z_p and Z_V are the values of the normalised density in the plane of the tube entrance where $X = 0$ and in the plane of the tube exit where $X = \Gamma$. It is assumed that there is no radial variation in density. The inclusion of Z_p and Z_V in the theoretical treatment is frequently referred to as the inclusion of end effects. The physical explanation of the dependence of Z_p and Z_V on Γ is as follows. When very few atoms flow through the tube, namely, when Γ is large, the atom density in the plane of the tube entrance is the same as that further away in the atom source, which we have designated V_p and so $Z_p = 1$. When this is the case the atom density in the plane of the tube exit can be assumed to be the same as that of the vacuum V_0 and so $Z_V = 0$. When Γ is small the flow through the tube is no longer negligible and Z_p is therefore less than 1 and Z_V greater than 0. However, as already implied by Equation 8.36 unlike the symmetry of $\eta(X)$, it does not necessarily follow that $Z_V = 1 - Z_p$.

Olander and Kruger (1970) produced the following universal Equations 8.37 to 8.39 for the angular distribution of a beam by retaining Z_V and Z_p in their equations. This means that they can be used to obtain the existing distributions such as those of Giordmaine and Wang (1960) and Zugenmaier (1966). They can also be used with any future values, including those obtained experimentally by Kurepa and Lucas (1981) and discussed by Davies and Lucas (1983) and also considered in Section 12.4.

However, this is mainly an academic exercise until the theory is developed to improve the agreement with experiment even when interatomic collisions are neglected. As introduced in Section 8.1 other factors such as surface diffusion need taking into account. The values of the transmission probability used in the calculations are discussed in Section 8.4.2. It does not appear explicitly in Equations 8.37 and 8.38 since Olander and Kruger combine the equation for the throughput N in terms of W and the requirement that the integral of the angular distribution over all forward angles must equal N. It is interesting to note that whereas the Clausing function is required for the equations that neglect interatomic collisions, it does not appear in the high pressure equations.

(i) For $0 \le \theta \le \theta_0$

$$f(\theta) = Z_V \cos\theta + [2/\pi^{1/2}][Z_V \cos\theta \exp(\delta'^2)/\delta']$$
$$\times \left\{ (R(p)/2)\left[\mathrm{erf}(Z_p\delta'/Z_V) - \mathrm{erf}(\delta') \right. \right.$$
$$\left. \left. + (2\delta'/Z_V\pi^{1/2})(1 - Z_p)\exp[-(\delta'Z_p/Z_V)^2] \right] + S(p) \right\} \qquad (8.37)$$

(ii) For $\theta_0 \le \theta \le \pi/2$

$$f(\theta) = Z_V \cos\theta + [2/\pi^{1/2}][Z_V \cos\theta \exp(\delta'^2)/\delta'] \, S(1) \qquad (8.38)$$

where $R(p) = \cos^{-1} p - p(1 - p^2)^{1/2}$ and $p = \Gamma \tan\theta$.

$$S(p) = \int_0^p (1 - z^2)^{1/2} \left\{ \mathrm{erf}\left\{ \delta'\left[1 + \frac{z}{\Gamma \tan\theta}(Z_p/Z_V - 1) \right] \right\} - \mathrm{erf}(\delta') \right\} dz$$

where $\delta' = \delta/(\cos\theta)^{1/2}$, $\delta = \left\{ (l/2\lambda)\left[Z_V^2/(Z_p - Z_V) \right] \right\}^{1/2}$ and λ is given by Equation 2.1. The error function of any variable x, namely, $\mathrm{erf}\,(x)$, used above, is discussed in Section 9.2.

The axial value $f(0)$ follows from Equation 8.37:

$$f(0) = Z_V + [Z_V\pi^{1/2}/2][\exp(\delta^2)/\delta]\left\{ \mathrm{erf}[Z_p\delta/Z_V] - \mathrm{erf}[\delta] \right\}$$
$$+ [(1 - Z_p)/Z_V)]\left[\exp\left\{ -\delta^2[(Z_p/Z_V)^2] - 1 \right\} \right] \qquad (8.39)$$

As discussed by Olander and Kruger (1970) the various theories differ in their assumed values of Z_V and Z_p. They give expressions for them in terms of Γ^{-1}, some of which are also shown graphically by Kurepa and Lucas (1981). Of particular interest to calculations of all the beam properties are those of Giordmaine and Wang (1960) and Zugenmaier (1966). The former assumed that $Z_p = 1$ and $Z_V = 0$. Zugenmaier retained $Z_p = 1$, but obtained $Z_V = W/2$, using his value of W given by Equation 8.24 and discussed in Section 8.4.2. Olander and Kruger (1970) give details of their derivation of values which (after correction of their original values for π) are $Z_p = 1.222$ and $Z_V = 0.519/\Gamma$. As they note a value of Z_p that is greater than unity is physically impossible. Nevertheless it is surprising that their derivation has been employed by others, so we mention it. At the time of these theories, as discussed in Section 8.4.2, the transmission probability W had not been calculated accurately over a wide range of Γ. However, the asymptotic value of $W = 4/(3\Gamma)$ given by Equation 8.15 was accepted, so tests of angular distributions could be checked against Equation 8.35. Giordmaine and Wang noted that this gives only half the correct value of W for large

Γ and so introduced an arbitrary correction factor. The best values of Z_V and Z_p to use in the above equations are discussed further in Section 12.4.

Lucas (1973a) carried out extensive numerical calculations over a wide range of parameters using Zugenmaier's equations. He obtained values for the throughput, axial intensity and beam halfwidth, the latter being obtained from the equations for the angular distribution by iteration. For $\lambda \gg l$ the results are the same as those obtained for the collision-free theory, Lucas finding close agreement when $\lambda > 100\ l$. However, when $\lambda = l\ H$ is 9% less when collisions are ignored and $f(0)$ is 17% greater.

We note that Zugenmaier's own equation for H (his equation 35) is incorrect since it is obtained by using the angular distribution for collision-free flow and the axial intensity from the theory including collisions. It therefore holds in neither regime. Lucas introduced reduced parameters to produce curves of universal application showing the variation of throughput, axial intensity and halfwidth for $\Gamma > 10$. These show clearly the change in behaviour in passing from low pressures where collisions in the tube are unimportant to those pressures where collisions are important, namely, that H changes from being almost independent of p at low pressure to a $\sqrt{p}$ dependence at higher pressure. Similarly $I(0)$ changes from a linear to a square root dependence on p. This is discussed further in Chapter 9. The curves and tables in Lucas (1973a) are superseded by the formulae in Chapter 9 now that spreadsheets are preferred to interpolation from tables and graphs. These formulae are used to compare theory with experiment in Chapter 11.

As well as Olander and Kruger's (1970) treatment, an earlier discussion of atomic beam production was made by Johnson et al. (1966) and later ones by Livshits et al. (1971) and Adamson and McGilp (1986). Livshits et al. obtained relative angular distributions for $\lambda > l$ up to $\Gamma = 100$ which include a probability of recombination following a single collision with the wall. A completely different approach to calculating angular distributions has been discussed by Malhotra et al. (2007), who proved the suitability of commercial ABAQUS radiation software by obtaining finite element simulations of the normalised angular distributions through three tubes for which Γ was less than 10.

We conclude that this section completes the theory of the flow of atoms through tubes at pressures such that $\lambda \geq d$. We note that no theoretical treatments have used the values of the Clausing function and transmission probability that are now known with higher accuracy than that needed for comparison of theoretical beam properties with experiment. There are now better estimates of the atom density within a tube, but is seems unlikely that the measured beam halfwidths will be predicted theoretically by including all these factors alone. A combination of allowing for interatomic collisions with interactions with the surface of a tube to include partial specular reflection and surface diffusion is needed. This is because of the failure of collisionless theory, as discussed in Sections 11.2.4 and 11.3.3. In Chapter 9 we follow the scheme we introduced in Section 1.1 to provide practical equations needed to design an atomic beam source. Before we compare theory with experiment it is preferable to discuss the design of collimators, which is in Chapter 10, so that Chapter 11 becomes the detailed comparison.

8.5.5 NEAR-FIELD EFFECT

In this section there are two effects that need to be considered. When the beam-forming impedance is either an aperture or a single tube, the angular distribution of the beam may depend upon the distance from the exit of the impedance to the observation region. This is because of its finite area, whereas the theoretical treatments normally assume it to be a point source. Hence it is important to establish at what distance from the source the theoretical distribution is obtained. More usually capillary arrays are used in preference to single tubes to form beams and the effect of their finite size on the beam properties is required. Before considering single tubes we first mention that Michalak (1993) calculated that the cosine distribution was only obtained from an orifice at distances of at least ten diameters.

We consider first the case when $\lambda > l$ so that there are no interatomic collisions. Neudachin et al. (1974) used the Maxwellian velocity distribution approach to calculate the angular distribution of atomic beams. They made calculations at various distances from tubes having $\Gamma \le 50$ and using both their approximations to the Clausing function discussed in Section 8.3.2. They showed that if the point of measurement is 40 tube diameters from the exit of the tube, the difference from Clausing's (1930) calculations for infinite distance does not exceed 5%.

The variation of the angular distribution with distance was also investigated by Adamson et al. (1988a,b) using a Monte Carlo simulation for tubes having values of Γ up to 54. At distances such that the tube can be approximated by a point source the Monte Carlo results are in close agreement with the theoretical profile. At smaller tube to detector distances the Monte Carlo simulation produces a profile that is broader than the theoretical curves. Adamson et al.'s preliminary conclusions were that the distance from the detector at which a tube behaves as a point source appears to depend on both its diameter and its length.

Hence for beams formed by single tubes, the near-field effect is negligible under conditions likely to be met in practice such that the beam is not employed very close to the tube exit. It would only appear to be of importance if high precision measurements are required. As discussed, particularly in Section 8.5.4, for designing an atomic beam experiment, there are other problems with the theoretical treatment that are more important to be resolved in order that the beam halfwidth can be more reliably calculated.

As described in Section 10.1 capillary arrays are normally used in preference to single tubes for forming atomic beams. It is therefore useful to know whether there is a distance from the front surface of a plane array to the point of observation where the angular distribution of the beam approaches that calculated for a single tube with the same dimensions as those in the array. We discuss two theoretical approaches. Barashkin et al. (1980) did not consider interatomic collisions within the array, whereas Brinkmann and Trajmar (1981) did. Barashkin et al. used a Maxwellian velocity distribution approach to calculate the angular distribution at different distances from an array and estimate that the beam narrows to that of a single tube at distances greater than 20 to 25 times the diameter of the array.

Brinkmann and Trajmar calculated the angular profile for an array with $\Gamma = 100$ at distances of from 2 mm to 20 mm from its 1.02-mm overall diameter. They used

Equations 8.37 and 8.38 for a single tube, although unfortunately with the Olander and Kruger (1970) values for Z_p and Z_v, and integrated over the array, assuming no significant interaction between the beams originating from the individual tubes. They present graphs at several input pressures of helium showing how the calculated profile narrows as the distance from the source increases, as is to be expected.

In conclusion we note that these near-field effects are normally quite small under typical experimental conditions. However, care might be needed when measurements are made on the beams formed by single tubes at a distance that is too close to the tube exit. These measurements are discussed in Section 11.2. In the typical case of measurements of the beam formed by a focussing capillary array discussed in Section 11.4, the near-field effect is unlikely to be sufficient to account for the anomalous halfwidth.

9 Designing an Atomic Beam

9.1 INTRODUCTION

In Chapter 8, the formal theory of the production of an atomic beam by a beam-forming impedance when it is a single tube has been presented. In Section 9.2 we provide easily applied relationships that allow the atomic beam properties to be predicted for any atom in practical circumstances. The limitations are that Γ should be sufficiently large and that the gas pressure should be sufficiently low that the beam shape is cusp shaped as shown in Figure 1.3 rather than approaching the cosine distribution of Figure 1.1. These limitations, which are normally achieved in practice for a well-collimated beam, mean that the equations are insensitive to tube end effects. Hence the gas density in the entrance plane of the tube Z_P is assumed to be the same as that of the ambient gas in V_P so that no allowance is made for the pressure drop due to the gas flow in the tube. However, the gas density in the tube exit plane Z_V is not assumed to be the same as the zero of V_0, since we are following Zugenmaier's (1966) approach that is discussed in Section 8.5.4. The theoretical tube end effects are discussed in Section 8.5.3. Further discussion of them is deferred to Section 12.4, since their existence could be a source of a disagreement between experiment and theories that do not take them adequately into account. For simplicity the shape of the distribution is also expressed in terms of the halfwidth. If the detailed shape of the distribution is required, this must also be calculated.

The assumption used throughout, namely, that of free-atomic flow, is not expected to be valid when the mean free path is less than the diameter of the tube. However, it is of interest to investigate where the following equations cease to be valid, in view of differing assumptions made by various authors. Hence in Chapter 11 we compare any measurements with them that continue upward in pressure from those in the transition region to hydrodynamic flow, in order to establish whether a general upper limit of validity exists.

The relationships in the next section can be combined with expressions already presented in Chapter 2 to obtain all the atomic beam information likely to be needed by the designer of an experiment. Thus we require the throughput of atoms through the beam-forming impedance so that the total rate of use of source material can be estimated using Equation 2.5 or 2.7. In addition, for gaseous beams the ambient pressure around the beam for a given pumping speed can be obtained from Equation 2.8. The axial intensity is required to estimate the beam density using Equation 2.2, and finally the halfwidth is needed to give the beam size where it is to be used and also to assist in designing orifices forming part of the differential pumping system that may be used to define the beam size.

In Section 9.3 we consider the question of the design of a collimator when variables are the tube diameter, the open area and thickness of an array and the input pressure to it. For example, thicker arrays may only be possible with larger-diameter tubes. We derive some expression of array quality, which we use in Chapter 10, to compare arrays. We also address the question of whether there is an optimum combination of tube diameter and length and open area that gives the best beam.

9.2 EXPRESSIONS NEEDED TO PREDICT BEAM PROPERTIES

With the advent of desktop computing facilities and particularly spreadsheets, the design equations given by Lucas (1973a) in tabular and graphical form can more usefully be replaced by algebraic equations. In order to simplify the application of these equations, tube diameters d are expressed in µm, and tube lengths l in mm, so that they are both close to values needed when they are applied. However, to avoid confusion, we retain $\Gamma = l/d$, with l and d in the same units, so that Γ represents the true dimensionless ratio of length to diameter, irrespective of the units of each, and we do not use Γ in the equations that follow. Atomic diameters σ are discussed in Section 2.4 and are given in picometre. As mentioned in Section 2.2, pressures p are expressed in Pascal, and temperatures T are in Kelvin. Halfwidths are in degrees, and are always the full width at half the peak height. This chapter assumes that the beam-forming impedances are either the right-circular cylindrical tubes defined in Section 8.2, or that they are less regular channels of equivalent diameters.

In order to quantify the statements in the opening paragraph of Section 9.1, the following expressions should only be employed for tubes where $\Gamma \geq 10$ and at input gas pressures such that $\lambda \geq d$. These conditions are normally well within the limits of what is required for a well-collimated beam. Under these conditions, the throughput N at all pressures and temperatures is given to three significant figures by

$$N = \frac{2.76 \times 10^9 d^3 p}{l M^{1/2} T^{1/2}} \text{ atoms s}^{-1}, \tag{9.1}$$

where M is the atomic weight ($M \sim 12$ for carbon).

At laboratory temperature ($T = 295$ K):

$$N_{295} = \frac{1.61 \times 10^8 d^3 p}{l M^{1/2}} \text{ atoms s}^{-1}. \tag{9.2}$$

We note that the throughput is independent of the atomic diameter.

At any temperature and at low pressures, the axial intensity $I(0)$ is given by

$$I(0) = \frac{6.57 \times 10^{11} d^2 p}{M^{1/2} T^{1/2}} \text{ atoms sr}^{-1} \text{ s}^{-1}. \tag{9.3}$$

At laboratory temperature

$$I(0)_{295} = \frac{3.82 \times 10^{10} d^2 p}{M^{1/2}} \text{ atoms sr}^{-1} \text{ s}^{-1}. \tag{9.4}$$

We note that the axial intensity is also independent of the atomic diameter at low pressures.

At high pressures, the axial intensity $I(0)$ is given by

$$I(0) = \frac{4.60 \times 10^{13} d^2 p^{1/2}}{\sigma M^{1/2} l^{1/2}} \text{ atoms sr}^{-1} \text{ s}^{-1}. \tag{9.5}$$

Note that this is temperature independent.

The transition between the two pressure regions is, unfortunately, frequently in the experimental range. However, if the error function can be readily calculated, a single expression for the axial intensity can be employed, namely

$$I(0) = \frac{4.60 \times 10^{13} d^2 p^{1/2} \text{erf}(1.26 \times 10^{-2} \sigma l^{1/2} p^{1/2} / T^{1/2})}{M^{1/2} l^{1/2} \sigma} \text{ atoms sr}^{-1} \text{ s}^{-1}, \tag{9.6}$$

where

$$\text{erf}(x) = \frac{2}{\pi^{1/2}} \int_0^x \exp(-x^2) dx. \tag{9.7}$$

Since erf $(x) = 2x/\pi^{1/2}$ when x is small at low pressures and erf $(x) = 1$ when x is large at high pressures, it follows that Equations 9.3 and 9.5 follow directly from Equations 9.6 and 9.7, which are consistent with the laboratory temperature data given by Lucas (1973a). The function erf (x) is now usually available on spreadsheets, which considerably simplify calculating axial intensities at any pressures likely to be needed. Otherwise it is tabulated, for example, by Jeffrey (2000).

At laboratory temperature

$$I(0)_{295} = \frac{4.60 \times 10^{13} d^2 p^{1/2} \text{erf}(7.36 \times 10^{-4} l^{1/2} \sigma p^{1/2})}{M^{1/2} l^{1/2} \sigma} \text{ atoms sr}^{-1} \text{ s}^{-1}. \tag{9.8}$$

At low pressures, the beam halfwidth H (degrees) depends only on the tube dimensions and is given by

$$H = 0.0962 \, d/l. \tag{9.9}$$

Under the same high-pressure conditions as for the axial intensity

$$H = \frac{2.32 \times 10^{-3} d \sigma p^{1/2}}{l^{1/2} T^{1/2}}. \tag{9.10}$$

At laboratory temperature,

$$H_{295} = \frac{1.35 \times 10^{-4} d \sigma p^{1/2}}{l^{1/2}}. \tag{9.11}$$

Note that at all pressures, the halfwidth is independent of the atomic weight.

In the same way as for the axial intensity, a general equation for the halfwidth may be obtained at any temperature using the error function.

We obtain

$$H = \frac{9.62 \times 10^{-2} d / l}{\mathrm{erf}\left(\dfrac{76.4 T^{1/2}}{\sigma l^{1/2} p^{1/2}}\right)}, \tag{9.12}$$

so that now erf $(x) = 1$ at low pressures.

At laboratory temperature

$$H_{295} = \frac{9.62 \times 10^{-2} d / l}{\mathrm{erf}\left(\dfrac{1311}{\sigma l^{1/2} p^{1/2}}\right)}. \tag{9.13}$$

These are believed to be the only simple expressions for the halfwidth that cover the entire pressure range from $\lambda > l$ to $\lambda \geq d$. Equations 9.1, 9.6 and 9.12 enable the entire theoretical beam-forming properties of any tube for any substance at any temperature to be determined by simple substitution and use of the error function. For beams at laboratory temperature, Equations 9.2, 9.8 and 9.13 may be employed. The equations specifically for high and low pressures are included in the event of the error function not being readily available.

Some authors, including Troitskii (1962), Ivanov and Troitskii (1963), Naumov (1963) and Barashkin et al. (1978), present their results in terms of the mean free path λ. Equation 2.1 may be used to substitute p for λ in those equations containing it to produce simpler expressions.

Note that from the low-pressure Equations 9.3 and 9.12, I (0) is proportional to p when H is approximately constant. From the high-pressure Equations 9.5 and 9.10, both I (0) and H are proportional to $p^{1/2}$. Since from Equation 9.1 N is proportional to p, I (0) H/N remains approximately constant at all pressures. Hence, improved beams are obtained when $\lambda < l$, since when some collisions occur within the tube,

I (0) increases with little increase in H. The relationship also forms a useful check on any errors in calculations.

In Chapter 11 detailed comparison of measurements of the properties of atomic beams with the above equations is made. Section 11.6 draws conclusions about possible disagreements between experiment and theory, particularly for the halfwidth of nonmetallic beams.

9.3   OPTIMISING THE BEAM-FORMING IMPEDANCE

The object of this section is to consider the optimum design of both a single tube, and a capillary array under the realistic assumption that there are practical limitations on both the maximum value of l, and the minimum value of d and also their combination. In addition, for a capillary array, its diameter D and its fractional open area f also have physical limitations. Careful consideration must be given to the design of a beam-forming impedance since the combination of these quantities must be taken into consideration. For example, it is likely that some capillary arrays containing tubes of particularly small diameters can only be constructed if they have both a small open area, and a small value of l. We therefore need to establish which combination of all these quantities produces the optimum atomic beam for any experiment.

We must, however, first carefully consider what we mean by optimum. The achievement of high axial intensities is the goal of developments in atomic beam techniques, and these need to be obtained both with low levels of background gas, and with low beam divergence.

If two beam-forming impedances produce the same axial intensity, but one does so with a smaller throughput, it is clearly superior, since less material is used to form the beam and less pumping speed is required to maintain the background pressure for a gaseous beam. More fundamentally, N produces a background that increases the time required to make a measurement that is proportional to the fluctuation $N^{1/2}$ in N. It follows that for a minimum measurement time to measure a quantity that is proportional to I (0), the beam-forming impedance should produce a beam with the largest value of I (0)$/N^{1/2}$.

If two beam-forming impedances produce beams with the same value of I (0)$/N^{1/2}$, that with the smallest beam divergence is in general superior for a number of reasons. It enables measurements with a higher angular resolution to be made, which means that differential pumping (Section 1.1) is more effective and, in general, a small beam is superior for crossed beam experiments when the second beam is charged or is a light beam and therefore more easily focussed. However, the beam size where an interaction takes place may be more dependent on the array diameter if a plane array is employed. In general we assume that an optimum atomic beam is obtained when I (0)$/(N^{1/2} H)$ is a maximum. For convenience we square the above expression to obtain I (0)$^2/(NH^2)$ as the quantity we seek to maximize. However, we emphasise that this is no universal expression. We cannot anticipate the ideal beam properties for every experiment. The presentation given here should be readily adaptable to other configurations, for example, to use I (0)$/N$ rather than I (0)$/N^{1/2}$.

To simplify the expressions as much as possible, we only retain the array dimensions in the discussion, since it is a trivial matter to use the equations in the previous section to add the other quantities that are relevant to the gas if these are required. They are not, however, parameters that are normally at the experimenter's disposal.

If we use Equations 9.1, 9.6 and 9.12 to calculate $I(0)^2/(NH^2)$ for a single tube subject to the usual conditions for the validity of these equations stated earlier, we find that $I(0)^2/(NH^2) \propto l^2/d$. Hence if it is required to select a single tube to produce a beam under these conditions, that with the largest ratio of the square of its length to its diameter should be selected.

We can now extend this discussion to cover the case of a capillary array. The effect of having as many tubes in the array as possible is that each will contribute to the atom density in the experimental region (otherwise they could be masked off with advantage, since they will contribute to N).

Consider for simplicity a circular array of diameter D containing a number h of capillaries of internal diameter d. It follows that the fractional open area f of this array is given by

$$f = d^2 h/D^2.$$

Hence

$$h = D^2 f/d^2$$

and the single tube values of both $I(0)$ and N must be multiplied by h in order to obtain the optimizing factor for an array. Hence

$$\frac{I(0)^2}{NH^2} \propto hl^2/d.$$

This requires that the angle subtended by a plane array at the experimental region is sufficiently small for the axial intensity to be the effective contribution. This is a good approximation in practice, and in any case, in a general treatment such as this one, no alternative approach is possible. For a focussing array, the assumption is effectively exact. In the few cases where an array is rectangular or square, we assume that D is the diameter of a circle of the same area.

Note that the above expression is more practically expressed as

$$\frac{I(0)^2}{NH^2} \propto D^2 fl^2/d^3,$$

since the array diameter and open area are more usually quoted than the total number of capillaries in it. Hence when designing an array, its individual capillaries should be selected to maximise l^2/d^3. The array should also have both as large an area and fractional open area as possible in it.

For a single tube in a capillary array we first define a single tube quality $Q_{CAP} = 10^6 \cdot l^2/d^3$, where l is measured in mm and d in μm. The factor 10^6 is introduced merely to obtain convenient values of Q_{CAP}. Although clearly Q_{CAP} is of limited use, it means that we are able to say something about the beam-forming quality of the array where the diameter and length of the individual channels are known, but not both f and D. In fact, it is not usually clear from the published details whether D is limited by the constructional technique. It is clearly important to include f if known. Hence we use

$$Q_{FRAC} = f \cdot Q_{CAP}$$

and

$$Q_{ARRAY} = D^2 \cdot Q_{FRAC}$$

with D also in mm, in order to compare the beam-forming qualities of the arrays discussed in Chapter 10.

We stress that it is highly important to consider exactly what the optimum atomic beam requirements are when an experiment is being planned, and then to use the equations in the previous section to establish the ideal array dimensions. The same equations are then used to predict the beam properties of the actual array that is used. It is important to bear in mind the conclusions we draw concerning the validity of the theoretical results for the halfwidth in Section 11.6.

10 Techniques of Multichannel Collimator Construction

10.1 INTRODUCTION

Since the introduction of the construction of arrays of long channels for the collimation of atomic beams by Zacharias and Haun (1954), many different techniques have been successfully employed for the construction of arrays of capillaries. In this chapter we include details of commercially available arrays as well as those constructed in the laboratory.

As shown in Section 9.3, the first quantity of importance when comparing arrays is Q_{CAP}, and this gives a measure of the atomic beam collimation produced by individual capillaries that form the array. In addition, Q_{FRAC} takes account of the parameter f, the fractional open area of a capillary array. Finally the overall quality of the array is given by Q_{ARRAY}, which takes its diameter D into consideration. Thus Q_{ARRAY} takes account of the number of capillaries contributing to the atomic beam.

Unfortunately it is not always possible to establish all three of these quantities from the publications. Although it is the only overall measure of array quality available, Q_{ARRAY} needs to be treated with some caution. First, unlike the other qualities, except sometimes the length of individual capillaries in the array, it is usually not clear whether the array diameter is limited by the construction technique. There are usually good experimental reasons for limiting the beam (and hence the array) diameter. These might include limiting the ambient gas pressure or pumping speed or either because of the desired size of the interaction region, or the requirement to minimise the consumption of beam-forming material. Hence the array diameter might be restricted by the experiment rather than the array constructional details. Focussing arrays may be less critical, but beam convergence might be important.

Note that the figures for Q_{CAP}, Q_{FRAC} and Q_{ARRAY} should only be considered as approximate, both because of the uncertainties in the authors' data and also because arrays tend to have minor defects that affect their quality. In addition to these quality conditions, other experimental requirements may affect the choice of collimator. For example, the vapour of the material to be formed into a beam may not be compatible with the material of the collimator or the desired atomic, molecular or radical state may not be produced or maintained on contact with it. Other considerations include compatibility with other requirements within the apparatus containing the collimator, such as a requirement that it should be electrically conducting or nonconducting, or nonmagnetic. The convenience of mounting, robustness, cost of purchase or ease of construction may be other factors of importance, as well as compatibility with the vacuum requirements, which might include the high temperatures required for baking an ultrahigh vacuum system. A further point is that continued development of some of the techniques discussed may lead to the production of collimators that are superior to the present best.

For each array we also give where possible the value of Γ, the ratio of the tube length to its diameter, that was introduced in Section 1.1. It is a dimensionless quantity, which gives a quick means of establishing the usefulness of an array, without needing to make any assumptions about the experimental conditions for which a beam is required.

It has been recognised for some time that, if other considerations allow, the performance of a plane array, in which the axes of the capillaries are mutually parallel, can be improved by the use of a cylindrically or spherically focussing array. In the former, the axes of the capillaries are parallel in one plane and convergent in a plane normal to it, so that a line focus of atoms is obtained. In the latter the focussing action is analogous to that of a spherical lens, compared with a cylindrical lens in the former case. We have not treated focussing arrays as a separate group, since they are either produced by the same initial construction techniques as a plane array of the same type, or the techniques used would also be suitable for producing an array of parallel capillaries. After discussion of the production of plane arrays, we consider those modifications that have been made to some of them to produce cylindrical or spherical focussing arrays in Section 10.7.

For the purposes of further discussion, we divide the constructional methods into several groups:

1. Arrays constructed of metal foils that are corrugated (Section 10.2); and
2. Arrays produced by embedding wires in a liquid plastic, and removing the wires after solidification of the plastic (Section 10.3); and
3. Arrays of glass capillaries (Section 10.4); and
4. Composite metal techniques (Section 10.5); and
5. Other techniques for producing collimating arrays in any material (Section 10.6).

We are attempting to make this section as comprehensive as possible, both by making reference to all the publications of which we are aware that describe construction techniques, and normally by giving the full details of the techniques that are contained in those publications. Since constructional details of arrays are frequently given in papers whose titles give no hint of this, we are unfortunately aware that it is unlikely that our coverage of the field has been complete. Unless we state otherwise, further details of the method of construction are not contained in the original papers.

10.2 METALLIC FOIL ARRAYS

We first describe the technique of stacking foils, followed by that of winding them on a mandrel. Finally in this subsection, other techniques for producing foil arrays are discussed in chronological order.

As mentioned in Section 10.1, the original method of constructing crinkly foils is described by Zacharias and Haun (1954). Their technique underwent development in several laboratories in the 15 years following the publication of their paper. Before reviewing this field, we note that the channels produced are rather irregular, so the comparisons of quality made in this section are probably less accurate than when channels of near circular cross section are being compared.

Zacharias and Haun produced long lengths of corrugated nickel foils about 13 mm wide. The corrugations were about 51 µm deep with a pitch of about 178 µm, and were made by rolling 25-µm-thick foil between interlocking bronze rollers. The crinkly sheet was cut in a direction parallel to the corrugations with household scissors and interleaved with plane sheet to build up a stack about 25 mm × 635 µm that contained about 300 channels 13 mm long. Hence $\Gamma = 107$, $Q_{CAP} = 93$, $Q_{FRAC} = 21$ and $Q_{ARRAY} = 416$. Some additional details of the rollers are given by King and Zacharias (1956). The rollers were grooved longitudinally in a milling machine equipped with a fly cutter so ground that repeated sharpening did not alter the groove profile, which is described as a 60° notch (Figure 10.1a). A dividing head was used to space the grooves uniformly around the circumference. After machining, the bronze rollers were hard chromium plated. Figure 10.1(b) shows their use. A schematic diagram of a completed array is given in Figure 10.1(c). The rollers had been used to produce about 30 m of crinkly foil at the time of their report. The authors claim 500 channels in the same area, leading to a proportional increase in the Q-values.

Minten and Osberghaus (1958) describe nickel crinkly foils that they obtained from the same laboratory with apparently the same dimensions of corrugations, but in strips of about double the width, namely, 25 mm wide. These authors used copper foil of unstated thickness as the interleaving material and built up an array containing 1300 channels in an area of 20 mm², with an open area of more than 50%. The values of $\Gamma = 205$ and at least $Q_{CAP} = 345$, $Q_{FRAC} = 172$ and $Q_{ARRAY} = 4.4$ k are a significant improvement over the array described above.

In the next two papers we describe a simplification of the stacking technique in which a corrugated and plain foil were wound together round a mandrel. Mitchell et al. (1960) made their ovens from foils 18 µm thick, using both nickel and stainless steel material. This foil is thinner than that used in earlier work. Mitchell et al. produced corrugations about 51 µm deep and 102 µm wide in foil strips that were 6 mm wide and about 760 mm long using a 2.5-mm-diameter mandrel. The authors neither state the overall diameter of the completed array nor discuss the effect of the mandrel on the atomic beam produced by it. They produced an array of 7000 channels with a Γ of 77 and Q_{CAP} of 75, with presumably much less difficulty than the stacking technique.

The same method of coil winding is briefly described by Giordmaine and Wang (1960), whose array (their Source 'C') consisted of 18,000 channels with a minimum diameter of 44 µm (calculated by dividing the perimeter of the approximately triangular-shaped channels by π). The authors state that the average diameter was probably up to 20% greater than the minimum. The channels were 9.5 mm long and the array was 11 mm in diameter. This gives $\Gamma = 177$, $Q_{CAP} = 613$, $Q_{FRAC} = 254$ and $Q_{ARRAY} = 31$ k (including the mandrel).

(a)

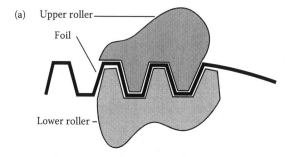

(b)

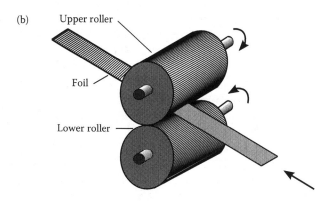

(c)

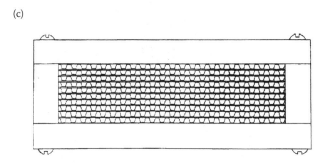

FIGURE 10.1   (a) The groove profile of King and Zacharias' rollers; (b) King and Zacharias' rollers for producing corrugated foils; (c) King and Zacharias' completed array. (Adapted from King JG and Zacharias JR, 1956. Some new applications and techniques of molecular beams. *Adv. Electron. Electron Phys.* **8** 1–88. With permission from Elsevier. Copyright 1956.)

Becker (1961a) gives details of four crinkly foil arrays that were constructed of nickel. The details are given in Table 10.1, where the capillary dimensions refer to the mean internal size. Becker's thinnest foil was over three times thinner than that used by Mitchell et al. (1960) described above but his arrays were comparable to theirs. A diagram of the array assembly is shown in Figure 10.2(a). Becker also describes a two-layer array of ground channels in brass and stainless steel as shown in Figure 10.2(b). No details are given of how they were constructed, but since the capillaries are quite large, they appear to offer no advantage over crinkly foil.

TABLE 10.1
Details of Becker's Crinkly Foil Arrays

Length (mm)	10	20	10	10
Width of capillary (μm)	112	270	122	122
Height of capillary (μm)	52	123	20	20
Number of channels	3200	580	7000	3500
Foil thickness (μm)	10	20	5	5
Dimensions width (mm) × height (mm)	1×26	1×26	1×26	0.5×26
Γ	119	99	125	125
Q_{CAP}	167	48	198	198
Q_{FRAC}	119	33	130	130
Q_{ARRAY}	3.9 k	1.1 k	4.3 k	2.2 k

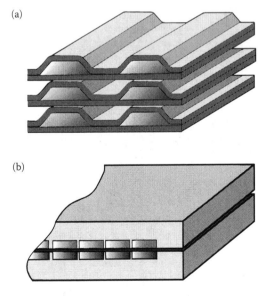

(a)

(b)

FIGURE 10.2   (a) The construction method of Becker's nickel array; (b) the two-layer array described by Becker. (Adapted from Becker G, 1961. Zur Erzeugung starker Molekularstrahlen im Hochvakuum mit Düsen. *Z. angew. Phys.* **13** 59–64. With permission from Springer Science & Business Media B V.)

A development of the technique of producing foils prior to stacking is reported by Kleinpoppen et al. (1962). Details of the method of construction do not appear to have been published. The following details were obtained from the then Director of the machine shop of the Physics Institute at the University of Tübingen. A lathe was used to cut triangular-shaped grooves in a silver–steel cylinder about 50 mm in diameter at a pitch of 10 mm⁻¹. Note that individual grooves were cut so that they did not form a thread. The grooved cylinder was hardened and mounted in a strong fork so that it was free to rotate. A mild-steel block was mounted on the table of a heavy-duty shaping machine. The block was machined and ground to a flat finish. The tool described above was then mounted in the tool holder and the machine was operated to generate a good impression of the grooves in the block. A tungsten strip, 20 μm thick and about 50 mm long and 10 mm wide was clamped on to the block with the grain of the metal perpendicular to its length. Grooves were formed in its entire length by about 5 passes of the tool. The strip was then cut to width on a guillotine.

The corrugated strips were interleaved with foil of 10 μm in thickness, and a package of about 25 pairs was eased into a tungsten tube with a rectangular bore. The aperture where the beam emerged retained the package in place at one end of the tube and the molybdenum holder into which it was frozen retained the package at the other end. As described in Section 6.2.3 this assembly was heated by electron bombardment (Hils et al. 1966) and similar tubes have been heated indirectly by tungsten heaters by Koschmieder et al. (1973) and Koschmieder and Raible (1975) to temperatures of up to 2600 K. Their purpose was to dissociate molecular hydrogen to form a beam of atomic hydrogen. The dimensions of the arrays are uncertain, but plausible values are $\Gamma = 485$, $Q_{CAP} = 318$, $Q_{FRAC} = 79$ and $Q_{ARRAY} = 758$.

Stanley (1966) briefly describes a different method of production of crinkly foils, which he used successfully to corrugate 18-μm-thick aluminium. The foil was laid over an assembly of 25-μm-diameter wires, and a lead block was placed over the foil. Pressure on the block produced corrugations in the foil that were 18 μm deep. Stanley produced his collimator by stacking not with alternate plain foils, but with foils of two different pitches, one twice that of the other. He does not discuss the advantage of this method, but it should be noted that the object was to obtain a fan-shaped array so that one-dimensional focussing of the atomic beam took place. We discuss these arrays in Section 10.7.

Stanley also briefly mentions the use of photoetching to produce channels in foils of beryllium copper that were under 13 μm thick, and so only about 70% of the thickness of his aluminium foils, and with channel widths of just over 6 μm, which is a considerable improvement over the widths produced in rolled foil. Stanley states that open areas of better than 50% were achievable with both his techniques. Channel dimensions produced by photoetching are given by Larson and Stanley (1967), but unfortunately the photoetching process is not described. In addition to the foils described above (Stanley 1966), individual channels that were 200 μm wide and 18 μm deep were produced in beryllium copper foil that was 3.2 mm wide, 25 μm thick and 14.3 mm long. A photograph of an individual foil is shown in Figure 10.3(a). The channels converge to a point about 20 mm from the foil. Figure 10.3(b) shows the cross section of a collimator assembly consisting of a stack of foils, of which about 40 made a complete array, with a Γ of 25, Q_{CAP} of 110, Q_{FRAC} of 46 and a Q_{ARRAY} of 586.

FIGURE 10.3   (a) One of Larson and Stanley's foils; (b) the schematic arrangement of Larson and Stanley's foils. (Reproduced from Larson HP and Stanley RW, 1967. Analysis of the He II 4686-Å (n = 4 to n = 3) line complex excited in an atomic-beam light source. *J. Opt. Soc. Am.* **57** 1439–49. Reprinted with permission from the publisher.)

Jones et al. (1969) describe an array used by Krakowski and Olander (1968) which was constructed of tantalum foils. The foil thickness is not stated, but the triangular channels were 50 μm high and 180 μm at the base, which the authors considered equivalent to a channel diameter of 76 μm. The array of stacked foils was 12 mm long and 3 mm high and produced channels 6 mm long. However, the array was stopped down to 1.56 mm in diameter, which contained a counted 104 channels with a Γ of 79, Q_{CAP} of 82, Q_{FRAC} of 20 and Q_{ARRAY} of 49. Krakowski and Olander operated this array at a temperature of up to 1800 K.

Mungall et al. (1981) describe an array for use in their caesium primary frequency standard. They employed 'a set of 20 fan-shaped collimator slots 12.5 mm long, 0.25 mm high, 0.25 mm wide at their entrance and 0.75 mm wide at their exit, spaced apart vertically by 0.125 mm.' Unfortunately they neither discuss the means of construction nor the rationale for the diverging channel shape.

In conclusion, the crinkly-foil technique can be used to produce multichannel arrays of any desired size in apparently any material that can be obtained in a foil. The foil-winding technique appears to reduce the skill and effort otherwise needed to assemble a stack of foils into an array, without serious loss of quality. Neither photographs of finished arrays nor measurements of channel uniformity appear to have been published. As we discuss in Section 11.3, performance appears to come up to expectations. The best crinkly foil arrays achieve values of Q_{CAP} and Q_{FRAC} of typically hundreds and Q_{ARRAY} of several thousands. The best reported array of this type is that of Giordmaine and Wang (1960). Foils of the metallic elements and alloys mentioned in this section at the thickness quoted, or frequently less, are readily available.

10.3   PLASTIC MATRIX ARRAYS

The techniques used for embedding wires in liquid plastic or resin, allowing this to harden, and then removing the wires to leave an array of capillaries are all similar, compared with the different approaches that have been made to construct arrays from metal foils. We discuss the plastic matrix method in chronological order of publication of the relevant papers.

The first technique that we are aware of that was used to construct capillary arrays in plastic was not, in fact, employed to produce an atomic beam. Huggill (1952) cast poly (methyl methacrylate) [poly $(C_5H_8O_2)$] (which is more commonly known by such trade names as 'lucite,' 'perspex,' or 'plexiglass') on to glass plates to produce sheets 400 ± 100 μm thick, which were cut into 6-mm-wide strips. These strips were wound uniformly with 42-μm-diameter copper wire, and each strip was well embedded by a coating cast from polymerised monomer between two plates. The strips were trimmed to identical size and were packed side by side into a mould and cast into a solid block by the polymerization of additional monomer. The block was milled at each end to remove the curved portions of the wire so that a plane parallel block remained, in which single wires ran between the milled end faces, which were then polished. The wires were etched out with a strong solution of ferric chloride, to leave an array of poly (methyl methacrylate) capillaries. Gold-walled capillaries were also produced by electroplating the copper wire with gold before winding, since ferric chloride does not attack gold.

Huggill gives the results of measurements of four of his arrays. We give in Table 10.2 his measurements of only two, since the others were deliberately made with fewer capillaries. Huggill attributed the slightly smaller diameter of the gold-walled capillaries to etching of the copper wire during the electroplating process.

Hanes (1960), who was apparently unaware of Huggill's work, describes a similar method of construction in detail. Hanes found enamelled copper wire superior to bare wire, since the enamel gave the final array more strength. The diameter of the copper in the wire was 39 μm, and so was approximately half that used by Huggill. The wire was wound on to the bobbin shown in Figure 10.4(a), which has a section 'A' removed. The wires in this region were wetted with epoxy resin and surrounded by the press shown in Figure 10.4(b), which was previously coated on its internal surfaces with silicone grease to act as a release. The press was next assembled and tightened, and the wires were cut at 'B' so that the wires and press could be placed

TABLE 10.2
Capillary Arrays Constructed by Huggill

Material	Plastic	Gold Wall
Array dimensions (mm)	16×11	17×10
No. of strips	20	10
No. of turns of wire/strip	35	70
Thickness (mm)	2.0	3.4
No. of capillaries	1400	1400
Diameter (μm)	83	78
Γ	24	44
Q_{CAP}	7	24
Q_{FRAC}	0.3	1
Q_{ARRAY}	67	200

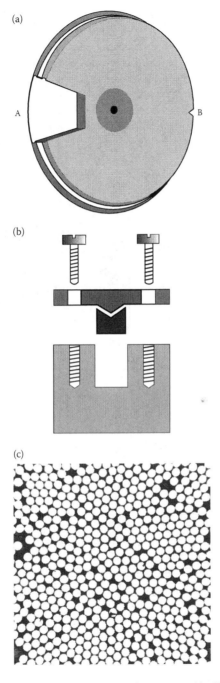

(a)

A

B

(b)

(c)

FIGURE 10.4 (a) The bobbin used by Hanes; (b) the press used by Hanes; (c) part of Hanes' completed array. (Photograph reproduced and diagrams adapted from Hanes GR, 1960. Multiple tube collimator for gas beams. *J. Appl. Phys.* **31** 2171–5. Reprinted with permission. Copyright 1960, American Institute of Physics.)

in an oven until the resin had set. The next stage was to cut off the surplus wire and to place the stack of wires and epoxy resin into a cylindrical mould into which more resin was poured so that the stack was cast into a rod with the wires parallel to its axis. This was then in a convenient form to be cut with a jeweller's saw into slices perpendicular to its axis. The end faces of the slices were ground smooth with fine emery paper which was placed on a flat surface.

Hanes removed the copper electrolytically, using, after numerous trials, an electrolyte consisting of about 100 ml of ammonium hydroxide (containing 30% NH_3 by weight) 50 g of glycine [$C_2H_5NO_2$] and 4 ml of hydrogen peroxide [H_2O_2] (containing 30% H_2O_2) in 1 l of water. The slice to be electrolysed was cemented to a glass tube which was placed in a beaker of electrolyte containing a platinum cathode. Electrolyte was poured into this tube in which a platinum anode was placed. The electrolyte was maintained at a temperature of 333 K and electrolysis was carried out at a current density of about 1 mA mm^{-2} of copper. Electrolysis took about two days for a 630-µm-thick wafer. A photograph of 1 mm^2 of the completed array is shown in Figure 10.4(c). Only a small percentage of the wires was not completely removed. Although Hanes achieved both smaller diameter capillaries and a larger open area in his arrays than Huggill, the approximately five times shorter length of the capillaries produces unimpressive values of $\Gamma = 16$, $Q_{CAP} = 7$, $Q_{FRAC} = 4$, and $Q_{ARRAY} = 1$. Hanes does not discuss whether the electrolytic technique is capable of making longer capillaries.

Aubert et al. (1971a) describe a technique that is essentially similar to that used by Hanes. They produced a uniform skein of copper wires 50 µm in diameter by winding the wires over two pegs as shown in Figure 10.5(a) with a 60 µm/turn feed. Simultaneously the skein was coated with a commercial resin from a drip feed. The bundle of wires was lightly compressed into the mould shown in Figure 10.5(b), and the resin was pre-cured for about 30 min at a temperature of 370 K. The bundle was then removed from the mould, lightly bound with fine wire, and coated with a commercial coating which served to protect the outer layers and increased its robustness. It also enabled the final wafer of material to be more deformable. The next stage of manufacture was to saw the bundle into wafers 800 µm thick, using the wire saw shown in Figure 10.5(c). A wafer was next sealed with paraffin wax into the recess of an aluminium alloy holder so that it could be polished against a similar flat surface on each face in turn, using polishing pastes of increasing fineness. The method of removing the copper electrolytically was similar to that described by Hanes, but Aubert et al. used 40 ml of ammonium hydroxide [NH_4OH] per litre of distilled water instead of 100 ml, and 5 ml of 110 volume hydrogen peroxide. They did not add glycine. Electrolysis took about twice as long as Hanes for a collimator thickness of about 800 µm compared with Hanes' 630 µm, but the current density at 400 µA mm^{-2} was less than one half that used by Hanes.

Aubert et al. obtained an unimpressive $\Gamma = 18$, $Q_{CAP} = 5$ and $Q_{FRAC} = 3$, but their Q_{ARRAY} of 162 is an improvement over Hanes' value. Unfortunately the authors do not provide a photograph of the completed array, so that it is not possible to conclude whether the differences are due to a difference in the enamelled copper wire, or a difference in the packing of the wires into the matrix. We discuss the work of Aubert et al. (1971b) to produce focussing arrays in Section 10.7.

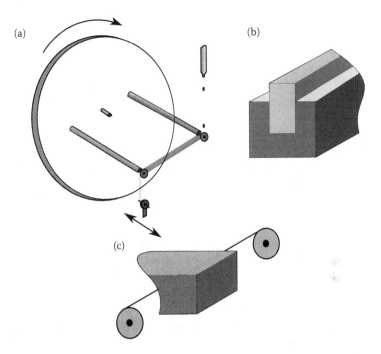

FIGURE 10.5   (a) Aubert et al.'s production of a skein of wires; (b) the mould used by Aubert et al.; (c) Aubert et al.'s wire saw. (From Aubert, D et al., *Production de jets molécu-laires de haute densité et excitation du spectre d'émission Entropie*, No. 42, 60–4, 1971.)

Fain and Brown (1974, 1977) used essentially the same technique of embedding wires in a plastic or resin, but they removed the wires by a much simpler technique. They employed stainless steel wires, down to 25 µm in diameter, which were washed in trichloroethylene (trichloroethene) [C_2HCl_3], sodium hydroxide [NaOH] solution and distilled water. The wire was wound into a skein in a similar way to that performed by the previous authors. It was coated with a monolayer of surfactant, type FX161, after it was deaerated in a centrifuge. After casting in the appropriate plastic or resin, the wires were pulled out of the plastic one at a time, using a microscope and a pair of tweezers, since the surfactant prevented the wires adhering to the matrix, and so made them relatively easy to remove. Fain and Brown reported that some wires that adhered very tightly did break, and so could not be removed. Capillaries between 25 µm and 26 µm in diameter were obtained in both poly (methyl methacrylate) and epoxy resin, of a length between 5 mm and nearly 8 mm. No information is given from which Q_{FRAC} and Q_{ARRAY} can be deduced, but the Q_{CAP} of 4.1 k obtained by Fain and Brown is superior to any other reported plastic matrix or crinkly foil array and the value of Γ of 320 is also superior to other plastic matrix arrays.

Experiments some years ago in the Physics Department at the then Royal Holloway and Bedford New College by Susan M Kay and her co-workers (private communication) were made to determine the best combination of surfactants and how the force required to remove a wire depends upon its size. They made enquiries of 3M UK plc concerning the use of surfactants to obtain the release of wires from plastics and resins. They were advised to use 'Fluorad' coating additives FC-430 or FC-431 in the resin or plastic before casting. 'Fluorad' fluorochemical surfactant FC-135 or 'Fluorad' antimigration coating FC-721 was recommended as a non-wetting coating for the wires, which can be applied simply by dipping. The force to remove a wire was found to be directly proportional to the wire diameter, but almost independent of its length. Hence, the method appears suitable for the preparation of much higher quality plastic collimators than has so far been achieved.

Kim et al. (1995) describe how films of polymer channels can be formed from a master that can, for example, be formed by photolithographic techniques. Channels can be formed that are typically 1.5 µm deep and 3 µm wide by placing the master on a support. Prepolymer enters by capillary attraction. Following curing of the prepolymer by thermal means or ultraviolet light the support can be dissolved, melted or vaporised, leaving the master to be reused. Although there is no interleaving with plane foils to build up an array of channels, this should readily be possible. Kim et al. give a valuable summary of advanced materials suitable for polymerization.

In conclusion, plastic matrix arrays are comparable both in open area and capillary diameter to the crinkly foil arrays, and recent developments have resulted in arrays that are superior in these respects to the best foil arrays. The main problem with plastic arrays is that it is technically difficult to obtain them with longer capillaries, and hence the quality figures appear disappointing. The capillaries closely approach the cylindrical shape used for the theoretical analysis, and they are very uniform in size. Though no studies appear to have been made of the parallelity of the axes of the capillaries in an array produced by the crinkly foil and the plastic matrix method, it seems likely that the latter will be superior. Wires of the materials mentioned in this section at the thickness quoted, or less, are readily available.

10.4   ARRAYS OF GLASS CAPILLARIES

10.4.1   INTRODUCTION

Capillary arrays have been successfully produced in glass by five main techniques:

1. Glass capillaries can be drawn from stock tubing and bundled together without being fused (Section 10.4.3).
2. Capillaries produced as in 1 can be fused together (Section 10.4.4).
3. Capillaries with a wire insert can be fused together, and the wire removed chemically after preparation of a wafer of the final thickness (Section 10.4.5).
4. Capillaries can be drawn, and fused together. The fused bundle of capillaries can then again be drawn and fused together in a one- or multiple-step process (Section 10.4.6).
5. Capillaries with a solid core of a readily etchable glass can be assembled as in 4, and the core is removed after preparation of a wafer of the final thickness (Section 10.4.7).

Any advantages and disadvantages of each technique will emerge from our subsequent discussion. Common to them is the method of drawing tubes into capillaries, and we discuss the principles before giving detailed treatment of individual papers.

10.4.2 CAPILLARY DRAWING TECHNIQUES

The starting point of many of the array construction techniques discussed in Section 10.4 use either small-bore tubes or tubes with a solid glass core which are then drawn into capillary tubes. Hence, we discuss methods of drawing them in this section.

The principles of drawing materials into fibres are discussed by Boys (1887). He obtained uniform glass fibres about 2.5 μm in diameter by firing molten glass attached to a straw arrow from a small crossbow made of pine. Fused silica fibres produced in the same way had a diameter estimated to be about one-tenth that possible with glass. Presumably glass capillaries could be produced in the same way, but it is now found that slow drawing is preferable.

Eschard and Polaert (1969) state that glass tubes can be drawn into fibres until their outside diameter is less than 100 μm, but this is close to the minimum practicable diameter, since the wall thickness increases because of the effect of the surface tension. The effect increases as the tube diameter decreases. However, by maintaining an excess gas pressure within the tube during drawing, open areas of 80% were obtained. They describe how the first drawing stage produces fibres with an internal diameter of less than 1 mm, which can be wound on a drum. The fibre is then cut into lengths, enamelled and a bundle of fibres placed in a hexagonal mould in a furnace to seal it. The second drawing operation uses two belts as described by Washington et al. (1971) and was repeated two or three times. The authors stress that the input and output velocities of the bundle and its temperature must be kept constant if a uniform array is to be achieved. They also describe an optical method of measuring the outer diameter of the fibres as they are being drawn to an accuracy of from 1% to 5%. The error signal from the reader was used to control the drawing speed. By combining the optical with a microwave measurement, it was then hoped to be able to monitor the wall thickness.

The principle of the drawing process is described in more detail by Washington et al. (1971), and is illustrated in Figure 10.6(a). Glass tubing is fed slowly into an oven with a velocity V_T and is drawn out of the softening zone at a much higher velocity V_C. Since the volume of glass entering and leaving the oven at any time must be constant, the reduction in cross sectional area of the glass of the tube is given by V_T/V_C. The wall thickness of the capillary tube can be kept closest to that of the initial tube by drawing the tube at the lowest possible oven temperature. Higher temperatures produce an increased wall thickness, and a reduction in both the bore and outside diameter. A slight overpressure of gas in the tube during drawing results in a decrease in wall thickness. This is because the resulting expansion of the capillary must result in a thinner tube wall, since the cross sectional area of the glass of the tube remains constant.

It is important that the oven temperature be controlled within close limits during the drawing process, since a 1-K change in ambient temperature causes a 1.5% variation in capillary diameter. It is therefore necessary to minimise convection currents by employing the smallest gaps between the glass and the oven walls. Since the drawing temperatures are kept low, it follows that the force on the take-up wheel is large, so slip can occur, which causes the capillary diameter to be non-uniform. Washington et al. passed the capillary through a pair of belts, as shown in Figure 10.6(b), which were rotated by small rollers. This arrangement reduced the variation in

FIGURE 10.6 (a) Washington et al.'s drawing process; (b) the drawing machine used by Washington et al. (From Washington, D et al., *Acta Electron.* 14, 201–24, 1971.)

capillary diameters to less than a few per cent. Capillary drawing machines are now available commercially with electronic control of the drawn diameter.

10.4.3 UNFUSED GLASS ARRAYS

Rudin et al. (1961) used a bundle of glass capillaries to produce a beam of atoms from deuterium that was dissociated in a discharge. About 50 tubes, 10 mm long, 1 mm in internal and 1.2 mm in external diameter were bundled together into an area of 60 mm². The open area was 60%. Hence $\Gamma = 10$ and both Q_{CAP} and Q_{FRAC} are less than unity, leading to a poor Q_{ARRAY} of 5.

Remick and Geankoplis (1973) constructed an array from 644 commercially available glass capillaries. Each tube was 39.1 ±1.2 μm in internal diameter and 9.6 mm long. They were sealed in holes drilled in an aluminium disk 35 mm in diameter giving $\Gamma = 246$, $Q_{CAP} = 1.5$ k, $Q_{FRAC} = 1$ and $Q_{ARRAY} = 1.5$ k.

10.4.4 SIMPLE FUSED ARRAYS

The earliest known description of the use of fused glass capillary arrays to form atomic beams is that of Craddock (1961). He describes briefly the production of a glass capillary array that was 8 mm in diameter and 2 mm thick that contained about

1000 capillaries. The open area was 35%. These figures indicate a mean capillary internal diameter of 37.4 μm. The collimator was produced by drawing hard-glass capillary tubes with a bore of 125 μm and outside diameter of 2 mm by hand, and then stacking them in a tube of unstated diameter and drawing this assembly down to the final diameter of 8 mm. Craddock states: 'This requires some practice.' The array was cut to the required thickness of 2 mm using a diamond saw. Hence $\Gamma = 53$. The irregular appearance of the finished array shown in Figure 10.7 is in marked contrast to the arrays currently manufactured commercially by other methods and shown in many later figures in this chapter. However, Craddock used it to obtain an atomic hydrogen beam halfwidth of about $10°$ at an unstated input pressure. Coating of the collimator with methyltrichlorosilane produced measured dissociations of hydrogen of typically 90%, depending upon the pressure and discharge conditions. His work indicates that a useful glass capillary array can be produced in the laboratory without the need for capillary drawing apparatus. Craddock's array had a $Q_{CAP} = 76$, $Q_{FRAC} = 27$ and Q_{ARRAY} of 1.7 k, but clearly much thicker arrays could be produced by his technique.

Fain and Brown (1974, 1977) describe an array of glass capillaries that had a bore of just under 10 μm, an outside diameter of 508 μm and a length of 4.4 mm. Two aligned rectangular wire meshes 19 mm × 38 mm were used to support 1054 capillaries, which were sealed in place with silica-filled epoxy resin, applied to only one of the meshes with a hypodermic syringe, one row of capillaries at a time. After the sealing, the other mesh was removed. The array had an impressive Γ of 440 and Q_{CAP} of 19.4 k, but of course a very poor Q_{FRAC} of 4 leading in combination to Q_{ARRAY} of 162, which presumably could be readily improved by using longer capillaries.

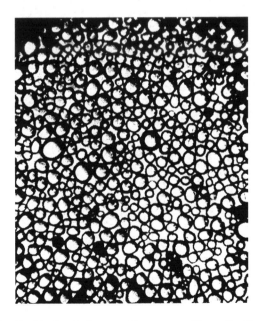

FIGURE 10.7   Craddock's fused glass capillary array. (From Craddock MK, 1961. The polarized proton source of the Harwell proton linear accelerator. *Helv. Phys. Acta Suppl.* VI 59–76. Reproduced with permission from Springer Science & Business Media B V.)

10.4.5 METAL CORE ARRAYS

We again refer to Eschard and Polaert (1969), whose paper is discussed in Section 10.4.2. They indicate that some of the advantages of tubes having a solid core leads to easier handling of less fragile fibres. In addition they can be compressed when hot without deformation to produce a fused array. Metal cores can be removed either by chemical or electrolytic means after the array has been cut to the desired thickness and glass cores are of a soluble glass. Details are given neither of suitable metals nor glasses. The capillaries in the arrays made from glass cores are stated to be not quite so uniform as those made with metal, but they are probably still adequate for producing atomic beams. A photograph of part of an array with capillaries that are 80 μm in diameter made by the metal core technique is shown in Figure 10.8, from which it can be seen that the open area is not particularly large. The thickness of the plate is not stated, but the authors indicate that it was desirable for their purposes of electron multiplication to have Γ in the range 40 to 60.

Washington et al. (1971) also discuss the preparation of glass capillaries that have a metal core. There are two basic methods. The first is due to Taylor (1924), who obtained cores in glass and fused silica of a number of metals and alloys. Initially glass tubing with a bore of 2 mm was sealed at one end so that a piece of metal could be retained in it. Repeated heating and drawing of the tube enabled the metal to be separated from its oxide, which was removed by drawing off that portion of the tube

FIGURE 10.8 Eschard and Polaert's array. (Photograph reproduced from Eschard G and Polaert R, 1969. The production of electron-multiplier channel plates. *Philips Tech. Rev.* **30** 252–5. Reprinted with permission. Copyright 1969, Koninklijke Philips Electronics N V.)

containing it. After drawing to a diameter of between 500 μm and 1 mm, the final drawing stages were similar in principle to those described in Section 10.4.2, but were performed by hand through a hole in a heated copper bar. Taylor gives details of the glasses that were suitable for drawing with nine metallic elements. Cobalt, gold, iron and silver wires could be drawn in fused silica.

The second method is to draw fine tubing over fine wire. Wire of the same diameter as the bore of the glass capillary to be produced is cleaned electrically or thermally and passed through a glass tube. This tube is then drawn by the method already described in Section 10.4.2, except that the temperature need not be kept to a minimum, because of the presence of the wire. Washington et al.'s octagonal take-up drum is shown in Figure 10.9(a). The drawn capillary was coated with a colloidon-based varnish on leaving the oven. The varnish bonded the fibres together on the drum after a drying period of about 10 hours. The octagonal coil was cut into eight similar plates, as shown in Figure 10.9(b), and placed in the press shown in Figure 10.9(c), which was heated in a furnace. After the varnish burnt off, the individual glass capillaries became fused together.

Washington et al. (1971) discuss the properties of metals and glasses that have been used to form capillary arrays by this method. The metal must be capable of being drawn without serious deformation of the wire drawing plate before the several kilometres of wire that are necessary have been produced. The breaking load of the wire at a temperature of about 1020 K must be not less than 25 g, because of the force exerted upon it by the glass drawing process. The wire must be removed by a process that does not attack the glass. If the expansion coefficients of the wire and the glass are not sufficiently matched, a helical break, such as shown in Figure 10.9(d) may occur. Successful combinations are shown in Table 10.3. It is not clear from Washington et al.'s discussion to what extent the values in the table are limiting values. The size of the finished arrays lay between 6 mm in diameter and 19 mm square but the fractional open area is not given. Hence Q_{FRAC} and Q_{ARRAY} are unknown.

Washington et al. (1971) mention other techniques that have been attempted to produce glass-cladded wire, without giving further details or references, namely, enamelling, crucible coating and electrophoresis. They also distinguish between sheathing, in which the wire remains loose within the capillary, and sealing, in which oxide of the metal of the wire interdiffuses with the glass, so some oxide may remain after solution of the metal. Washington et al. suggest aqua regia at a temperature of 350 K to remove ferro-nickel wires and a mixture of sulphuric and nitric acids at the same temperature to remove molybdenum. They found that the metal removal time was reduced by the application of alternating current across the plate, but they give no details of the voltage applied. A photograph of a completed array made by this technique is shown in Figure 10.9(e). Unfortunately the size of the capillaries is not given.

10.4.6 DRAWING OF FUSED CAPILLARIES

The method of drawing a fused bundle of capillaries to form a bundle of finer capillaries is described by a number of authors. Collins et al. (1963) describe the production of an 8-mm-diameter array which is very similar to the same size array that is

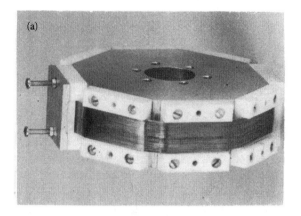

FIGURE 10.9   (a) Washington et al.'s take-up drum; (b) the plates produced by Washington et al.'s process; (c) Washington et al.'s press; (d) a helical break described by Washington et al.; (e) an array produced by Washington et al. with the metal core process. (From Washington, D et al., *Acta Electron.* 14, 201–24, 1971.)

(d)

(e)

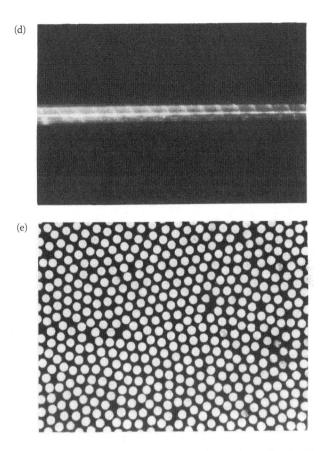

FIGURE 10.9 (Continued) (a) Washington et al.'s take-up drum; (b) the plates produced by Washington et al.'s process; (c) Washington et al.'s press; (d) a helical break described by Washington et al.; (e) an array produced by Washington et al. with the metal core process. (From Washington, D et al., *Acta Electron.* 14, 201–24, 1971.)

TABLE 10.3

Combinations of Satisfactory Glasses and Metals for Capillary Array Production

Glass	Metal	Diameter (μm)	Channel Length (mm)	Γ	Q_{CAP}
Soda	Ferro-chrome	60	1.5	25	10
Lead	Molybdenum	40	2.0	50	63
Soda	Cupro-nickel	80	4.0	50	31
Lead	Ferro-nickel	80	4.0	50	31

briefly described by Stafford et al. (1962). Its use to obtain beams of atomic hydrogen is discussed in Section 6.3.4. Collins et al.'s array contained slightly more tubes, namely, 1200, but of the same length (2 mm), and with internal diameters quoted as ranging from 140 μm to 150 μm, hence Γ was approximately 14, which compares with Stafford et al.'s Γ of about 20, based on approximately 100-μm-diameter tubes. Collins et al. started with Pyrex glass tubing which was 20 mm in diameter and with 1-mm-thick walls. This was drawn in a capillary drawing machine for which no further published details are given to form a continuous length which was cut into pieces 80 mm long. The pieces were packed uniformly into a tube which was 40 mm long and 200 μm greater in internal diameter than the final size of the finished array. This tube served as a temporary holder and enabled the bundle of capillaries to be tied with tungsten wire about 100 μm in diameter and sealed to a rod. This bundle was withdrawn from the holder using the rod as a handle and wound evenly with tungsten wire, which was tied securely.

The fusing was carried out with the assembly held by the rod in a clutch which was rotated at 15 min⁻¹. Firing was at a temperature of 853 K for 10 min followed by 883 K for 5 min. The temperature was then lowered quickly to 853 K again, followed by reduction to below 670 K at a rate between 5 K min⁻¹ and 10 K min⁻¹. Disks were cut from the fused capillary bundle, after removal of the oxidised tungsten wire, by means of a diamond-edged wheel 25 mm in diameter, rotating at 12,000 min⁻¹ under a jet of water. A photograph of the finished array is shown in Figure 10.10. It reveals some departure from ideal packing, quite a wide variation in the bore of the tubes, estimated at about double that given by the authors and some edge distortion. Direct comparison with Figure 10.9(e) is, however, likely to be misleading, since we do not know the size of Washington et al.'s capillaries. From their data estimates Stafford et al.'s (1962) array with 35% open area had Q_{CAP} of 4, Q_{FRAC} of 1 and Q_{ARRAY} of 90.

A variation on this method is described by Ad'yasevich and Antonenko (1963), in that instead of the tubes being fused together after the initial drawing, they are fused before any drawing takes place. This was possible by using smaller tubes initially, with outside diameters in the range 1 mm to 5 mm. The glass used was a molybdenum type which was prepared in a vacuum furnace. Figure 10.11(a) shows how both vertical and horizontal compressive forces were applied to the square cross-section assembly of tubes in a muffle furnace. The finished bundle of tubes was approximately square in cross section, with a side of up to 80 mm. It was stretched by the apparatus shown in Figure 10.11(b), from which it can be seen that the principle is exactly the same as that already described. The direction of motion is upwards, so that the lower clamp moves slowly. The oven temperature was controlled by the commutator which sensed the drawing speed. The drawn bundles were constant in cross sectional dimensions along the length to ±2.8%. The stretched bundles were fused together in the same way as the first bundle was produced, and then stretched again. The wall thickness of individual tubes needed to be at least 10% of their outside diameter, otherwise the square cross section of the initial bundle became deformed into a diamond shape after stretching, so that further fusing could not take place.

When tubes that were initially 5 mm in outside diameter were used, with a wall thickness of 30% of this, an open area of 30% was achieved. This is rather low, but Ad'yasevich and Antonenko etched the completed array in a weak aqueous solution

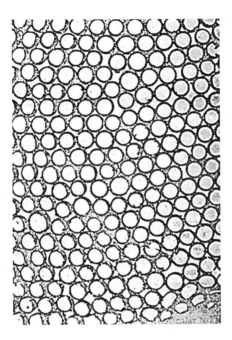

FIGURE 10.10 The array produced by Collins et al. (Photograph reproduced from Collins ER, Glavish HF and Whineray S, 1963. An ion source for polarized protons and deuterons. *Nucl. Instrum. Methods* **25** 67–76. Reprinted with permission from Elsevier. Copyright 1963.)

of hydrofluoric acid, to improve the open area to 80%, which is approaching the theoretical maximum of 91% for touching cylinders of zero wall thickness with a solid interstitial space. A photograph of a completed array is shown in Figure 10.11(c), from which the capillary diameter can be estimated to be 25 μm. The authors claim to have produced arrays down to 10-μm-diameter capillaries with 80% open area.

It is clear from the work of Ad'yasevich et al. (1965) that the authors thought that thin plates were desirable, since they believed that the mean free path of the gas atoms should exceed the length of the capillary. The etching technique is not believed to be applicable to the production of plates much thicker than the 100-μm maximum quoted by the authors. Nevertheless the wall thickness of 10% of the outer diameter that they quote should allow arrays to be manufactured with an open area approaching 30%, without apparently any limitation on their thickness. The overall area of an array did not exceed 20 mm^2. Due to the extreme thinness of the completed arrays, with Γ only 10, their quality appears unimpressive, namely, $Q_{CAP} = 1$, $Q_{FRAC} = 1$, and $Q_{ARRAY} = 13$. Ad'yasevich et al. show a photograph of an array produced by the same technique. The array consists of 40-μm-internal-diameter capillaries 400 μm long, with a fractional open area of 80%. The array was used to obtain a beam of atomic hydrogen from a type of Wood's tube as described in Section 6.3.2. With the same value of Γ as before it gives a slightly improved Q_{CAP} of 3 and Q_{FRAC} of 2, but since no overall diameter is given, we are unable to estimate Q_{ARRAY}.

Washington et al. (1971) employed glass tubing that was initially between 500 μm and 2 mm in diameter and fed it into their oven at a rate between 10 μm s^{-1} and

(a)

Weights

Glass tubes

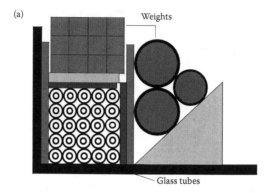

(b)

Weight

Carriage

Heater

Glass
billet

Thermo-
couple

Stepped pulley   Commutator

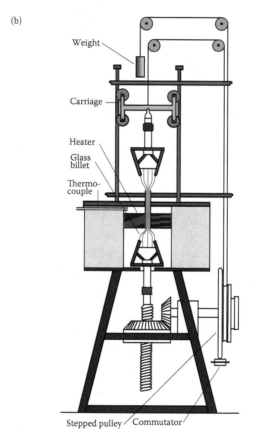

FIGURE 10.11   (a) Ad'yasevich and Antonenko's method of applying compressive forces; (b) the apparatus used by Ad'yasevich and Antonenko to stretch the bundle of tubes; (c) Ad'yasevich and Antonenko's completed array. (Adapted from Ad'yasevich BP and Antonenko VG, 1963. Preparation of glass collimators. *Instrum. Exp. Technol.* **2** 308–10 [*Prib. Tekh. Éksp.* **2** 126–8 1963]. With permission from Springer Science & Business Media B V.)

(c)

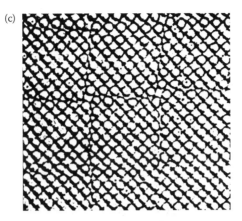

FIGURE 10.11 (Continued) (a) Ad'yasevich and Antonenko's method of applying compressive forces; (b) the apparatus used by Ad'yasevich and Antonenko to stretch the bundle of tubes; (c) Ad'yasevich and Antonenko's completed array. (Adapted from Ad'yasevich BP and Antonenko VG, 1963. Preparation of glass collimators. *Instrum. Exp. Technol.* **2** 308–10 [*Prib. Tekh. Éksp.* **2** 126–8 1963]. With permission from Springer Science & Business Media B V.)

100 µm s⁻¹. Capillaries were drawn out of the oven at a rate between 1 mm s⁻¹ and 100 mm s⁻¹. A bundle of tubes drawn in this way was placed in a heatproof hexagonal mould and fused in an oven at temperatures in the range of about 800 K to 900 K. The fused bundle, which was typically 30 mm in diameter and 500 mm long was drawn in exactly the same way as the initial tube, after cementing supporting tubes to it. Its appearance is then as shown in Figure 10.12(a). It was again an advantage for the drawing to be at the lowest possible oven temperatures. The fusing of the hexagonal bundles of capillaries produced by the second drawing process into the final capillary array was performed in the same way as the production of bundles from individual tubes. Distortion during the fusing processes could be reduced by coating the tubes or bundles before drawing with the powder of a glass that has a lower softening point than the glass of the tube. The powder was applied with a suitable binder such a nitro-cellulose, which burnt away during the drawing operations to leave a glass tube coated with lower softening-point glass. During the next fusing process, the glass coating softened and so fusion took place without the glass of the capillaries becoming softened.

This method is stated to be suitable for finished arrays up to about 25 mm in diameter. Millar et al. (1972) describe how the construction technique has been extended to manufacture plates up to 124 mm in diameter. The initial stages are identical, but the bundles of fused fibres are finally packed into a flexible-walled hexagonal metal cage, which is compressed hydraulically on all six faces equally with a pressure of 2.6 MPa, when the assembly is heated to the softening temperature. Grinding the fused bundle to a cylindrical shape removes the distorted regions at the edges of the hexagonal structure. Finally the core glass is removed as before by etching.

Washington et al. do not give any details of the slicing and subsequent polishing of the fused bundle to produce capillary arrays of the required thickness, but

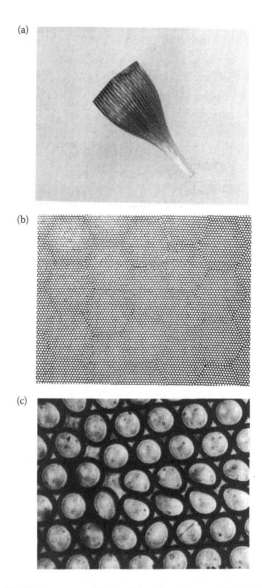

FIGURE 10.12 (a) Washington et al.'s fused bundle; (b) part of Washington et al.'s array; (c) deformation at the boundaries of hexagonal bundles of an array produced by Washington et al.

they advise filling the bundle with wax before cutting, to prevent debris entering the channels. The wax is dissolved away after all machining operations are complete. Part of a completed array produced by this technique is shown in Figure 10.12(b). Unfortunately the size of the capillaries is unstated. The hexagonal sub-structure is evident due to slight deformation of individual capillaries at the boundaries of the hexagonal bundles as can be seen in Figure 10.12(c), which shows, at higher magnification, the deformation at the junction of three hexagons.

It should be borne in mind that Washington et al. are concerned with the quality of optical images produced by channel image intensifiers, for which the processes we describe here are the first stages of manufacture. Small nonuniformities in the capillaries and distortion of a small proportion of the total number of the capillaries in an array are believed to be unimportant for the purposes of atomic beam formation, since glass arrays are in general of high quality compared with crinkly-foil arrays, which produce satisfactory beams.

10.4.7 DRAWING OF SOLID-CORED CAPILLARIES

The fifth type of glass capillary array is that produced initially with a solid glass core which is etched away. The method has been used in the production of commercially available fused glass capillary arrays. For example, Johnson et al. (1966) used arrays of this type for the formation of helium atomic beams. They were available with capillaries as small as 3 µm in diameter with 50% open area and up to 500 µm thick. A plate 2.4 mm in diameter gives a Γ of 167, Q_{CAP} of 9.3 k, Q_{FRAC} of 4.6 k and Q_{ARRAY} of 27 k.

Gardner (1968) gives the composition of glasses that are suitable for drawing by the solid core technique. The rods are formed of lanthanum silicate glass, which consists of 10% lanthanum oxide (probably [La_2O_3] but he states [LaO_3]) by weight. The principal ingredient is barium oxide [BaO], which makes up 47% of the mixture, which also contains 18% boron oxide [B_2O_3], 12% silicon dioxide [SiO_2] and 10% thorium oxide [ThO_2]. The balance of 3% consists of iron and aluminium oxides. This glass is etchable in 0.5 molar nitric acid. The glass for the tubes is made of 80.6% silicon dioxide by weight, with the balance of the materials being 13% boron oxide, 3.8% sodium oxide [Na_2O], 2.2% aluminium oxide [Al_2O_3] and the 0.4% remainder being potassium oxide [K_2O].

Commercially available arrays are also discussed by Jones et al. (1969). 'Mosaic A' was 2.5 mm in diameter. It had capillaries 5.5 µm in diameter but only 250 µm long, giving $\Gamma = 45$, $Q_{CAP} = 376$, $Q_{FRAC} = 195$ and $Q_{ARRAY} = 1.2$ k and 'Mosaic B' was 2.2 mm in diameter and had capillaries 4.5 µm in diameter and 280 µm long giving $\Gamma = 62$, $Q_{CAP} = 860$, $Q_{FRAC} = 293$ and $Q_{ARRAY} = 1.4$ k. These are shown in Figure 10.13(a) and 10.13(b), respectively.

Measurements of the throughput through these arrays, which are discussed in Section 11.3.1, and beam parameters produced by them led Jones et al. to conclude that the channels were partially blocked to an extent greater than that shown in the photographs, which show open channels by the lighter regions in Figure 10.13(a) and dark spots in Figure 10.13(b). These authors presumably received faulty samples of arrays in which the solid core has not been completely etched away, since Wiza (1979) confirms that the etchable core technique is employed by the same company to make microchannel plate detectors and presumably therefore the fused glass capillary arrays described by Johnson et al. (1966) and Jones et al. (1969). Figure 10.14 shows a scanning electron microscope photograph of a small portion of an array described by Wiza. The phenomenon of blocked capillaries in glass arrays does not appear to have been reported by any other authors than Jones et al.

We return to Washington et al.'s (1971) paper for further details of the construction of arrays produced initially with a solid glass core. The drawing process is

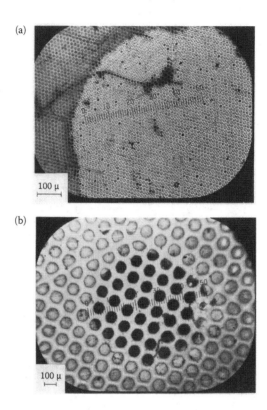

FIGURE 10.13   (a) Mosaic 'A' of Jones et al.; (b) Jones et al.'s Mosaic 'B'. (Photographs reproduced from Jones RH, Olander DR and Kruger VR, 1969. Molecular-beam sources fabricated from multichannel arrays. I. Angular distributions and peaking factors. *J. Appl. Phys.* **40** 4641–9. Reprinted with permission. Copyright 1969, American Institute of Physics.)

essentially the same as we have already described, following the cementing of a core rod into the glass tubing. The presence of a core again means that oven temperatures are less critical than when tubes without a core are drawn, but convection currents can cause an uneven cross section. The second drawing process is performed as before, but the final fusing can take place at a much higher pressure when solid cores are present. A suitable press is shown in Figure 10.15(a), but a preferred alternative, especially for smaller subassemblies of tubes which tend to suffer dislocations, is to pack the hexagonal subassemblies into a hexagonal glass cylinder, closed at one end, shown in Figure 10.15(b). After sealing on a lid, pressure can either be applied uniformly externally, using high pressure gas, or by evacuating the interior. A completed array made by this method is shown in Figure 10.15(c). Comparison with Figure 10.12(b), which shows an array with the same capillary bore, indicates that the hexagon structure is no longer apparent.

   Lieber et al. (1975a) report on studies of fused glass capillary arrays of unstated diameter from another manufacturer. The capillary diameter was 9 μm, and the thickness of the plate they studied was 8 mm, though the authors state that 10-mm-thick plates could be constructed. The plates were made by the etched core technique, and

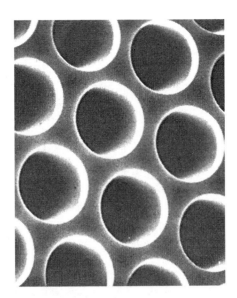

FIGURE 10.14 A scanning electron microscope photograph of Wiza's array. (Photograph reproduced from Wiza JL, 1979. Microchannel plate detectors. *Nucl. Instrum. Methods* **162** 587–601. Reprinted with permission from Elsevier. Copyright 1979.)

had an open area of 60%. The measured optical transmission was only 4%, which the authors attributed to curvature of the individual capillaries that is comparable to their diameter, and to non-uniform etching. This array has $\Gamma = 889$, $Q_{CAP} = 88$ k and $Q_{FRAC} = 53$ k. Lieber et al. (1975b) made measurements on the electron transmission at 10 keV of plates containing 7.9-µm capillaries that were up to 840 µm thick. The measured 39% transmission was close to the expected 41%, calculated from the dimensions of the capillaries. Hence these plates offer satisfactory beam collimation provided the plates are not too thick. The thinner plate has $\Gamma = 106$, $Q_{CAP} = 1.4$ k and $Q_{FRAC} = 558$.

Lampton (1981) indicates that not every type of glass retains its cross-sectional shape on being drawn. This property appears to be restricted to glasses that can be worked over a wide range of temperatures. For channel electron multipliers, where the electrical properties of the walls after final processing are important, glass containing about 50% lead oxide, 40% silicon dioxide and a number of alkali oxides is employed. Figure 10.16 shows part of one of Lampton's arrays containing 40-µm-diameter capillaries.

The importance of optical fibre for data communications has led to further developments in the production of capillaries with a solid core. Kuijt and Teunissen (1988) describe the plasma-activated chemical vapour deposition method. A fused silica tube is first chemically cleaned, and flame polished under internal pressure so that it does not collapse. A microwave discharge in the halides of the material to be deposited mixed with oxygen is generated in the rotating tube, which moves through the resonator at a rate of about 8 m min⁻¹. The next process is to collapse the tube into a solid rod, which can be drawn as described above. The authors discuss this method

(a)

(b)

(c)

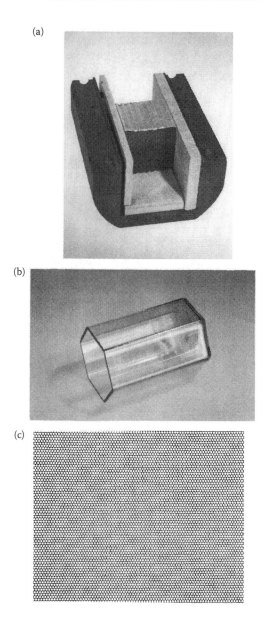

FIGURE 10.15   (a) The press used by Washington et al.; (b) Washington et al.'s hexagonal glass cylinder; (c) Washington et al.'s array produced from solid-drawn capillaries.

in relation to the very pure uniform fibres needed for good optical transmission, so it is not known whether it offers any advantages for producing capillary arrays, such as more readily etchable cores or smaller capillaries.

Laprade and Reinhart (1989) show photomicrographs of six microchannel plates used for electron multiplication with pore sizes from 32 μm down to 4 μm in diameter. Based

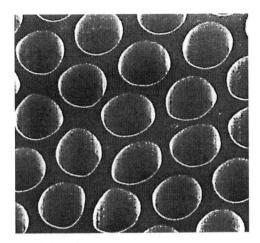

FIGURE 10.16   Lampton's array produced by the same method. (The photograph is reproduced with permission. Copyright 1981, Scientific American, Inc. All rights reserved.)

on their data for centre to centre spacing and assuming a hexagonal structure their smallest array had an open area of 58%. The array with 6-μm-diameter pores had an open area
of 67% and a thickness of 720 μm Hence $\Gamma = 120$, $Q_{CAP} = 2.4$ k and $Q_{FRAC} = 1.6$ k. The
diameter of the arrays is not given so Q_{ARRAY} cannot be determined.

Horton and Tasker (1992) indicate that the core and cladding glasses must be
chosen to have suitable temperature dependence of their viscosities. They state that
multi-component alkali lead silicate glasses are used for the cladding with barium
borosilicate glasses used for the core.

Tonucci et al. (1992) briefly describe the technique used to produce arrays containing much smaller diameter tubes. Their starting point was clear potash lead
glass tubing in which was inserted a boron-rich acid-etchable glass rod, the rate of
etching of which was several thousand times that of the tube. The main departure
from the methods we have described above is the drawing at a high (but unspecified) temperature in vacuum. However, they state that interdiffusion of the glasses
at elevated temperatures may be a problem and annealing may be necessary if work
hardening can occur. After several stages of drawing, stacking in a bundle, refusing
and redrawing, the final assembly was given a glass cladding similar in composition
to the rod and cut with a diamond saw in slices between 500 μm and 5 mm thick.
The end faces were ground with diamond compound and polished with 500 nm
diamond powder. After cleaning in methanol [CH₃OH], and annealing to reduce the
fragility of the disks, the etching followed in dilute hydrochloric acid. Finally three
rinses in distilled water were followed by one in methanol and then the disks were
dried. Following annealing, arrays described as 'very durable' were produced with
450-nm-diameter capillaries with an estimated open area of 33%, 95-nm-diameter
capillaries with an open area of about 27% and 33-nm-diameter capillaries with an
open area of about 35%. A scanning electron micrograph of the former array is shown
in Figure 10.17. If plates are achievable which are 5 mm thick, in the latter case this
leads to $\Gamma = 152$ k, $Q_{CAP} = 7 \times 10^{11}$ and $Q_{FRAC} = 2 \times 10^{11}$. These arrays are obviously

FIGURE 10.17   An example of one of Tonucci et al.'s arrays with 450-nm-diameter capillaries. (From Tonucci, RJ et al., *Science* 258, 783–5, 1992.)

of particular interest for their use in collimating atomic beams, but their use for this purpose has not yet been recorded.

10.4.8 OTHER METHODS OF PRODUCING GLASS ARRAYS

Dimensional details of glass collimators are also given by Meshcheryakov et al. (1967) and Slabospitskii et al. (1970). No constructional details are given, and since both arrays have rather large cross sectional areas for the individual channels, they are not considered further.

In Section 10.4.2 Eschard and Polaert's (1969) methods of monitoring the capillary dimensions during drawing are mentioned. In an appendix to their paper Washington et al. (1971) describe methods of continuous measurement of capillary diameters by three methods. In the first the displacement of position sensors in contact with the capillary is measured electrically. The second method is optical, and photocells in a differential circuit measure the difference in light intensity produced by the capillary. The third employs X-band microwave cavities to sense the volume of the glass. Employment of this method in conjunction with one of the other methods that senses the outer diameter of the capillary enables the bore of a hollow capillary to be determined. Some further details of all these methods are given in Washington et al.'s paper.

A sixth technique for producing capillary arrays in glass is described by Lampton (1981). Photolithographic etching is used to produce fine parallel grooves on both sides of thin glass plates. A stack of plates is fused into a block from which plates are sliced. Lampton states that the controlling of the width and depth of the grooves during etching and fusion proved to be difficult.

The reviews by Washington et al. (1971), Wiza (1979) and Lampton (1981) which we have considered deal with the production of microchannel plates. These are electrically conducting fused glass arrays with a high secondary electron coefficient. From the point of view of atomic beam collimation, the only difference between non-conducting fused glass capillary arrays and micro-channel plates is that the latter have axes of the capillaries that are not parallel to the normal to the surface of the plate. The angle between the axes and the normal is usually between 5° and 20°. These plates are now produced by a number of manufacturers of electron devices. The capillaries have diameters in the range 20 μm to 50 μm with open areas of the plate of at least 50%, and lengths of about 1.8 mm, giving up to $\Gamma = 90$, $Q_{CAP} = 405$, $Q_{FRAC} = 203$ and $Q_{ARRAY} = 127$ k for a typical 25-mm-diameter plate. The tilt of the plate would not be a serious inconvenience for most atomic beam applications, but unfortunately enquiries indicated that very few of the plates that would be suitable for atomic beam collimation fail to meet their specification for charged particle performance, so as to be available as an inexpensive byproduct.

Bellman and Raj (1994, 1997) briefly describe the use of Fotoform orifice plates for use in chemical beam epitaxy (CBE). In the first experiment a plate which was 45 mm in diameter was used and in the second 21.6 mm. Each contained about 3000 holes with a Γ of about 50 with a very small open area. Since there are inconsistencies in the data given in the two papers, it would unfortunately be misleading to quote values for Q_{CAP}, Q_{FRAC} and Q_{ARRAY}.

Though no details of the capillary drawing technique are given, Konijn et al. (1998) indicate that 45-mm-diameter arrays which are from 100 mm thick with capillary diameters in the range from 10 μm to 30 μm containing one million capillaries have been constructed by two manufacturers. They are presumably made by drawing hollow tubes since for tubes with a core, etching depth appears to be the limitation of plate thickness. The authors describe how the plates can be cut to size using a fine granularity diamond saw using high pressure pure water to remove the debris and blowing nitrogen through the tubes from the opposite end to the cutting. Though these arrays are designed for liquid scintillation detectors, it is useful to know that plane capillary arrays thicker than those hitherto used for beam collimation, with Γ up to 10 k are possible. From the photograph of a section of an array, an open area of 62% is estimated. This leads to $Q_{CAP} = 10^7$, $Q_{FRAC} = 6.2 \times 10^6$ and $Q_{ARRAY} = 1.3 \times 10^{10}$.

In addition to the 'slumped' microchannel plates, which are described in Section 10.7, Brunton et al. (1999) give details of some commercially available plates of which the smallest capillaries were 8 μm square and 800 μm long, so that $\Gamma = 90$, based on a circle of the same area. These were contained in a 21-mm × 21-mm array with an open area of 44%. This leads to $Q_{CAP} = 870$, $Q_{FRAC} = 383$ and $Q_{ARRAY} = 214$ k.

10.4.9 Section Conclusion

The processes which have been used commercially for producing microchannel image intensifiers have been critically reviewed by Lampton (1981). The metal-core technique has two major disadvantages. The first is technical, namely, the difficulty of producing a uniform winding of the glass-coated wire. When Figure 10.9(e) is compared with figures showing glass arrays produced by other techniques, the more irregular distribution of the capillaries, with the consequential reduction in open area, is apparent. The second is economic since almost all handling takes place when the tube is at its final diameter.

No discussion of the minimum internal diameter of the capillaries that can be fused into an array has apparently been attempted. We mentioned in Section 10.4.2 that the minimum external diameter is just below 100 µm. With the wire insert method the limiting factors are presumably either the smallest wire diameter that can be obtained, or the thinnest wire with sufficient breaking strength to withstand the forces exerted upon it during drawing. With the solid core technique both the smallest diameter and the largest value of Γ are presumably limited by the difficulty of obtaining fresh etchant where needed and of removing waste products. With the drawing of tubes without any core these limitations no longer apply. It is not clear whether finer tubes are difficult to draw, because of, for example, inhomogeneities in the glass, or temperature gradients during the drawing process, or whether they are too fragile to be fused together without excessive distortion.

Glass capillary arrays are much superior in quality to any other type of array that has been considered so far. They have the advantage of both being readily available commercially and not requiring construction techniques beyond those available to a laboratory. However, as pointed out by Leavitt and Thiel (1990), care must be taken if there is any possibility of a chemical reaction between the beam-forming substance and the glass of the array, because of the large total surface area of the channel walls. For example, the array used by Lucas (1973b) and referred to in Section 10.7 is fairly typical and has a channel surface area of 4×10^5 mm^2, or about the size of a coffee table top. Leavitt and Thiel suspected decomposition of perfluorodiethyl ether (perfluoroethyl ethyl ether) [$C_4H_5F_5O$] on the channel walls, despite its being generally unreactive. They thought that other constituents in the predominantly silicon dioxide array may promote chemical reactions.

10.5   COMPOSITE METAL TECHNIQUES

The construction methods of klystron grids is described by Varian and Varian (1952). Wires of a soft malleable metal, such as annealed zinc, aluminium or iron, or a fibre, such as hemp or cotton, are coated with the material which is to form the grid structure. Suitable materials must be both malleable and ductile, and suggested metals are copper, gold, molybdenum, nickel, platinum, rhenium, silver or tungsten. Colloidal graphite is also proposed. The means of coating may either be electroplating, dipping in molten metal, anodising, metallising or painting with a fluid mixture of powder and binder, or even winding a spiral ribbon round the wire and spot-welding the ends to it.

A bundle of coated wires or fibres is placed in a sheath of a readily drawable material, such as copper, but complete filling of the tube is not necessary. Repeated drawing or swaging of the assembly produces sufficient pressure to deform the elements into a hexagonal cross section. The drawn assembly can be sawn into sections of the desired thickness. The core material is then removed by etching. For example, aluminium is removed with a sodium hydroxide solution and zinc with hydrochloric acid. The sheath may also be removed by etching if required, or it may be bonded to the honeycomb structure by sintering in a hydrogen furnace at temperatures of 1100 K to 1200 K for copper, 1600 K for nickel and 800 K for silver. This sintering is in any case necessary to prevent individual cells of the structure from being separated. For certain materials such as iron, Varian and Varian state that it is necessary to perform the sintering before the etching. They state that the wall thickness can be as small as 2.5 µm, but the maximum possible length of the collimating channels is not discussed.

Markus (1957) gives some further details of the construction of the microscopic grids used in miniature klystron tubes. All sizes have been converted from inches and rounded. Lengths of aluminium wire 127 µm in diameter and 16 mm long were coated in an unspecified manner with copper to a thickness of 5 µm. Then 23 were inserted into a copper tube that was 750 µm in internal diameter and 24 mm long in a three-stage process using initially tubes of slightly larger diameter. The tube was threaded internally to a depth of 5 mm so that it could be fitted to a copper rod to facilitate successive drawing through 24 dies whose internal diameters decreased from 3.9 mm to 3.3 mm. A slitting saw was used to cut 76-µm-thick slices which were next polished until they were one-half this thickness. The aluminium wires were etched out with hydrogen peroxide and this was followed by furnacing in a hydrogen atmosphere to bond the copper structure. Ammonia solution was used to etch the structure until the wall thickness was about 1 µm. Photographs of the grid described show a rather irregular structure, with an octagonal central hole surrounded by a ring of eight hexagons or pentagons with an outer ring of 14 four- or five-sided holes. A slightly thicker 800-µm-diameter grid containing 48 irregular hexagons is shown in Figure 10.18. These grids were not designed for atomic beam collimation, so the limit of extension to thicker, larger diameter structures was not considered.

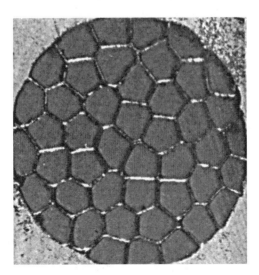

FIGURE 10.18 A klystron grid described by Markus. (The photograph is reproduced with permission from Varian Medical Systems. All rights reserved.)

Giordmaine and Wang (1960) investigated the beam-forming properties of their source 'A' which was made of what they called 'extended' klystron grid. The collimators were of unstated composition. The hexagonal channels had a distance between opposite flats of 299 µm and a wall thickness of about 25 µm. A plate with an overall diameter of 5.1 mm contained 224 individual capillaries. The plate was 6.6 mm thick, and if we follow Giordmaine and Wang in designating the effective diameter of an irregular shape to be the perimeter divided by π, the tubes had a Γ of 24. Both Q_{CAP} and Q_{FRAC} are about 2 and $Q_{ARRAY} = 46$.

Helmer et al. (1960) indicate that klystron grid material is also made by plating aluminium wires with copper, and the deformation is performed by packing the plated wires in a copper tube, which is swaged. The authors claim that the smallest wire which can be used is 25 µm in diameter but they offer no reason. Apparently 127-µm-diameter wire is more practical, as mentioned by others discussed in this section. Since the authors believed that thickness of their array had to be limited so that the mean free path of their ammonia molecules was greater than this, their array was only 3.8 mm thick. It was 3 mm in diameter with an open area of 85%. Hence $\Gamma = 30$, $Q_{CAP} = 7$, $Q_{FRAC} = 6$ and $Q_{ARRAY} = 27$.

Mitchell et al. (1960) describe a collimator made commercially by a similar technique to that already described. Aluminium wires approximately 127 µm in diameter and 100 mm long were used. They were plated with nickel and then packed into a tube of unstated composition that was 3.2 mm in internal diameter. The fusion process was carried out in their case by applying heat at high pressure. Etching of the aluminium was by means of potassium hydroxide solution. The solid rod was sliced into sections 3.2 mm thick, yielding $\Gamma = 25$ and $Q_{CAP} = 5$.

The grids currently in use in reflex klystrons contain hexagon cells of the size described by Giordmaine and Wang (1960), but the plates are only 300 µm thick, or 22 times thinner than used by Giordmaine and Wang, so that they are unattractive to use for atomic beam collimation, even if they could be stacked to form longer channels.

We discuss Beijerinck and Verster's (1975) measurements on atomic beams produced by stainless steel collimated holes structures which were then commercially available in Section 11.3. Details of the method of construction are given by Roberts and Surprenant (1969). Monel metal rods are sheathed in stainless steel tubes which are placed in an outer stainless steel tube. Reduction in diameter is performed by repeated drawing through a die. As stated by Roberts and King (1971) the drawing causes diffusion bonding of the assembly, though this is later (Roberts and King 1972) described as fusion bonding. Roberts and Roberts (1970) state that the construction process causes atom migration which results in 'a common indistinguishable crystalline structure creating a monolithic structure'. The composite rods can again be assembled into a stainless steel tube after drawing and this assembly can be drawn again in the same way that glass capillary arrays are produced by a repeated process. After the drawn composite rod is cut into disks the monel rods are dissolved in nitric acid. Roberts and Surprenant (1969) state that the method is suitable for producing capillaries that are less than 1 µm in diameter. The uniformity of cross section is better than ±5% and the open area is about 50%. Roberts and Roberts (1970) found that there may be a 10% variation in cross sectional area along a tube, as well as from tube to tube.

These authors state that suitable materials other than stainless steel include aluminium, copper, iron and carbon steel, lead, nickel, niobium, tantalum, titanium or tungsten, but thermoplastic materials other than metals might be used. They specifically suggest the combination of copper tubes and lead wires, since construction can be performed at workshop temperature, and the lead can be removed by melting. With this combination, an array of tubes was produced that was 88.9 mm long containing capillaries that were just over 150 μm in diameter giving $\Gamma = 593$, $Q_{CAP} = 2.3$ k and $Q_{FRAC} = 1.2$ k.

The drawing process eliminates the interstitial spaces. Other construction methods suggested include rolling, extruding, swaging and forging. The drawing of stainless steel assemblies is carried out at a high temperature, but at a temperature that is well below the melting point of either tube or core material. Annealing is advantageous before drawing. The tubes may also be drawn on to the wires before packing into the larger tube. In one process a hexagonal array of 91 cored stainless steel tubes was packed into a mild steel tube of circular cross section. The gaps around the periphery were filled with mild steel spacers. After the drawing process the mild steel was etched away with dilute nitric acid. The assemblies were then cut into 76.2-mm lengths and 91 of these were packed into a second tube, which was processed in the same way. A further drawing stage produced over 7.5×10^5 capillaries, each 4 μm in diameter in a 6.86-mm-diameter array, yielding an impressive $\Gamma = 19$ k, $Q_{CAP} = 9 \times 10^7$, $Q_{FRAC} = 5 \times 10^7$ and $Q_{ARRAY} = 2 \times 10^9$.

Production samples were prepared using wire cores of a nickel copper alloy that were 635 μm in diameter, which were threaded through type 304 stainless steel tubes that were 889 μm in diameter. Approximately 5600 cored tubes were placed in a stainless steel can that was 71.1 mm in internal diameter. The can was evacuated, sealed and heated to a temperature of 1283 K and extruded to give an outside diameter of 16.5 mm. The can was then drawn until the diameter was reduced by a further 635 μm. The diameters of the individual capillaries were only reduced by about a factor of five at this stage. However, the sample was again reduced in diameter, namely, to 6.35 mm, which produced a further approximate reduction factor of 2.5 to give 51-μm-diameter capillaries contained in a 3.2-mm-thick plate. By annealing between drawing stages it was possible to reduce the sample further to 5.18 mm in diameter. The composite rods were cut into approximately 150-mm lengths, and 61 of these were packed into a stainless steel can with an internal diameter of just over 54 mm. A hexagonal array of rods was obtained by filling the peripheral spaces with strips of stainless steel 19 mm wide and 380 μm thick. The extrusion reduced the diameter to 16.5 mm as before, and the final drawing again reduced this diameter by 635 μm. The final array contained 335,600 capillaries, 6000 less than the theoretical number, with a mean internal diameter of 12 μm and stated wall thickness of 5 μm. The initial tubes had a 635-μm bore and an 889-μm outer diameter giving a wall thickness of 127 μm. If the reduction in diameter of the tubes takes place without change in the ratio of the diameters, it appears that the authors are quoting the wall thickness as the difference in diameters rather than the more usual difference in radii! Hence the wall thickness is probably one-half that quoted.

The plate was 508 μm thick, giving $\Gamma = 42$, $Q_{CAP} = 149$, $Q_{FRAC} = 102$ and $Q_{ARRAY} = 26$ k. Finally further reduction in diameter of the composite rod to 2.5 mm in diameter enabled collimating plates to be produced that were 127 μm thick containing capillaries that were 1.9 μm in diameter, giving $\Gamma = 67$ and $Q_{CAP} = 2.4$ k. It is not clear whether only thinner plates could be manufactured as the capillary diameter is reduced, since only slightly improved values of Γ and Q_{CAP} are reported as the diameter of the capillaries is reduced by over an order of magnitude. The Q_{CAP} of the stainless steel etched plates is considerably less than the copper array in which the zinc was removed by melting.

Beijerinck and Verster (1975) have made measurements on stainless steel arrays produced by this means, but which do not correspond with the dimensions of those discussed above, even allowing for the fact that Beijerinck and Verster claim that their measurements differ 'appreciably' from the values stated by the manufacturer. They state that the channel cross section was 'irregular' in one sample and the tubes exhibited a 'slight waviness' in a different sample. Since these arrays are no longer available we see no point in giving their dimensions, which are not superior to those quoted above.

10.6   OTHER COLLIMATOR CONSTRUCTION TECHNIQUES

10.6.1   INTRODUCTION

In the last four subsections we have discussed four principal and different techniques which have been successfully employed in the production of atomic-beam collimators. In this subsection we have grouped all other techniques together. The constructional techniques to be described do not fall into one of the previous categories and the collimators have not yet been developed to the stage where they produce beams that are competitive with collimators already described. It is, however, important to discuss any techniques that might lead to the construction of improved collimators, or enable the production of collimated beams of some substance that is incompatible with the collimator materials so far described.

Note that porous materials are not, in general, likely to have any beam-forming properties, as was pointed out by Hanes (1960). Sintered materials are normally produced by heat and pressure on small spheres of material, so no beam-forming properties are to be expected.

10.6.2   STACKS OF METAL WIRES AND TUBES

A means of producing an array of small parallel channels of non-circular cross section is briefly described by Allen and Misener (1939). Up to 2000 stainless steel wires of 61 μm in diameter were bound in a parallel bundle and placed in a thick-walled nickel–silver tube. This tube was drawn through a succession of steel dies of decreasing diameter. As a result the wires were deformed into an approximately hexagonal shape, leaving channels between them of mean width as small as 120 nm and 150 mm long. They were not used to form an atomic beam, which would of course be of very low quality because of the very small open area of the array.

Eminyan et al. (1974) mention the mounting of commercially available fine metal tubes of unstated size and composition in a stainless steel cylinder 25 mm long and 1.0 mm in diameter. No further details are given, but the smallest tubes listed by the manufacturer have an internal diameter of 50 μm with a small wall thickness. Tubes are available in a large number of metals and alloys, including stainless steels. A stack of metal tubes has also been employed by Nguyen et al. (1975). Each tube of unknown origin and constructed of 'inox' (stainless steel) was 100 μm in diameter (or less if the diameter quoted is the external one and not the internal) and 25 mm long, yielding $\Gamma = 250$. The bundle consisted of 20 tubes mounted in an 'inox' cylinder of 1-mm bore giving $Q_{CAP} = 625$, $Q_{FRAC} = 125$ and $Q_{ARRAY} = 125$. Hollywood et al. (1979) formed a beam with nineteen stainless steel tubes which were 100 μm in diameter and 24 mm long which formed an array about 1 mm in diameter. This leads to $\Gamma = 240$, $Q_{CAP} = 576$, $Q_{FRAC} = 109$ and $Q_{ARRAY} = 109$. Gerginov and Tanner (2003) packed type 304 stainless steel tubes having an internal diameter of 584 μm and an external diameter of 813 μm into a hole of dimensions 5 mm × 15 mm. The tubes were cut to length by grinding and after mounting in an ultrahigh vacuum flange and sealing with vacuum epoxy resin, the ends of the array were ground and polished to give a collimator that was about 10 mm long. The completed array contained well over 100 tubes and the interstitial spaces also assisted the collimation. The collimator was used to produce a caesium beam as mentioned in Section 4.5. A second collimator further limited the beam divergence. It was made from an assembly of glass microscope cover slips that were 25 mm square and 170 μm thick and spaced with pieces of cover slip.

A bundle of tubes clearly has the advantage that it is usually quite simple to obtain a large value of Γ and Q_{CAP} but it will always remain a difficult process to obtain high values of Q_{FRAC} and Q_{ARRAY} when it is remembered that the best glass capillary arrays may easily contain 10^7 tubes!

10.6.3   HOLE BORING TECHNIQUES

In Section 3.4.2 we discussed techniques that have been developed for drilling single small diameter holes in materials in order to produce near ideal orifices. In this section we are interested in methods that could produce arrays of holes.

Fleischer et al. (1963) report on the etching of radiation-damaged mica and quartz. The material has to be thin enough to enable complete penetration, which, in their case was only 7 μm of mica. Yet holes could be obtained as small as 2.5 nm after etching in a 20% solution of hydrofluoric acid for a few seconds, which gives an impressive Γ of nearly 3000. The open area of the array of irregularly distributed approximately square cross sectioned holes was only 0.5%, but 10^9 channels mm^{-2} were obtained. The channels produced were uniform in cross sectional area along their length. Parallel channels were obtained from a collimated particle beam. No reports are known of particle track collimators having been used for atomic beam production, but a Q_{CAP} of 3×10^9 and a Q_{FRAC} of 1.6×10^7 indicate considerable potential if the irregular arrangement of the capillaries can be tolerated, which is caused by the random distribution of the incident radiation beam.

Jones et al. (1969) describe an attempt to produce a capillary array by the electron-beam milling technique in fused silica of unstated thickness. However, in a later paper, Jones et al. (1972) state that this was 300 μm and that it contained 70 capillaries which were 50 μm in diameter. As shown in Figure 10.19, taken from their earlier

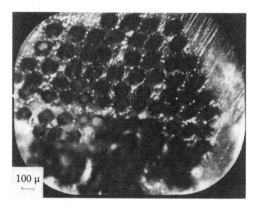

FIGURE 10.19   An array produced by electron beam milling described by Jones et al. (Photographs reproduced from Jones RH, Olander DR and Kruger VR, 1969. Molecular-beam sources fabricated from multichannel arrays. I. Angular distributions and peaking factors. *J. Appl. Phys.* **40** 4641–9. Reprinted with permission. Copyright 1969, American Institute of Physics.)

paper, nonuniform holes were produced, some of which are greater than 100 μm in diameter. The material ejected by the electron beam not only resolidified in some of the capillaries, causing partial blocking, but also produced fingers of silica which produced the 'nondescript' region at the bottom of the figure. The array when tested with helium produced a beam that was little better than that obtained from an orifice, indicating that the array was about the same thickness as the hole diameter. The plate did offer some collimation at temperatures up to 1500 K.

Some of the techniques discussed by Hedley et al. (1977) for producing single small-diameter holes and referred to in Section 3.4.2 might prove suitable for producing many holes. Mechanical twist drills with a diameter of 100 μm can drill holes with a Γ of 20. Pulsed electron beams have been used to produce 40-μm-diameter holes of 'almost unlimited' depth. Fischer and Spohr (1988a,b) review the construction and uses of etched radiation damaged materials. The finest channels so far obtained have diameters of 10 nm and lengths of 100 μm, giving $\Gamma = 10$ k and $Q_{CAP} = 1 \times 10^{10}$.

10.6.4   ALIGNED BORED FOILS

Several experimenters have overcome the problem of drilling deep holes by aligned stacking of a number of identically drilled thin plates. Giordmaine and Wang's (1960) Source 'B' used metal foils of unstated thickness, containing 12,800 holes, 46 μm in diameter in a 13-mm-diameter foil. An unstated number of foils was assembled into a stack 3.1 mm thick. The authors state that the optical transparency of the stack was about 85% of that of a single foil. Hence the walls of each capillary have a corrugated structure, which is important to bear in mind when comparing its beam-forming properties with other sources, and with theory. The tube diameter quoted was measured with a magnifying optical projector, which measures the minimum diameter along the tube. Giordmaine and Wang estimate that non-uniformities mean that the quoted diameter is up to 10% smaller than the average. Assuming this, $\Gamma = 75$, $Q_{CAP} = 135$, $Q_{FRAC} = 18$ and $Q_{ARRAY} = 3$ k.

Kapitanskii et al. (1971) also employed a collimator produced by the stacking technique. They used 1-mm-diameter holes drilled in PTFE disks that were 1 mm thick. The thickness of the stacked disks was 22 mm and the area of each disk was 64 mm² containing 60 holes so the open area was nearly 74%. The large diameter of the holes means that Γ is only 22, both Q_{CAP} and Q_{FRAC} are less than 1 and $Q_{ARRAY} = 29$.

Tunna et al. (2006) describe in detail their experience of obtaining 50-μm-diameter circular apertures in 100-μm-thick tungsten sheet. Lasers were employed to trepan the holes by cutting round their perimeters. Trials with a 15-ns laser pulse revealed problems with the ejection of the material but an ultrafast 20-fs laser operating at a wavelength of 780 nm was used to produce an array of 16 × 16 holes with a pitch distance of 300 μm. It was also found beneficial to direct a jet of compressed air across the sheet surface to reduce the build-up of debris from the drilling. They employed a stack of 11 plates with an overall length of 330 mm to form an X-ray collimator, so it has not been used to form an atomic beam. but the formation of an atomic-beam collimator by this means is clearly feasible. The technique obviously does not limit the production of a higher density of holes, but whether holes can be obtained in much thicker material is uncertain. Their alignment would allow a stack to form a collimator with a larger value of Γ.

10.6.5 WOODEN COLLIMATORS

The research worker, calculating the cost of manufacturing or purchasing a high-performance atomic beam collimator might well declare, colloquially, that they do not grow on trees. Surprising enough, however, they do. Miles (1978) has published photomicrographs of woods at magnifications of either ×25 or ×60 in cross section, and either ×25, ×60 or ×250 in tangential longitudinal section or radial longitudinal section. Examination of the photographs of the softwoods shows that the cells are very approximately circular in cross section, with diameters of about 35 μm and with open areas that are remarkably high. Only just under 1.4 mm of the longitudinal section is normally shown, but a conifer such as *Podocarpus spicatus* shows only a few cells that do not traverse this length. As can be seen from Figure 10.20 the straightness is such that there are rectilinear paths through the majority of cells. The hardwoods show vessels that are typically 100 μm in diameter, but which only form a fraction of the total area. The non-botanist will find Jane's (1970) book invaluable to the understanding of the structure of wood, as well as providing the common names of the botanical names used by Miles. BS EN 13556 (2003) also includes French and German names. Jane states that the vessels in hardwoods may extend up to metres in length.

Lucas et al. (1980) studied the beam-forming properties of a few sample woods, using the mass-spectroscopic method of measurement described by Hanes (1960). Initially cylindrical samples of several woods were turned and faced to size on a lathe, so that the axes of the cylinders were approximately parallel to the axes of the cells. Microscopic examination of the cross sections of the cylinders at this stage

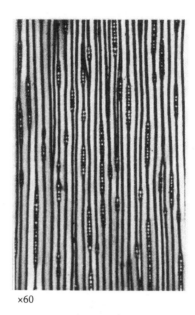

×60

FIGURE 10.20   The structure of *Podocarpus spicatus*.

revealed an appearance similar to a sawn log rather than the cross section shown by Miles. The next stage in the preparation of collimators is familiar to the botanist, and is described by Jane (1970). The samples were placed in cold water which was brought to the boil. The boiling water was then replaced by cold water, which was also brought to the boil. This process was repeated from six to ten times until the wood sank. The samples were stored until required in ethanol [C_2H_5OH]. They were then soft enough to be sliced on a sledge microtome. Its use is also described by Jane, but normally the microtome is used to produce thin slides from a small specimen, so an additional support was provided for the wooden cylinders in the form of a double vee-shaped holder. The microtome was used to slice thin sections from one end of the cylindrical sample until they had the appearance under the microscope of the cross section shown by Miles. The sample was reversed in the holder and sections were cut in the same way from the other end. The cylinder could then be examined under the microscope by transmitted light to reveal the open channel structure. Before a sample was mounted in the vacuum system so that its beam-forming properties could be measured, it was dried at a temperature of 330 K for about 20 hours.

Exley et al. (1977) recommend that samples cut from the living tree should be stored in formalin aceto-alcohol, 30% ethanol, rather than being left to dry out. This mixture also contains formaldehyde [CH_2O] and acetic acid [$C_2H_4O_2$]. Dry samples of wood should be soaked for at least 24 h in distilled water. Hand held single-edged hard backed razor blades were found preferable to a microtome for slicing the samples. The final cut was made with a new blade.

Trial measurements by Lucas et al. (1980) on some hardwoods in 11-mm-long cylinders and on a 2.5-mm-thick softwood sample showed that beam formation was comparable to that obtained by a metal collimated holes structure that had 51-μm-diameter channels 3 mm long. However, more work would be necessary to obtain quantitative results and to establish both the optimum woods and the optimum thickness of the sample.

The wooden collimator will certainly form an inexpensive means of producing an atomic beam from chemically inert gases, that is unlikely to be surpassed in quality by any but the better collimators described in this chapter. In fact the apparatus used to make physical measurements that are of the highest precision may well employ collimators for caesium that are inferior to wooden gas collimators, since Glaze et al. (1974) in describing the realisation of primary frequency standards, state that a collimator of only 500 channels was then used.

10.6.6 Dry Etching Techniques

Horton and Tasker (1992) propose the etching of suitable materials, such as semiconductors, ceramics and glasses, using an ion beam in conjunction with a mask of, for example, photosensitive polymer which is prepared to match the required channels.

Snider et al. (1994) state that the lower limit of capillary diameter when a tube has to be activated for electron multiplication is around 10 μm. They describe the development of a dry etching technique for producing capillary arrays in both gallium arsenide [GaAs] and fused silica, with the object both of making arrays with smaller capillary diameters and of reducing costs. However, the aim was for values of Γ only in the range from 10 to 50. The gallium arsenide was described as showing 'more promise' as a substrate than silica. Etching of the former was undertaken both with a commercial chemically assisted ion beam etcher and an electron cyclotron resonance etcher, whereas silicon was etched with a commercially available magnetron reactive ion etcher. The etch masks tested were of metals, dielectrics and photoresist, and details are given of the preparation of these.

Shank et al. (1995) give further details of the extensive steps in the production of masks for the dry etching process. Their aim was to obtain channel dimensions of from 1 μm to 2 μm having values of Γ up to 60. They investigated various etching techniques using commercially available equipment, including reactive ion, magnetron reactive ion, chemically assisted ion beam or streaming electron cyclotron resonance. However, the disadvantage of the best methods which did not exhibit aspect-ratio dependent etch effects, namely, electron cyclotron resonance and reactive ion etching, was the unacceptable decrease in etch rate with increasing Γ. The etch rate increased to hours as the depth increased. An alternative method, which led to some improvement, was therefore tried in which etching was carried out on a silicon template to leave pillars rather than pores. Low pressure chemical vapour deposition was then used to fill the interstitial space with silicon nitride [Si_3N_4] followed by etching away the template.

10.6.7 Arrays in Silica, Silicon and Alumina

Some of the techniques described in this section have been employed to produce capillary arrays used to form an atomic beam. Other arrays are similar to arrays used for beam formation that have been described in previous sections.

Olt et al. (1961) describe how silica tubes may be drawn, using a machine similar to that already described in Section 10.4.2, but with one important addition. Since the tube was heated with an oxy-propane flame from one side, uniform heating was obtained by rotating the tubes during drawing. Since silica has a defined melting point, unlike glass, which softens, the problem of requiring special glasses, mentioned by Lampton (1981) and discussed in Section 10.4.7, in order to obtain round drawn capillaries, seems difficult to understand. Although no details of the construction of fused silica capillary arrays have been given, we assume that they can be prepared using a silica-sealing glass to bind the individual capillaries together during fusion. Wilmoth (1972) briefly reported on the use of a fused silica capillary array that was claimed to be 1 mm in diameter and 1.5 mm thick and containing about 5×10^5 capillaries of 5 μm in diameter. The areas are inconsistent but possibly $\Gamma = 300$. His array was used to form beams of argon and molecular nitrogen.

Wainer et al. (1971) describe in detail the construction of microchannel plates for use as electron multipliers by anodising aluminium film in a bath of oxalic acid $[C_2H_2O_4]$. This was followed by removal of the remaining aluminium in a saturated aqueous solution of mercury chloride $[HgCl_2]$, followed by washing in 5% nitric acid to remove mercury salts, rinsing and drying. The channels so produced were typically 20 nm in diameter and 10 μm long, leading to a Γ of 500, but with an open area of only about 5%. However, etching was accomplished to increase the pore diameter in a two-stage process. Initially the sample was soaked in a saturated solution of sodium hydroxide in ethyl alcohol for an hour. This did not cause any etching until the sample was placed in distilled water, when open areas in the range 20% to 70% were achieved with a Γ in the range 50 to 80. Plate diameters up to 25 mm were obtained, leading to $Q_{CAP} = 2 \times 10^4$, $Q_{FRAC} = 8 \times 10^3$ and $Q_{ARRAY} = 5 \times 10^6$.

The method used to produce a capillary array in fused silica employed by Knight et al. (1996, 1997) was to start initially with a silica rod which was 30 mm in diameter. A 16-mm-diameter hole was drilled along its length and the outside of the rod was milled to give a hexagonal cross section. This was then drawn at a temperature of about 2300 K in a fibre drawing tower to 800 μm in diameter. After cutting and bundling this fibre it was drawn again to yield a spacing between the capillary centres of about 50 μm. A third drawing stage reduced this spacing to about 2 μm. Since the authors' aim was to construct an optical fibre, its length was many metres. The distance across the flats of the finished structure was 38 μm, but the method appears suitable for producing capillary arrays in fused silica with much larger overall diameters. From the dimensions of the original hexagonal hollow rod, the open area can be calculated to be 34%, but the authors state that surface tension effects reduce the final capillary diameter. By varying the furnace temperature during drawing, a further reduction of the diameter of the capillaries was possible. Though the authors' interest was not forming an atomic beam, their method is of importance if a capillary array in silica is required that is capable of producing large values of Q_{CAP} combined with good values of Q_{FRAC} that should in principle be capable of producing much larger values of Q_{ARRAY} than was required for the authors' optical fibre.

Woznyj et al. (1981) claim that fused silica capillary arrays are commercially available. The array they used had 25-μm-diameter capillaries 1.9 mm long in a

3-mm-diameter array with an open area of 50%, leading to $\Gamma = 76$, $Q_{CAP} = 231$, $Q_{FRAC} = 116$ and $Q_{ARRAY} = 1.04$ k. Their measurements obtained with this array are discussed in Section 11.5.

Govyadinov et al. (1998) give examples of microchannel plates made from anodic aluminium oxide by a development of the technique described by Wainer et al. (1971) and mentioned above. They claim that it is theoretically possible to obtain 10^8 channels mm^{-2} with values of Γ up to 300. An electron microscope photograph of one with channel diameters in the range 100 nm to 300 nm with a Γ in the range 100 to 500 is shown in Figure 10.21, from which it can be seen that rather irregular channels are produced with a relatively small open area estimated from the figure to be about 33%. A directional etching technique using a photolithographic mask enabled more regular channels to be produced, but with values of Γ not exceeding 24.

Taking $d = 100$ nm and $\Gamma = 500$ leads to $Q_{CAP} = 2.5 \times 10^6$ and $Q_{FRAC} = 8 \times 10^5$. Q_{ARRAY} is unknown since the overall diameter of these plates, which incidentally are limited to 100-μm thickness, is unknown.

The production of an alumina microchannel plate is described by Yi et al. (2000). They first prepared a slurry, which consisted of 2-μm alumina particles with a binder, dispersant and solvents. A ball mill was then used to disperse the alumina particles. The slurry was formed into alumina tape, which was cast to form what is referred to as a 'greensheet'. Holes about 100 μm were next punched in the sheets using a computer programmed puncher. They were stacked and aligned to form a laminate which was heated in a furnace to a temperature of about 1900 K. The resulting array contained 170-μm-diameter channels that were 2 mm long in a square array with

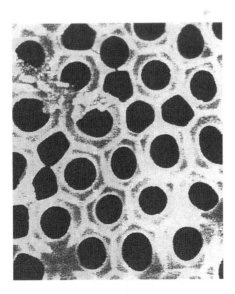

FIGURE 10.21   Govyadinov et al.'s anodized aluminium array. (Photograph reproduced from Govyadinov A, Emeliantchik I and Kurilin A, 1998. Anodic aluminum oxide microchannel plates. *Nucl. Instrum. Methods A* **419** 667–75. Reprinted with permission from Elsevier. Copyright 1998.)

220-µm centres in a plate 50 mm square. Thus channels had a Γ of about 12 and the plate had an open area of 47%. Hence $Q_{CAP} = 0.81$, $Q_{FRAC} = 0.38$ and $Q_{ARRAY} = 1.2$ k.

A micromachining process for silicon wafers based upon photolithography is mentioned by Beetz et al. (2000). It can produce channel diameters between 500 nm and 25 µm with values of Γ up to 300 for 3-µm channels. The theoretical limit for Γ is stated to be greater than 2000. The scanning electron microscope image they show (Figure 10.22) has channels which are 3 µm square with a 6-µm spacing, leading to a calculated open area of 25%, which the authors claim to be approximately 95%. From the image it appears that the open area is about 50%. It is claimed that open areas of greater than 90% are achievable. It is possible to produce plates up to the 300-mm diameter of the silicon wafers but the largest described are 100 mm in diameter. They can be used at temperatures up to about 1500 K. For 3-µm channels $Q_{CAP} = 3 \times 10^4$, $Q_{FRAC} = 1.5 \times 10^4$ and $Q_{ARRAY} = 1.5 \times 10^8$.

Unlike some of the construction techniques described in this chapter Rossi et al.'s (2000) aim was to produce an atomic beam collimator. Their array was constructed from a 25-mm-square piece of n type <100> single polished silicon which was 300 µm thick. It had a resistivity of 60 Ω mm and it was first cleaned in a 10% solution of hydrofluoric acid [HF] for 10 min to remove native oxide. It was then photo-electrochemically etched using a circulating solution of 5% hydrogen fluoride, ethanol and deionized water when illuminated with a 240-W halogen lamp through an infrared-rejecting filter. A double electrochemical cell was employed, so that etching was carried out from both surfaces of the silicon simultaneously, using a voltage of about 4 V applied to platinum electrodes, which provided a current density of about 200 µA mm^{-2} over the 158-mm^2 etched area. Etching times of four hours yielded an array of randomly distributed 5-µm-diameter pores with a Γ of 60. The open area was not estimated. After etching it was necessary to remove crystalline silicon with potassium hydroxide [KOH] solution, followed by rinsing in deionized water. The array should be capable of operation at a temperature of 1300 K. Because of the

FIGURE 10.22 The micromachined array produced by Beetz et al. (Photograph reproduced from Beetz CP, Boerstler R, Steinbeck J, Lemieux B and Winn DR, 2000. Silicon-micromachined microchannel plates. *Nucl. Instrum. Methods A* **442** 443–51. Reprinted with permission from Elsevier. Copyright 2000.)

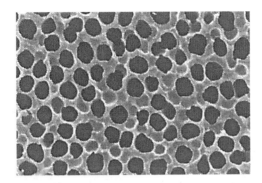

FIGURE 10.23 An Anodisc 13 alumina array described by Roy et al. (Photograph reproduced from Roy S, Raju R, Chuang HF, Cruden BA and Meyyappan M, 2003. Modeling gas flow through microchannels and nanopores. *J. Appl. Phys.* **93** 4870–9. Reprinted with permission. Copyright 2003, American Institute of Physics.)

relatively thin size of the plates a Q_{CAP} of only 720 was obtained. Without knowing the open area, we are unable to estimate Q_{FRAC} and then Q_{ARRAY}.

Figure 10.23 shows a scanning electron microscope image of a commercial Anodisc 13 alumina filter. It was stated by Roy et al. (2003) to be 60 μm thick and with irregular pores having measured average pore diameters of 212 nm. The open area was 25% ±5%. From the manufacturer's data, the diameter of the filter is 13 mm. Hence $\Gamma = 283$, $Q_{CAP} = 4 \times 10^5$, $Q_{FRAC} = 9 \times 10^4$ and $Q_{ARRAY} = 1.6 \times 10^7$. The authors state that the cylindrical pores are well defined and aligned with their axes perpendicular to the surface. They were used for throughput measurements at higher pressures than those considered in Chapter 11 and unfortunately their beam-forming properties were not studied.

10.6.8 Possibilities for the Future

We include in this sub-section other possible methods of array construction which on further development have the potential to be useful for the formation of atomic beams, but as far as we are aware, have not yet been employed for this purpose.

Arrays have been produced by the electrodeposition of metals such as gold and platinum on to glass fibres by Slayter (1960). The coated fibres were welded together by being heated under pressure in a radiofrequency field. The glass was subsequently removed thermally. The capillary arrays were not intended for gas beam collimation and although no detailed dimensional details are given, we assume that high quality arrays of some metals with an open area of at least 50% should be capable of construction by this technique.

The fibre reinforcement of metals is reviewed by Kelly and Davies (1965) and by Cratchley (1965). Aligned drawn rods of, for example, metal wire or glass can be incorporated into pure metals and alloys, in order to increase their strength. There are several methods of introducing the rods into the metal, some of which are appropriate for a metal that has a higher melting point than the rod. One method is to pack the rods into a fused silica tube which dips into a crucible containing molten metal, and to draw the metal up, using a vacuum pump. Powder-metallurgical methods can also be employed, and deposition of metal on to the glass by a number of techniques is possible. These include electroplating and plasma coating. Fused silica can also be coated during the drawing process by passing it through a bead of molten metal before it reaches the take-up mechanism. The coated rods are bonded in each case by applying pressure and heat in a mould for typically one hour. After incorporation of the fibres in the metal, size reduction by extrusion and rolling is possible, with consequential improvement in the alignment of the rods. These authors do not, of course, discuss the removal of the rods to leave an array of metal capillaries, but clearly etching could be used. Melting under differential pressure is probably more effective for capillaries with a large Γ when the difference in melting points of core and metal allow this.

Cline (1981, 1982) describes the production of alternate thin strips of two metals that form a eutectic mixture, for example Pb–Sn and Cd–Pb on a glass substrate. One metal is evaporated on to the substrate, and the second is evaporated on to the first in films that are 1 μm thick. A Nd:YAG laser is used to produce a large temperature gradient to melt the metals as they pass under it. When they solidify they do so in strips about 1 μm wide parallel to the direction of motion. It is suggested that the completed structure can be used as the mask for X-ray lithography. There are obvious possibilities for making corrugated foils by this technique that are much finer than those made by mechanical means.

Williams and Giordano (1984) describe in detail the production of 8-nm-diameter gold wires by electroplating holes produced in etched particle tracks in mica of near circular cross section. Hence it might be possible to use such very fine wires to produce capillary arrays by a variation of any of the techniques we have described that employ wires, so overcoming the poor open area of the radiation damaged plate. The handling problems appear formidable, however.

Becker et al. (1986) have used the synchrotron radiation lithography, galvoforming and plastic moulding (LIGA) technique to produce honeycomb structures. Those in poly (methyl methacrylate) have a cell size of 80 μm and a wall thickness of 4 μm with a length of 350 μm. Honeycomb was also produced in nickel with the same cell size but with a length of 330 μm. The authors do not discuss whether much larger values of Γ could be produced by this technique, which is obviously not yet producing useful collimators, though the high quality of the structures suggests that this is possible. A development of the LIGA process is described in detail by Liu et al. (1999), who also review other methods of producing capillary arrays for use as electron multipliers. Spin-on glass is combined with the LIGA process to produce arrays which are limited to a Γ of 20 in a 150-μm-thick plate with a large open area, but the authors stated that they are hoping to improve upon the aspect ratio. Figure 10.24 shows a photograph of a section of an array with square channels with a side of 18 μm.

In view of the importance of nanotechnology the amount of published work in this area is enormous. Nanotubes are usually produced as single tubes which have one end closed, but Ghadiri et al. (1993) describe the stages in the production of self-assembly arrays by controlled acetification of alkaline peptide solutions. Individual channels are uniform, under 1 nm in diameter, and with values of Γ of some hundreds, so these arrays are likely to be useful for atomic beam collimation. Xu (private

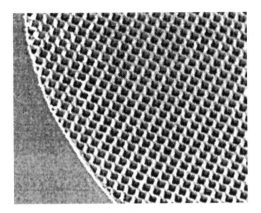

FIGURE 10.24   An array produced by Liu et al. using the LIGA technique. (From Liu, RH et al., *J. Microelectromech. Syst.* 8, 146–51, 1999.)

communication) indicates that aligned carbon nanotube arrays are now being manufactured commercially, but references to their application to atomic beam formation have not yet been located.

Porous membranes can be prepared in the laboratory as discussed by Itaya et al. (1984), Masuda et al. (1997) and Elam et al. (2003). Itaya et al. have made throughput measurements at several temperatures for various gases through the membranes they prepared. They obtained the expected linearity with gas pressure. The potential for use as collimators can be seen to be enormous since Elam et al. report $\Gamma = 2.5 \times 10^4$ and straight parallel channels with a large open area are possible. Both Hulteen and Martin (1997) and Huczko (2000) give details of commercial sources of both track etch and porous alumina membranes as well as discussing nanochannel arrays in many other materials in their review articles.

10.7   FOCUSSING CAPILLARY ARRAYS

As mentioned in Section 1.1 both cylindrical and focussing arrays have been used to form atomic beams. We treat these chronologically without distinguishing those reported to be available commercially from those constructed by the experimenters.

Varian and Varian (1952) describe how klystron grid structures in several metals and graphite can be made into spherical focussing arrays. Their method is an extension of that described in Section 10.5 and commences after the structure has been sawn into sections, and before etching of the core material. The section is plated or coated on both plane surfaces with, for example, zinc. The plating on the surface that will become convex after curvature is made thicker than that on the concave side. The curvature is then produced by the use of a press. Varian and Varian state that in this manner the wafer itself is kept under compression in a transverse direction during the shaping operation, so that there is no tendency for the individual elements to separate as they are displaced.

Cylindrical focussing arrays made from metal foils are described by Stanley (1966) and Larson and Stanley (1967) Their method is different from the others in that they only produced cylindrical arrays, but we have largely described their technique under plane arrays, since the method is also suitable for producing these. It appears from Figure 10.3(a) that about 36 channels are directed to a point on a line 20 mm in front of the aperture.

Aubert et al. (1971a,b) describe the preparation of both cylindrical and spherical focussing arrays made from an epoxy resin. Their constructional technique has already been described in Section 10.3 for the production of plane arrays, where the quality details are given. When cylindrical or spherical focussing arrays were required, the plane array was bent after the polishing process. The design of Aubert et al. for a cylindrical focussing array is shown in Figure 10.25(a) and for a spherical one in Figure 10.25(b). The array was placed between moulds with suitably curved

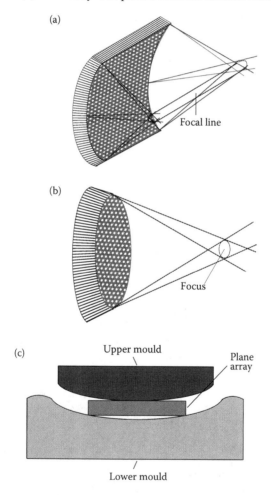

FIGURE 10.25   (a) A cylindrical focussing array described by Aubert et al.; (b) Aubert et al.'s design of a spherical focussing array; (c) Aubert et al.'s moulds used to produce a spherical focussing array. (From Aubert, D et al., *Procédé de fabrication de collimateurs multicanaux focalisants Entropie*, No. 42, 64–5, 1971.)

surfaces, as shown in Figure 10.25(c), and heated to a temperature of 500 K, when the resin softened and bent apparently under the weight of the upper mould. After this bending, the array was polished again, using the moulds instead of the plane surfaces used for polishing plane arrays. The etching of the wires was identical to that for plane arrays. About 100 channels were directed to a point on the line focus, which was 30 mm from the cylindrical surface. The spherical focussing array had the same radius of curvature and contained 10^4 channels directed at the focus.

Roberts and Surprenant (1972) state that the technique they describe for producing plane collimated holes structures can also be used to obtain focussing arrays of capillaries, but they give no further details.

Lucas (1973a,b) reports that spherical focussing arrays are commercially available in glass. They contained 5-μm-diameter capillaries in a 2.5-mm-thick plate curved to a radius of curvature of 50 mm, which had a usable diameter of 23 mm. The plate had an open area of 50%, so 10^7 channels were directed to the focal point giving $\Gamma = 500$, $Q_{CAP} = 5 \times 10^4$, $Q_{FRAC} = 2.5 \times 10^4$ and $Q_{ARRAY} = 1.3 \times 10^7$. Measurement of the beams produced by them is discussed in Section 11.4.

Brunton et al. (1999) were interested in the X-ray optical properties of what they called 'slumped' microchannel plates, which had a radius of curvature of nominally 500 mm. The channels were 8.5 μm square and the plate size was 21 mm square and was 340 μm thick, leading to $\Gamma = 40$. The plate was made from lead silicate glass with the composition $Si_6O_{17}Pb_2K$ and had an open area of 50%. This leads to $Q_{CAP} = 91$, $Q_{FRAC} = 46$ and $Q_{ARRAY} = 2.5 \times 10^4$. Profile measurements showed that a nominal 500-mm radius of curvature was in fact 537 mm in one direction and 549 mm in the other.

An unusual design of collimator which is a development of that described by Ross and West (1998) and discussed in Section 5.2, is described by Seccombe et al. (2001). In each case an interacting photon beam was required to pass through an aperture in a focussing array so that the direction of the atom beam and the interacting beam were coaxial rather than mutually perpendicular, as is usually the case. Seccombe

FIGURE 10.26 Seccombe et al.'s design of a focussed beam source. (Photograph reproduced from Seccombe DP, Collins SA and Reddish TJ, 2001. The design and performance of an effusive gas source of conical geometry for photoionization studies. *Rev. Sci. Instrum.* **72** 2550–7. Reprinted with permission. Copyright 2001, American Institute of Physics.)

et al.'s innermost hollow truncated cone of the source had a semiangle of 35° and an internal diameter of 4 mm through which their photon beam passed. This contained 20 straight 250-μm-wide and 250-μm-deep grooves along the conical surface each of which was 25 mm long. A second hollow truncated cone with an outer semiangle of 45° fitted over this, thus completing the innermost channels. It was provided with 30 channels of the same size and a third truncated cone with an outer semiangle of 55° and 40 grooves completed those channels in the same way. This outermost row of channels was completed by an ungrooved cone thus making 90 grooves in all which were all directed toward the same point on the photon beam axis 4 mm from the front surface of the array. The entire source was made of aluminium. A photograph of the front of the source is shown in Figure 10.26. The channels were 25 mm long, leading to a Γ of 100 and a Q_{CAP} of 40. In view of the novel type of source, we do not attempt to estimate Q_{FRAC} nor Q_{ARRAY}.

10.8   CONCLUSION

In conclusion the crinkly foil arrays of the 1950s and 1960s are now normally superseded by other construction techniques. In particular arrays of fused glass offer especially good quality, provided glass is compatible with the beam-forming substance. The plastic matrix array may be competitive if further development is carried out. Metallic arrays produced by the drawing technique are rugged and suitable for a wide variety of materials. Construction is hardly a straightforward workshop process, however, and industrial production has unfortunately ceased. However, it can only be hoped that the lapsing of the patents will stimulate commercial interest. Some novel techniques have the potential to surpass the quality of arrays that have hitherto been used for beam formation. As discussed by Lucas (1973a) focussing arrays should always be used when experimental conditions allow and we have shown that many of the techniques used for constructing plane arrays can be adapted to produce focussing systems.

11 Comparison of Theory with Measurements

11.1 INTRODUCTION

The main purpose of this chapter is to collect together in a logical sequence all measurements of beams formed both by single capillary tubes in Section 11.2 and by plane capillary arrays in Section 11.3 and to compare these with the theoretical evaluation presented in Section 9.2. These two groups are further divided into comparison of theory with experiment in turn for the throughput, the axial intensity and the angular distribution of the beam, which we use to obtain the halfwidth if this is not stated specifically. This covers the majority of measurements, but in Section 11.4 we treat focussing arrays separately from plane arrays. Section 11.5 discusses cases where only an indirect comparison of theory with experiment is possible. Unfortunately some results are not presented before reduction of the data from our preferred quantities, but we include them in appropriate sections when it has been possible to make comparison with the equations given in Chapter 9. In Section 11.6, conclusions are drawn from the previous sections and finally in Section 11.7 we make what comparison is possible with beams produced by free jet sources. The aim in all cases is to establish primarily the reliability of the application of the equations given in Section 9.2 but also the range of their validity as well as to look critically at measurements.

Mention is made of collimators which have values of Γ that are less than 10, since they are useful in evaluating the end effects discussed in Section 8.5.4 which are more important for small values of Γ. However, no detailed comparison is made since channels with small values of Γ do not form useful collimators, so the extra effort in evaluating the expressions that are valid for $\Gamma < 10$ does not seem justified. For orifices $\Gamma \to 0$ and good agreement is obtained between experiments and the theory presented in Section 2.3.

We have not knowingly excluded any measurements from our discussion, even if they are now shown to be unreliable, since it is believed to be useful to indicate possible pitfalls that unwary experimenters may face. However, the most detailed discussion is given of the more accurate measurements. In some cases only a limited comparison with theoretical predictions is possible. Whereas input gas pressures in the range required are readily determined absolutely, it is unfortunate that the conversion of temperatures to vapour pressures is less certain, as is well known from the variation between experimental results. Where vapour pressures are not stated by the authors, conversion from temperatures has been made using Yaws (1994, 1995a). Throughout this chapter, temperatures that are not stated are assumed to be at a laboratory temperature of 295 K.

The equations given in Section 9.2 are usually easier to apply than those tabulated by Lucas (1973a), since many measurements fall in the intermediate pressure region where neither the low pressure nor the high pressure equations are valid for the beam properties. Whereas some authors use various theoretical predictions with which to compare their own measurements, for consistency we use those from Section 9.2, which usually yield similar, though not necessarily identical results.

11.2 BEAM FORMED BY SINGLE TUBES

11.2.1 INTRODUCTION

We will now discuss over 20 papers which present measurements of parameters of the atomic beam formed by a single capillary tube that can be compared with the equations in Section 9.2. Unfortunately, the majority of authors do not measure all the parameters, but measurements of the beam formed by a single tube are particularly important since they eliminate the uncertainties introduced by capillary arrays. In particular, the measured halfwidth can normally be compared directly with theory, without the necessity of allowing for the finite size of an array. However, caution is needed if the aperture of the beam detector subtends a significant angle at the tube exit, but fortunately the sensitive detectors usually employed in recent measurements enable very small detecting apertures to be used. As discussed in Section 8.5.5, measurements made very close to a tube exit may also be unreliable because of a near field effect.

We firstly discuss the results of throughput measurements in Section 11.2.2, followed by those for the axial intensity in Section 11.2.3, and then the beam shape in Section 11.2.4.

11.2.2 MEASUREMENTS OF THROUGHPUT

We preface the following individual measurements of throughput by noting that many authors make measurements over a particularly wide range of input pressures. Measurements of throughput have often been extended to pressures such that the mean free path λ is less than the tube diameter d, and they therefore give some information on the highest pressures at which Equation 9.1 is valid. Measurements therefore range from very low pressures where λ is much greater than the tube length l to very high pressures where $\lambda \ll d$.

Results are sometimes plotted in terms of the conductance C, which was introduced in Section 2.2 and is the throughput N divided by the pressure p. The conductance is clearly a more sensitive representation of small departures from linearity of the throughput than the throughput itself when both are plotted against pressure. It is generally observed that until $\lambda \sim d$ the measurements exhibit a constant value of conductance which, however, increases rapidly at higher pressures. Between the change, as discussed in Section 8.4.5, a slight minimum in conductance is usually noted and this has become known as the Knudsen minimum. Both because of its insignificant amount and the fact that it is at a pressure which is normally higher than used in atomic beam formation we do not draw attention to it in the following discussion. If the conductance results as presented enable us to compare the throughput measurements as a function of pressure with Equation 9.1 we have done so. These remarks also apply of course to the results presented in Section 11.3.1 for the throughput through capillary arrays.

Carlson et al. (1963) measured the flow of mercury vapour through an orifice and several capillary tubes of accurately known dimensions by measuring the weight loss of their oven and assembly by means of a vacuum balance. Their 'series 6' and 'series 7' tubes, had diameters of approximately 600 μm. The 'series 6' tube was approximately 30 mm long, so $\Gamma = 49.8$, and the 'series 7' 20 mm long, so $\Gamma = 33.2$. Measurements were made at pressures from those such that $\lambda > l$ to those where $\lambda < d$. At pressures such that $\lambda > d$, the throughput for these tubes calculated from Equation 9.1 averaged to about 8% greater than that measured. The flow through the orifice was close to that expected from Section 2.3 when $\lambda > d$. At higher pressures, the measured throughput in all cases became much larger than calculated, indicating that the formulae cease to be valid when $\lambda < d$.

Davis et al. (1964) made measurements of the throughput of several gases through cylinders of glass and brass which had a maximum value of Γ of only just over 2.5, which is unfortunately outside the scope of this comparison. Some of the latter were machined to produce corrugations in the bore, so providing what the authors describe as 'rougher than rough' capillary tubes. They found the throughput decreases slightly for rough tubes compared with glass. Hence the smoothness of the tube surface may also be an important factor in its beam-forming properties. Further evidence of this is given by Hobson (1969) who studied the thermal transpiration of helium through a pyrex tube that was 22 mm in diameter and about 750 mm long so $\Gamma \sim 34$. This tube was leached for 100 h at a temperature of 373 K in 0.05 N hydrochloric acid in order to produce a porous surface, followed by washing in distilled water and air drying. The results were compared with the earlier measurements of Edmonds and Hobson (1965) in unleached tubes, from which it was deduced that specular reflection produced erroneous results in the earlier work.

Jones et al. (1969) studied a single tube that was 380 μm in diameter and 4.8 mm long, and hence $\Gamma = 12.6$. The throughput they obtained for helium is equivalent to 88% of that obtained from Equation 9.2.

Barashkin et al. (1978) made measurements of the throughput of carbon dioxide and hydrogen at a temperature of 293 K through a number of glass tubes, but only one, with $\Gamma = 24.1$ had $\Gamma > 10$. The ratio of their measurements to the theoretical value is close to unity at low pressures but exhibits an increase when the mean free path is less than the tube diameter. Hedley et al. (1978) have studied the flow of air and helium at near atmospheric pressure through single capillary tubes with values of Γ of at least 100. The mean free paths for all measurements were less than the tube diameters and there is considerable scatter in the ratio of their measurements to those predicted by Equation 9.1 but those considered by the authors to be most reliable show throughputs larger than predicted which is consistent with the fact that Equation 9.1 is not valid when $\lambda < d$.

For completeness we include Jackson et al.'s (1985) measurements on the mass flow of zinc vapour through single tubes at temperatures in a range of about 150 K up to about 900 K. The one with the largest Γ had $\Gamma = 9.94$, which makes use of

Equation 9.1 uncertain. The pressure of measurements was also in the range where Equation 9.1 is not expected to be valid. At the lowest temperature, the experimental throughput was about 25% more than that given by Equation 9.1, but this increased with temperature so that at the highest pressure the measured throughput was about three times that predicted. The measurements are therefore consistent with the high pressure measurements of Carlson et al. (1963), discussed earlier in this section.

We also mention Tison's (1993) results for the conductance of helium through a single capillary tube that was approximately 2.16 μm in diameter and 2 mm long, hence with a Γ of 926. Unfortunately his considerable scaling of the results makes comparison with Equation 9.1 uncertain, but a rapid increase in scaled conductance is apparent at pressures such that the mean free path is comparable to or less than the diameter of the tube. At lower pressures the near pressure independence of the conductance is consistent with Equation 9.1.

Rugamas et al. (2000) measured the throughput of helium, neon, argon, xenon, hydrogen, nitrogen, carbon dioxide and acetylene through a single capillary tube that was 800 μm in internal diameter and 20 mm long. Thus $\Gamma = 25$ and the gas pressures used were such that the range of λ was usually from greater than the tube length to less than the tube diameter. The authors were careful to correct the input pressure to the tube for the pressure drop from the measurement position of the capacitance manometer to the tube inlet. The throughput at the laboratory temperature of 300 K was derived from measurements at 1-s intervals of the pressure rise in a calibrated volume of the gas inlet system for a period of about 1 min. The ratio of their measured throughput to that predicted by Equation 9.1 tends to increase with pressure, which, as discussed at the start of this section, is unexpected. This makes detailed analysis unmeaningful, but we note that the average of all the ratios for which $\lambda > d$ is close to unity. This means that averaged experimental results agree with theoretical predictions despite the pressure dependence.

We conclude this section by stating that, although there is a scatter among the measurements of the throughput through a single capillary tube, there are no strong indications that the theory is in error. As we shall see later, discrepancies between theory and experiment are much greater for the axial intensity and angular distribution than for the throughput. The pressure range over which the theoretical relationship based on molecular flow is valid is an additional factor of interest. It appears that at a pressure such that $\lambda = d$, it is certainly valid, but it is unlikely to be valid at higher pressures.

11.2.3 Measurements of Axial Intensity

Lahmam-Bennani and Duguet (1979) made measurements in carbon dioxide at a fixed distance of 560 μm from a single duralumin tube with $\Gamma = 10$, namely, 100 μm in diameter and 1 mm long at input pressures from 8.5 kPa to 133 kPa. The density they derived was proportional to the pressure over this entire pressure range, hence the axial intensity is also linear with pressure. This is an interesting result in this case where $\lambda \ll d$ ($\lambda = 0.005\ d$ at their lowest pressure of 8.5 kPa), since the linear dependence of the theoretical axial intensity upon p at pressures such that $\lambda \gg l$ changes to a square root dependence up to pressures such that $\lambda = d$, so it is unfortunate that these authors were unable to continue their measurements on a single capillary tube downward in pressure to check the theoretical predictions.

Drullinger et al.'s (1985) caesium recirculating oven is described in Section 4.6.3. They measured the axial intensity of the caesium beam produced by a collimating tube that was 2 mm in diameter and 45 mm long, hence $\Gamma = 22.5$. The authors' measurements of the axial intensity at a temperature of about 330 K agree closely with that obtained from Equation 9.6, but as the temperature was increased to about 460 K, the measured intensity fell gradually to about a quarter of the theoretical value. The authors believe that these differences are not significant in view of the difficulty of knowing the atom temperature in an oven of this type, which has a temperature gradient along the collimating tube.

Van Zyl and Gealy (1986) measured the axial intensity of molecular hydrogen at laboratory temperature that was produced by a tube 2.57 mm in internal diameter and about 76 mm long. Hence $\Gamma \approx 30$. At a pressure such that λ was just larger than d, their axial intensity was 77% of the theoretical value, which is within their estimated uncertainty of ±30%. They also made measurements on oxygen flowing through a different tube, but since they do not state its length, we avoid comparing their results with theory because of this uncertainty.

We have been unable to locate any other measurements of the axial intensity of the beam produced by a single capillary tube with $\Gamma \geq 10$ where all quantities are known that enable comparison to be made with theory. Hence the situation is unsatisfactory, but fortunately measurements of the relative angular distributions are more common, and we discuss these to complete this section.

11.2.4 MEASUREMENTS OF HALFWIDTHS

Angular distribution measurements are preferably made by rotating either the beam detector or the beam source at a constant source to detector distance so that the complete profile of the beam can be determined. Uncertainties in the position of the baseline when $\theta = \pm 90°$ are then eliminated. Many measurements have been made using linear motion drives, since they are used to adjust the position of the beam. For example this method was used by Adamson and McGilp (1986), which we discuss in more detail later in this section. This means not only that only part of the angular distribution can be sampled, but as pointed out by Kuhl and Tobin (1995), a $\cos^3 \theta$ correction has to be applied to such measurements. A $\cos^2 \theta$ term arises because the linear sampling results in the sampling distance being increased by a factor of sec θ and hence the solid angle is smaller by $\cos^2 \theta$. This factor has to be combined with a further cos θ term which is due to the angle subtended by the detector decreasing with increasing θ. This effect has also been discussed by Rugamas et al. (2000). As they state, it is uncertain the extent to which this correction has been applied to earlier measurements where it is required.

Troitskii (1962) gives no details of his experiments but he quotes the mean free paths at which his angular distributions were measured. In cases where unambiguous details are given and $\Gamma \geq 10$ we have derived his halfwidths and compared these with Equation 9.12 by substituting Equation 2.1. A tube with $\Gamma = 10$ produced a beam with a halfwidth 1.8 times that given by Equation 9.12, whereas a tube with $\Gamma = 66$ produced a halfwidth at a higher pressure that was 3.8 times the theoretical.

Ivanov and Troitskii (1963) present some measurements that were apparently made by Naumov (1963), but not quoted in the paper they cite. The measurements of the relative angular distributions of ammonia through two tubes with $\Gamma = 15$ and $\Gamma = 50$ were made at such low pressures that $\lambda = 10 \, l$, so that Equation 9.9 can be applied. There is unfortunately insufficient data to apply Equation 9.12. In each case, the measured halfwidths were about a factor 2.4 greater than predicted theoretically.

We have already referred to Barashkin et al.'s (1978) measurements of the throughput of carbon dioxide and hydrogen through single capillary tubes in Section 11.2.2. Measurements of the angular distribution of the carbon dioxide beam were presented for the single tube with $\Gamma = 24.1$ at a range of input pressures of which two may have $\lambda > d$. At these two pressures the halfwidths obtained from their angular distribution measurements are about twice the theoretical. In another plot the same factor is obtained for that tube if we accept that the pressure is so low that the limiting halfwidth is obtained theoretically. Barashkin et al. also plot the pressure dependence of the halfwidth of both carbon dioxide and hydrogen for the tube with $\Gamma = 24.1$. The values increase slowly from about twice that expected from Equation 9.12 at pressures such that $\lambda \approx l$ and then more rapidly up to pressures such that $\lambda < 0.1 \, d$, where a limiting halfwidth of about $75°$ is obtained.

Harvey and Fehrenbach (1983) present one measurement of the shape of a beam of atomic hydrogen produced by a radiofrequency discharge at an estimated input pressure of 1.0 Pa and collimated by a tube that was 1 mm in diameter and 20 mm long, hence $\Gamma = 20$. If we use Allison and Smith's (1971) calculated value of the viscosity of atomic hydrogen at a temperature of 300 K, we obtain $\sigma = 294$ pm and a mean free path which is about half the length of the tube. The beam halfwidth is less than 30% of that predicted theoretically which the authors attribute to recombination on the walls of the tube. A further possibility is an error in the pressure estimate. However, the near Gaussian beam profile is quite different from the cusp shape that is expected from a beam that has a halfwidth of about $2°$.

Drullinger et al.'s (1985) axial intensity measurements made with their caesium recirculating oven are discussed in Section 11.2.3. Their design of necessity uses a temperature gradient along their collimating tube from about 300 K at the source increasing to about 360 K at the exit. This clearly makes the temperature of the emitted atoms uncertain but their measured beam halfwidth of $3°$ is less than expected from Equation 9.12, which predicts over $4°$. The authors suggest that the narrower distribution is because atoms striking the walls of the porous tungsten collimator are absorbed rather than being isotropically reemitted, which is the assumption used in calculations.

We have already mentioned Jackson et al.'s (1985) measurements of the throughput in Section 11.2.2. At a temperature of 855 K they obtained a halfwidth of $66°$ for a zinc beam flowing through a tube that was 3.18 mm in diameter and 31.6 mm long, hence $\Gamma = 9.9$. At this temperature, the vapour pressure of zinc is 1.1 kPa, so $\lambda \ll d$. It is therefore surprising that the halfwidth of $66°$ has only reached 55% of the $120°$ of a cosine distribution, when Equation 9.12 predicts an impossibly broad distribution since it is being applied to a pressure where it is no longer applicable.

Adamson and McGilp (1986) studied the flow of hydrogen and carbon monoxide through two glass tubes. They used an enclosed ionisation gauge to sample the beam through an aperture 1.5 mm in diameter in an end cap about 26 mm in diameter that was normally mounted 12 mm from the end of the tube. Measurements were made by a linear scan which allowed an equivalent angular scan of under ±60°, so that only part of the angular distribution of the beam could be sampled. Only one of the tubes, namely, with l = 8.6 mm and d = 500 µm had Γ > 10, namely, Γ = 17.2. For each gas, three different pressures were used, and the highest in each case was still such that $\lambda > d$. Unfortunately, the pressures were uncertain by ±50%, but the uncertainty is insufficient to account for our data reduction leading to measured halfwidths of factors of 3.10 ± 0.4 greater than given by Equation 9.13, when no account is taken of the finite resolution of the detector of ±3°, as they discuss. Adamson and McGilp varied the source to detector distance and obtained consistent results, so eliminating the effect of detector resolution. They claim good agreement between theory and experiment, due to their adjustment of the pressure of comparison to give good agreement. In fact, the actual halfwidths are probably broader than those quoted, which are measured at the half height of their curves, since these do not tend to zero asymptotically at large angles, so the true zero of the curves is below that indicated.

The same apparatus was used by Adamson et al. (1988a,b), who in addition to making further measurements of the angular distribution of nitrogen from the same tubes, extended their measurements to two further glass and six stainless steel tubes. One of the glass and three of the metal ones had Γ > 10. The authors state that they have removed the uncertainties of the pressure measurement reported by Adamson and McGilp (1986), but unfortunately they do not state the pressures of any of the measurements, so we are unable to compare their results directly with Equation 9.12. For five tubes where Γ > 10, the measured halfwidths were an average of 1.5 times that given by their quoted values of Zugenmaier's (1966) theory after this has been corrected for the finite aperture of the detector and normalised to the axial intensity. They attribute the discrepancy to the fact that both the size of the detector orifice and its distance from the tube are respectively comparable to the tube diameter and length.

Adamson and his co-workers have indicated an important possible source of error in measurements of angular distributions, especially when a single tube is used as a source, if the detector aperture is comparable to the tube diameter. It is important to realise, however, that the ionisation gauge is rarely used as a detector, because its insensitivity makes the use of a large aperture necessary compared with that of a mass spectrometer detector.

We have already commented upon van Zyl and Gealy's (1986) measurement of the axial intensity in Section 11.2.3. They also presented six measurements of the angular distributions in hydrogen, of which one was at a temperature of 2300 K when the molecules were partially dissociated. At pressures and temperatures such that the mean free path was greater than or comparable with the tube diameter, the halfwidths were about twice those expected from Equation 9.12. At shorter mean free paths, the theory is not expected to be valid, but we note that the ratio of measurements of the angular distribution to those expected theoretically decrease from the above.

Buckman et al. (1993) present measurements of the angular distribution of helium and nitrogen flowing through a single glass tube which was 15 mm long and 1 mm in diameter and hence with a Γ of 15. They used a miniature ionisation gauge detector which was scanned linearly across the beam for up to 5 mm each side of the intensity maximum. Their measurements were made at distances of 1.5 mm, 2.5 mm and 4.5 mm from the source exit. The true zero of the angular distribution as mentioned above (Adamson and McGilp 1986) was also uncertain. The sensitivity of the detector was insufficient to enable measurements to be made in helium under conditions such that the mean free path was greater than the tube diameter. However, it is interesting to note that when they measured the beam profile for nitrogen at only 1.5 diameters from the end of the tube from pressures such that the mean free path was longer than the tube length to pressures where it was less than the tube diameter, no change in beam profile was observed. However, when measurements were made at a distance of several tube diameters from the end of the tube, a pressure variation of the beam profile was measured. This is important since it shows that the expected pressure dependence of the beam is not observed close to the tube exit. The possibility of near field effects is discussed in Section 8.5.5. Clearly further measurements are needed to determine where the angular dependence of the axial beam intensity becomes independent of the distance from the source. Their one measurement we can compare with Equation 9.13 is for a nitrogen beam and it is over 40% wider than expected, but this percentage would increase if allowance could be made for the fact that their measurements for this case only extended to ±60°.

Eibl et al. (1998) measured the angular distribution of the atomic hydrogen beam produced in a tungsten tube with $\Gamma = 75$, namely that was 600 μm in internal diameter and 45 mm long. When the last 4.5 mm of the tube were heated to a temperature of 1960 K, the beam was completely dissociated with a measured halfwidth of 40°. Unfortunately experimental uncertainties make comparison of their data with what is expected theoretically unreliable but their halfwidth is over three times that expected from the low input pressures to the tube.

Tschersich and von Bonin's (1998) tube source of atomic hydrogen with $\Gamma = 64$ was described in Section 6.2.3. Their beam was over 80% dissociated but their quadrupole mass analyser measured only the angular distribution of the atomic component of the beam. This was measured at four gas pressures from 0.47 Pa to 140 Pa over an angular range of ±15°. This range was unfortunately insufficient to determine the beam halfwidth at the highest pressure at which they made measurements. We used Allison and Smith's (1971) calculations of the viscosity of atomic hydrogen to obtain the atomic diameter at the temperature of 2600 K used for the measurements. The average measured halfwidth at the pressures other than their highest was over five times that expected from Equation 9.12, despite the mean free path being greater than the tube diameter for these cases.

We also mentioned Schwarz-Selinger et al.'s (2000) hydrogen beam source in Section 6.2.3. They made measurements at four different pressures of their deuterium beam formed by their 50 mm long tube, which was 1 mm in internal diameter, so having $\Gamma = 50$. The measurements average to about 1.9 times what is expected from Equation 9.12 when we use Allison and Smith's (1971) calculations of the viscosity of atomic hydrogen scaled by the ratios of the viscosities for molecular deuterium and hydrogen.

We have already commented upon Rugamas et al.'s (2000) throughput measurements for several gases through a single tube with $\Gamma = 25$ in Section 11.2.2. They also made detailed measurements of the angular distribution of the beam formed by a single tube over the same wide range of pressures. To sample the beam, the authors used a Bayert–Alpert type ionization gauge which could be rotated up to an angle of 100° either side of the tube axis, which is important, since uncertainties in the position of the base line mentioned earlier (Adamson and McGilp 1986, Buckman et al. 1993) are removed. In addition, measurements were made of the distribution produced by an aperture that was 200 μm in diameter and 50 μm thick and the fact that a cosine distribution was obtained was a useful test of the apparatus. The beam was sampled through apertures that were either 500 μm or 800 μm in diameter at a distance usually of 25 mm or 30 mm from the tube exit. This resulted in angular resolutions stated to be 1° to 2°. In addition to tabulated beam halfwidths, the complete angular distribution is shown for a few pressures of argon from which the expected near-hyperbolic shape of the beam can be recognised. Some further details of the apparatus for these measurements are included in their paper.

We have compared Rugamas et al.'s (2000) tabulated evaluations of the beam halfwidths with those to be expected from Equation 9.12 and the atomic and molecular diameters tabulated in Section 2.4 at all tube inlet pressures where $\lambda > d$. Unlike the throughput comparison, no significant variation in the ratio of measured to theoretical halfwidths was found either with gas species or gas pressure. We have therefore combined all 68 measurements to yield a factor 2.1 ± 0.1 by which the measured halfwidths are greater than those calculated. Rugamas et al. also combined their measured halfwidths in plots against mean free path, but unfortunately without including any theoretical curves. In Figure 11.1 we show their data for argon which start from the lowest pressure of all the gases they employed, namely, where $\lambda = 4.4\ l$ and then we continued up to $\lambda = d$. The theoretical curve is multiplied by the mean factor by which these argon measurements are greater, namely, 2.07. The graph shows the expected near independence of the halfwidth upon pressure at low pressures and the square root proportionality at higher pressures. The close match of the shape of the experimental and theoretical curves over the whole pressure region is remarkable. It is unfortunate that more data were not available in the low pressure region. We conclude that whatever the reason is for the discrepancy between measured and theoretical curves, it is not due to interatomic collisions, since it also exists when $\lambda > l$.

We have already mentioned Cvejanovic and Murray's (2002) production of a beam of calcium in Section 5.2. The beam was emitted from a single tube 20 mm long and 900 μm in internal diameter, namely, with $\Gamma = 22.2$ and was found to diverge at about 3°. This is less than expected from Equation 9.12 in the limit of low pressure and it is about six times less than we calculate on the basis of their oven temperature of about 950 K. This suggests that the beam halfwidth was measured at a much lower calcium pressure than quoted for the operation of their oven.

We have already referred to Ahmed et al.'s (2005) oven in Sections 5.12 and 5.16. They also measured the halfwidth of a beam of uranium with the oven operated at a temperature of about 2500 K. The shape of their beam was not like the cusp shape to be expected from a single tube, so their measured halfwidth, which is about 70% of that expected from Equation 9.12, is unfortunately unreliable.

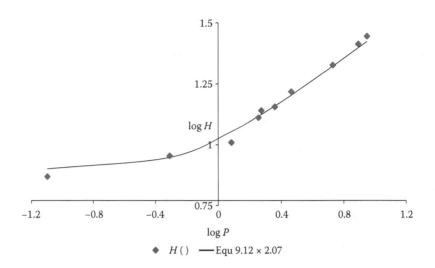

FIGURE 11.1   Rugamas et al.'s measurements in argon compared with theory × 2.07.

Hence we conclude that measurements of the halfwidth of the beam formed by single capillary tubes in general show both a relatively large scatter among the results presented in each paper and among the results presented by different authors. They point, however, to a tendency for the halfwidths obtained by the most reliable measurements to be greater than that predicted theoretically by typically about a factor of two. Although the large spread of factors indicates experimental difficulties in making even a relative measurement, it should be noted that the systematic discrepancy with theory is much greater than can be explained by experimental uncertainties. The most likely of these are knowledge of the gas pressure at the tube entrance and the elimination of the background gas when measuring the angular distribution. There is a tendency to overestimate the gas pressure at the tube entrance, since allowance is not made for the pressure drop between the position of measurement and the tube entrance. This should, however, lead to a measurement that is narrower, not wider than predicted. If the background pressure is not correctly subtracted, the point of measurement of the halfwidth is in fact too near the peak of the curve, hence again an underestimate is made.

Measurements on single capillaries tubes are particularly important, since measurements on plane arrays of capillaries have the added uncertainty that it is difficult to deconvolute the measured angular distribution for an array to that for a single tube. This is necessary so that measurements may be compared with single tube theory. It is unfortunate that very few measurements on the angular distribution of vapour beams from single tubes have been reported, since, as we shall see in Section 11.3.3, they might have provided more evidence for a possible explanation of the discrepancy between theory and experiment.

We will discuss all results further after presenting results for capillary arrays in the same order of parameters as for single capillary tubes, but considering plane arrays first, followed by cylindrical and then spherical focusing arrays.

11.3 BEAMS FORMED BY PLANE CAPILLARY ARRAYS

11.3.1 MEASUREMENTS OF THROUGHPUT

Huggill (1952) was not interested in the atomic beam formed by an array, but his throughout measurements are of interest. We have already discussed the construction of his plastic matrix arrays in Section 10.3. Huggill measured the flow of hydrogen, helium, nitrogen, argon and carbon dioxide through both poly (methyl methacrylate) and gold coated arrays with $\Gamma = 24$ and $\Gamma = 44$, respectively, and in addition measured the flow of ethylene through the former. In each case, the pressure was varied from very low values to those such that $\lambda \approx d$. Graphical results are presented which are essentially plots of quantities proportional to the conductance as ordinate and p as abscissa for carbon dioxide, nitrogen, argon and helium. This ratio increases by an average of $(22 \pm 4)\%$ from very low pressures up to those such that $\lambda = d$. As discussed at the beginning of Section 11.2.2, such a variation is not expected, so we average the 51 tabulated results to give a mean of 0.97 ± 0.07 for Huggill's quotient of the measured flow and that at very low pressure.

Giordmaine and Wang (1960) measured the flow of carbon dioxide through three capillary arrays. That called Source 'A' contained channels with a hexagonal cross section with $\Gamma = 20$. Source 'B' contained circular channels with $\Gamma = 66$ and Source 'C' was constructed of crinkly foil with $\Gamma = 177$. The arrays have been fully described in Sections 10.5 for 'A', 10.6.4 for 'B', and 10.2 for 'C'. The measured throughput through Source 'A' was 1.1 ± 0.3 times the theoretical. With Source 'B' the factor was 6.2 ± 1.0 which we disregard since it is assumed to be in error and with Source 'C' the factor was 1.7 ± 0.2. The authors quote errors of $\pm 20\%$ for the throughput measurements and ± 4 Pa for the measurements of pressure which was in the range 3 Pa to 250 Pa. The scatter in the ratio of the measured to the throughput calculated from Equation 9.1 for arrays 'A' and 'C' is consistent with these rather large errors. At some of the highest pressures of measurement $\lambda < d$, but we have not been able to establish a departure from theory due to this cause, because of the large scatter in the results. In order to make a useful comparison of the axial intensity and beam halfwidth in the presence of the large uncertainties in the pressure, Giordmaine and Wang presented their results in terms of the throughput. When we come to discuss these quantities, however, we shall remain consistent and keep the dependent and independent variables separate.

Hanes (1960) measured the conductance of helium flowing through the array he made with $\Gamma = 16$ which is described in Section 10.3. He obtained almost constant values at low pressures which is equivalent to the expected linear increase of throughput with pressure. His results are in satisfactory agreement with Equation 9.1 when his allowance for the finite size of the array is taken into account. However, the conductance already begins to decrease at higher pressures as soon as the mean free path is less than the capillary length, which has not been observed by any others.

Johnson et al. (1966) measured the throughput of helium through fused glass capillary arrays having values of Γ of 110, 51, 20 and 13, and capillary diameters of respectively 3 μm, 10 μm, 25 μm and 40 μm. For each array they made measurements of the rate of fall of pressure from a known volume at pressures of 3.3 Pa, 27 Pa, 130 Pa,

670 Pa and 2 kPa. When compared with Equation 9.1, the theoretical throughput was found to be about 44% more than that measured for tubes of 3 μm and 25 μm in diameter and 63% more for those of 10 μm and 40 μm in diameter. There is only a small scatter between the results at five different pressures for each array.

In Section 11.2.2 we discussed Jones et al.'s (1969) measurements of the throughput through a single tube, which they expressed in terms of a conductance. They made the same measurements both using air and helium on the arrays which we have already described in Sections 10.2, 10.4.7 and 10.6.3. For the crinkly foil array (Section 10.2), their measured throughput was about 21 times that calculated from Equation 9.1, but their remaining measurements were about 0.6 ± 0.3 times the theoretical. The authors attribute the large factor of the crinkly foil measurements to a poorly constructed mounting, which allowed gas leakage. The fused silica membrane (Section 10.6.3) had poorly defined channels, so the theoretical estimate was subject to a large error. Both glass mosaic sources were partially blocked, and as discussed in Section 10.4.7 this appears to be an exceptional condition. Hence none of these measurements is useful for comparison with theory, but they do point to the fact that measurements of throughput, once the theory has been confirmed, or a correction established, are a useful pointer to the correct behaviour of an array.

Our objective of detailed coverage requires the inclusion of a series of measurements of the throughput of gases through capillary arrays that were originally published between 1972 and 1974, and carried out at the Ural Polytechnic Institute, Sverdlovsk, USSR. Whereas our interest is in the throughput as a function of gas pressure, which is basically what was measured, the extensive reduction of data from the measurements makes comparison with Equation 9.1 uncertain. However, some useful information can be deduced, such as the low pressure linearity of the throughput and an indication of the shortest mean free path in terms of the tube diameter at which Equation 9.1 is valid. The authors plot a quantity which is proportional to the conductance against pressure. Fortunately an advantage of the way the data is presented means that the uncertainty in the diameter of the capillary tubes in all but the first paper does not prevent knowing the gas pressure at which the mean free path is equal to their diameter.

Borisov et al. (1972) made a series of very precise measurements of the rate of flow of helium, neon and argon through an array of 100 'smooth' glass capillary tubes having a mean diameter of (99.72 ± 0.6) μm and a length of 29.76 mm $\pm 0.1\%$, so $\Gamma = 298$. They obtain values effectively for the ratio of their measured throughput to that given by Equation 9.1 of 1.126 ± 0.007 for helium, 1.139 ± 0.006 for neon, and 1.037 ± 0.010 for argon at temperatures of 293.5 K, 292.8 K and 293 K, respectively. These are obtained by reference to Berman's (1965) theoretical transmission probabilities which within their errors can be considered to be those obtained by Zugenmaier (1966). From Equations 8.23 and 8.24, we namely find that, at their value of Γ, Zugenmaier's values are larger than Berman's by only about 3 parts in 10^5. The pressure range of measurements extended from that where $\lambda = 103\ d$ up to those where $\lambda \ll d$. At the lower pressures, the results are consistent with the linear dependence of throughput on pressure up to a pressure such that $\lambda \approx d$. Borisov et al. interpret their results in terms of the incomplete accommodation of the tangential components of momentum flux of the atoms with the walls of the tubes.

Porodnov et al. (1974a) made further measurements in the same apparatus of the flow of argon, helium, neon and nitrogen through a different array of about 640 capillaries, each stated to be about 51 mm long, but apparently with a diameter of 270 ± 1 μm, hence $\Gamma = 190$. Their results are also consistent with the throughput being proportional to pressure at low pressures but increasing more rapidly when the pressure is such that the mean free path is less than the diameter of the capillaries.

Porodnov et al. (1974b) continued these measurements using apparently the same array of about 640 capillaries, each now stated to be 13.5 μm in diameter, hence $\Gamma = 3800$. Measurements were made at laboratory temperature over a wide range of mean free paths for the inert gases and some common molecules, but only those for helium, hydrogen and xenon are presented graphically, from which it can be seen that these exhibit the same dependence on pressure as found by Borisov et al. (1972) and Porodnov et al. (1974a). The measurements were also used to derive the free-molecular reduced flow rate, which they define and discuss.

Akin'shin et al. (1975) extended this series of measurements using apparently the same array as described by Porodnov et al. (1974b), but stated to contain 135 ± 1-μm-diameter tubes. They made measurements at temperatures of 194.7 K and 77.2 K as well as 293 K with more atoms and molecules. The graphical results presented for the relative flow rate of argon, deuterium and helium at 194.7 K exhibit the now familiar pressure dependence.

Continuing now with other measurements, we note that Beijerinck and Verster (1975) measured the throughput of oxygen at pressures from $\lambda > d$ through the collimated holes structures described in Section 10.5. They presented their results in terms of the ratio to the theoretical throughput equivalent to that given by Equation 9.2. For the array containing 16-μm-diameter channels, with $\Gamma = 28$, the measured throughput is only $(66 \pm 3)\%$ of the theoretical. With the arrays containing 50 μm ($\Gamma = 54$) and 140-μm-diameter channels ($\Gamma = 8.35$), the throughput was respectively $(11 \pm 2)\%$ and $(2 \pm 6)\%$ greater than theoretical. The authors attribute the discrepancy for the array with the smallest channels both to their irregular cross section and to the resulting error in their measured diameter.

Detailed studies of the flow of argon through 2-mm-thick glass plane capillary arrays were made by Guevremont et al. (2000). Individual tubes were either 10 μm or 50 μm in diameter. The arrays had a useable diameter of 11.4 mm but this could be reduced by the use of 100-μm-thick apertures that were either 1 mm or 170 μm in diameter and placed in contact with the array on the gas inlet side. Measurements were made with a mass spectrometer that sampled the beam through a 170-μm aperture. Linear motion drives enabled the arrays to be moved in all three dimensions relative to this aperture. Experimental problems led to many of the curves being less smooth than is usual but the linear dependence of the throughput on input pressure was indicated over a large range of argon pressures such that the mean free path of the atoms was greater than the diameter of individual tubes.

We have already commented in Section 11.2.2 upon Rugamas et al.'s (2000) throughput measurements for several gases through a single tube. With the exception of carbon dioxide, they made the measurements on the same gases through an array of 91 tubes that were 50.8 μm in diameter and 5.08 mm long, so that $\Gamma = 100$. We note the same pressure variation of the ratio of their measurements to theory as before that makes the analysis of their results uncertain, but note that their average is exceptionally about 2.5 times that expected.

Marino (2009) measured the throughput of an unstated gas through an array of 20,000 'inox' (stainless steel) tubes having $d = (110 \pm 1)$ μm and $l = (150 \pm 0.05)$ mm so $\Gamma = 1364$. The mean free path ranged from about 0.7% of the tube lengths to much less than their diameter. He compares his results with the single tube measurements of Tison (1993) that are discussed in Section 11.2.2 and finds close agreement.

In conclusion, the throughput measurements through arrays of capillaries are generally in satisfactory agreement with theoretical predictions. Where discrepancies occur, they are explicable in terms of experimental uncertainties. Throughput measurements through arrays present no particular problems compared with single capillary tubes, but discrepancies between theory and experiment might indicate an uneven pressure distribution across the diameter of the input side of the array.

11.3.2 Measurements of Axial Intensity

Zacharias and Haun (1954) quote their axial intensities of caesium as 10^5 atoms s^{-1} at a temperature of 341 K and 10^7 atoms s^{-1} at 417 K into a solid angle cone of 0.01 r for each of their 300 oven channels, the construction of which is described in Section 10.2. As mentioned in Section 4.2, the oven was filled with caesium carbonate and potassium. If we assume that the partial pressure of caesium in the heated mixture corresponds to the vapour pressure of caesium at the oven temperature, and that each channel of triangular cross section is equivalent to one of circular cross section 121.8 μm in diameter and 12.7 mm long, we obtain theoretical intensities of 3.6×10^{11} and 2.6×10^{13} atoms s^{-1} from Equation 9.6, which are considerably more than those measured. This is probably due to the above assumption of the vapour pressure of caesium being governed by the oven temperature rather than its rate of production from the chemical reaction.

We have already discussed Giordmaine and Wang's (1960) results for the throughput in the previous section. As explained previously, we disregard all their results for Source 'B'. Their measured axial intensity for the other sources at the same pressures is about 2.6 times that predicted theoretically at low pressures and the ratio of measured to theoretical intensity increases with pressure by a factor of about two even though $\lambda > d$ for all their measurements.

Kotlikov and Khryashchev (1986) formed a beam of sodium with a multichannel array with an overall diameter of 1.2 mm having an unstated number of 20-μm-diameter tubes which were 6 mm long. They measured the absolute intensity in the beam and used the data given by Lucas (1973a) to compare with Zugenmaier's (1966) calculations. They obtained good agreement at sodium pressures from about 1 mPa up to about 31 mPa beyond which at pressures up to 43 mPa the intensity appeared to reach a maximum which the authors attributed to collisions of the beam atoms with the background sodium vapour. Note that Equation 9.6 agrees quite closely with their measurements up to pressures of 31 mPa. The measurements agree with those they have obtained from Lucas (1973a) when the temperature dependence in that paper is corrected by using Equation 9.6.

Khryashchev et al. (1988) used a fluorescence method to measure the axial intensity of beams of sodium atoms formed by two glass capillary arrays. One array consisted of 20-μm-diameter capillaries 6 mm long, so that $\Gamma = 300$. The second had $\Gamma = 800$, with 10-μm-diameter and 8-mm-long capillaries. The method was not absolute, so the authors normalised effectively to Equation 9.6. Measurements were made at sodium temperatures from about 425 K to 545 K which represents a pressure from 940 μPa to 565 mPa, namely, a range of a factor 600. The pressures were, however, so low that even the highest pressures were such that $\lambda > l$. Good agreement is obtained with the shape of the intensity against temperature curves for both arrays at the lower pressures. The theoretical values are not reached experimentally at higher pressures. They attribute the discrepancies to degrading vacuum at the higher pressures, as also noted by Kotlikov and Khryashchev (1986), mentioned above.

Hisadome and Kihara (1992) made three twelve-channel collimators of unknown material and shape in which each channel had a cross section of 6×10^4 μm². They had thicknesses of 17, 34 and 50 mm leading to values of Γ of respectively 61.5, 123 and 181. The overall size of the collimator is not stated. The collimators were tested using caesium at oven temperatures in 10-K intervals from 343 K to 423 K. The atomic beam was detected using a tantalum hot wire. Measurements were made from which $I(0)$ and the beam halfwidth could be deduced as a function of input pressure, but angular measurements were only made within about 1.5° of the beam axis. Hisadome and Kihara neither state how they treat the finite size of their collimator nor how they converted their oven temperature to vapour pressure in order to compare their experimental data with theory. We derived the atomic diameter of caesium from Yaws (1995b) viscosity data at his lowest temperature for data for this atom of 950 K rather than extrapolating a non-linear curve. This leads to an underestimate of the diameter. We find that their average axial intensities were about 84% of the values given by Equation 9.6. The discrepancies are probably due to problems in determining the oven temperature and possibly collisions of the beam atoms with background vapour on the rather long route to the detector of up to 300 mm.

We note that this conclusion is not consistent with the theoretical curves presented by the authors, whose measurements are shown to agree with their theoretical predictions at temperatures between about 350 K and 370 K but exceed them by typically up to a factor of two at the highest temperatures. The discrepancy appears to be caused by the unusually large estimate of the caesium atomic diameter of 2.4 nm that they used, which is abut four times larger than our value.

Ikegami (1994) studied the beam of caesium formed by both an array of seven stainless steel tubes and a glass capillary array. The metal tubes were 700 μm in internal diameter and 20 mm long, hence with $\Gamma = 28.6$. Unfortunately no details of the open area are given. The individual capillaries in the glass array were 50 μm in diameter in a plate that was 2 mm thick, so $\Gamma = 40$. The effective diameter of the plate was 2 mm and its open area was 50%. Ikegami measured the relative axial intensity of the beam and plotted this against the reciprocal of the temperature of his oven. When the intensity is plotted against the caesium vapour pressure, the intensity of the beam formed by the stainless steel tubes follows a linear law with pressure at the lower pressures, and changes to the anticipated square root dependence at the higher pressures. The relative intensity of the beam formed by the glass array remained linear with pressure, which is consistent with the measurements being made at a mean free path always longer than the capillary length.

Libuda et al. (2000) used an ionization gauge to measure the axial intensity of an argon beam that was produced by a 12-mm-diameter glass capillary array. It was 1 mm thick and contained 50-μm-diameter capillaries. Over a range of two orders of magnitude of input pressure to the array up to pressures where the mean free path was less than the length of individual tubes, the expected linear relationship between axial intensity and input pressure was found. At higher pressures there is some indication that the axial intensity is following the expected square root dependence on pressure before a decrease in axial intensity with increasing pressure was observed. The authors suggest this reduction is caused both by collisions of atoms in the beam with the background gas over the 240-mm path length and by collisions of atoms within the beam. Although the authors present absolute measurements, they are unfortunately not presented in a manner which makes comparison with Equation 9.6 possible.

We find, in conclusion, that we have been able to make very few comparisons of $I(0)$ for a plane array with theory. Other authors do present measurements, but they either contain insufficient geometrical or pressure details for a comparison with Equation 9.6 to be made, or more frequently they present the atom density in front of an array without attempting to reduce their results so that comparison with the theory for a single capillary tube can be made.

11.3.3 MEASUREMENTS OF HALFWIDTHS

We now make comparisons of the beam halfwidths produced by plane arrays with theory. The comments on the corrections to measurements made with linear sampling compared with angular at the beginning of Section 11.2.4 also apply. If the angular acceptance of the measurement probe is small, as it usually is, another advantage of the angular measurement is that the finite size of the array requires less correction. The uncertainty of all angular distribution measurements of the beam formed by arrays of capillaries is whether these have been reduced to single tubes.

We referred to Zacharias and Haun's (1954) intensity measurements in the previous section. They also measured the angular distribution of their caesium beam at a pressure such that λ was greater than the channel length. The halfwidth was almost independent of the caesium vapour pressure and about 0.6°, which is about 43% of the theoretical value. The shape of their angular distribution curves, however, changed much more rapidly from a narrow peak to high wings at larger angles than is expected theoretically.

Minten and Osberghaus (1958) measured the flow of nitrogen and hydrogen through their crinkly foil array described in Section 10.2. Comparison of their data with theory is slightly uncertain because it is not clear whether the measurements relate to nitrogen or hydrogen, and because of the uncertain geometry of the cross section of the channels. As they noted, the halfwidth at the lowest pressure is much broader than expected from low-pressure theory when corrected for the finite size of their array and detector, since the pressure is quite close to that where the limiting halfwidth should be expected. We deduce a halfwidth about a factor 2.5 greater than theory at the two lowest pressures, with the measurement at the highest pressure in slightly better agreement.

Giordmaine and Wang (1960) are among the few authors who have measured all parameters of interest and we have already commented upon their throughput and axial intensity measurements. Source 'A', discussed in Section 10.5, gives a halfwidth of 1.7 ± 0.4 times the theoretical. For Source 'C' (Section 10.2) the corresponding factor is 3.0 ± 0.7. We note that the authors' own discussion is incorrect since they assume that the halfwidth is proportional to $N^{1/2}$ (and hence to $p^{1/2}$) without any constant term, but of course at vanishingly low pressure Equation 9.9 applies so the halfwidth still remains finite, due to the finite acceptance angle of each capillary tube. Hence their forced fits through the origin of their plots of halfwidths against $N^{1/2}$ are an incorrect analysis of their data.

We compared Hanes (1960) conductance measurements for helium flowing through his collimator described in Section 10.3 with theory in Section 11.3.1. When his halfwidth data are compared with theory, we find that his measurements are 1.8 ± 0.2 times greater. However, Hanes states that the tubes in the collimator are not exactly parallel, which leads to an increase in his measured halfwidths by an unknown amount.

Naumov (1963) measured the flow of ammonia through a few single channels (but unfortunately no data are presented for $\Gamma \geq 10$) and arrays with both 7 and 26 channels. One with $\Gamma = 51.7$ produced a beam with a halfwidth that appears to be about 4° at low pressure, whereas Equation 9.9 predicts under 2°, but it appears that correction has been made neither for the 6.3-mm diameter of the array nor for the resolution of the detector. The array subtends an angle of nearly 2° at the ionization gauge detector which subtends an angle of about 0.6° at the array. Naumov quotes his angular distribution curves in terms of mean free path, but, as indicated in Section 9.2, Equation 2.1 can be substituted in Equation 9.12 to yield theoretical halfwidths without knowing the ammonia temperature, pressure or molecular diameter. The measured halfwidth at low pressure is over twice that expected before making the uncertain angular corrections and at pressures ten or more times higher, experiment is greater than theory, but by a lesser extent. Naumov also plots his halfwidths for arrays of various dimensions that show the expected near pressure independence at low pressures which should correspond to the limiting halfwidth given by Equation 9.9, when corrected for the finite size of the array. The array diameters for which Equation 9.12 applies are all within 0.5 mm of 6.5 mm. Hence the variation, from about twice to three times this theoretical low pressure limit, is not attributable to the finite array size.

Ivanov and Troitskii (1963) compare their calculations for a rectangular array with Naumov's (1963) experiments, referring to the paper discussed above, but apparently using other results. The halfwidths of beams obtained from sources with $\Gamma = 15$ and $\Gamma = 50$ at low pressure are about 2.4 times the low pressure limit of Equation 9.9.

We have already referred to Johnson et al.'s (1966) throughput measurements for helium through a plane capillary array in Section 11.3.1. Unfortunately their angular distribution measurements are not absolute, so we have been unable to compare their axial intensity measurements with theory. The authors' presentation of combined halfwidth data for all arrays over a pressure range of about two orders of magnitude show the halfwidth to be proportional to the square root of the pressure, as is to be expected from the high pressure Equation 9.10. Since the dimensions of the tubes in the array are known and the points are labelled according to the pressure of measurement, λ can be calculated for each point and so the corresponding array can apparently be unambiguously identified, enabling comparison with Equation 9.12. However, whereas the text indicates that 25-μm-diameter tubes are included, we did not find any by this means. The two results at a pressure of 27 Pa give halfwidths that are $(85 \pm 12)\%$ of the theoretical prediction. The remaining eight measurements of the halfwidth are 1.76 ± 0.41 times the theoretical. It therefore seems likely that the angular measurements at the lowest pressure of 27 Pa are in error, although the throughput discussed earlier shows no such anomaly. The throughput was anomalous for the 3-μm-diameter tube, but the halfwidths do not exhibit any deviation.

We have already discussed Beijerinck and Verster's (1975) measurement of the throughput in Section 11.3.1. Unfortunately their angular distribution measurements were not made at angles less than 1°, so the axial intensity cannot be compared with theory. The authors do, however, present peaking factors, which we compare with theory in Section 11.5. Beijerinck and Verster's measured halfwidths are 1.9 ± 0.3 times that expected from Equation 9.13. Despite the measured throughput of one of their arrays being significantly different from the others, when compared with theory, there is no significant difference in the measured halfwidths.

Shen (1978) measured the angular distribution of silver atoms which were collimated by his array that is described in Section 5.9, having $\Gamma = 10$. They occupied a 17.5-mm-diameter circle. The silver was deposited on a plate which was 102 mm distant from the collimator and the thickness of the deposited film was measured optically. The measured halfwidth of the beam from the whole array was 10° at a temperature of about 1240 K, which corresponds to a silver vapour pressure of 262 mPa. The value calculated from Equation 9.12 is uncertain because Yaws (1997) only gives viscosity values at temperatures between 2485 K and 2885 K, but the near linear dependence of viscosity on temperature in this range means that linear extrapolation to 1240 K in order to obtain the diameter of silver atoms at that temperature is acceptable. This leads to a theoretical value that closely agrees with the measurement.

Pine and Nill (1979) show the angular distribution of helium through a commercial fused glass capillary array which was masked to an active area of 5 mm by 20 mm. It contained 5-μm-diameter capillaries that were 2 mm long, so $\Gamma = 400$. Their halfwidth is found to be a factor 29 times the theoretical. However, their measurements at large angles are much greater than the usual cusp shape. Hence either their measurement or their reduction of the data to allow for the finite source area is presumably in error, and the authors state that no correction has been made for the angular resolution of the detector. From the information they give, we expect both to be small compared with their measured halfwidth of 18°.

Barashkin et al. (1980) present results for the angular distribution of low pressure hydrogen flowing through a 5.6-mm-diameter capillary array containing 50-μm-diameter capillaries with $\Gamma = 141.5$, which indicate a halfwidth of over four times the theoretical low pressure limit of Equation 9.9. The curve appears to have been corrected for the finite size of the array.

Brinkmann and Trajmar (1981) present some measurements of angular distributions of helium and nitrogen produced by an array of 300 capillaries each 50.8 μm in diameter and 5.08 mm long, hence $\Gamma = 100$. Since similar results are obtained for each gas, which is not expected theoretically, it appears that they have not reduced their data to that of a single tube, so they obtain much wider halfwidths than predicted.

The halfwidth of an ammonia beam was measured indirectly and hence relatively by Baldacchini et al. (1982) when it was collimated by a capillary array which was 10 mm long, 3 mm wide and 600 μm thick. It contained 'almost circular' tubes that were about 12 μm in diameter, so $\Gamma = 50$. The open area of the array was about 40%. A square root dependence on pressure was measured up to pressures where the mean free path was about 60% of the tube diameters. At higher pressures no further increase in halfwidth was observed, which is probably due to the presence of collisions of the beam with the background gas. Their measurements on helium with 10% ammonia showed the same dependence up to a mean free path equal to the tube diameter, based on the helium atomic diameter alone.

We have already mentioned Kotlikov and Khryashchev's (1986) measurements of the axial intensity of a beam of sodium atoms in Section 11.3.2. At a pressure of 11.4 mPa they measured a beam halfwidth of $0.5° \pm 0.05°$ when $0.32°$ was expected theoretically, so it is just over 50% greater than predicted theoretically.

Hisadome and Kihara's (1992) measurements of the axial intensity of the caesium beams formed by their three collimators were discussed in Section 11.3.2. From their curves we are able to calculate the beam halfwidths for each collimator at oven temperatures of 343 K and 383 K. Using the same atomic diameter as before, we find that their measured halfwidths average 1.8 ± 0.7 more than the values given by Equation 9.12.

We have already referred to the measurements of Buckman et al. (1993) of beam profiles obtained from a single glass tube in Section 11.2.4. They also measured relative angular distributions of the beams of helium, neon, argon, krypton and nitrogen with limited data in hydrogen that were produced by a glass capillary array which consisted of 40-μm-diameter tubes 1 mm long. Hence $\Gamma = 25$ and the active diameter of the array was 1 mm with an open area of 50%. It was estimated to contain about 290 capillaries. Measurements of the beam profiles were made at three distances from the array which were comparable with its overall diameter. The limited angular range of measurements mentioned in Section 11.2.4 also apply to these measurements, so we limit our discussion to stating that the halfwidths show the expected change from being only slightly pressure dependent when $\lambda \leq l$ to being increasingly pressure dependent up to and beyond $\lambda = d$. Buckman et al. also made measurements of the profile of helium, argon and nitrogen beams formed by a stainless steel array with an active diameter of 890 μm which contained 50-μm-diameter capillaries 3 mm long, so $\Gamma = 60$. The measurements are subject to the same criticism already mentioned above and also lead to similar conclusions. The authors note that their measured halfwidths are greater than those expected from calculations but they do not present a detailed comparison. They follow Adamson et al. (1988a) in blaming the discrepancies on the finite diameter of the sources.

Ikegami's (1994) measurements of the axial intensities of a caesium beam have already been discussed in Section 11.3.2. He measured angular distributions over a range of 20° either side of the normal. Unfortunately the angular measurement was limited by an aperture which reduced caesium contamination of the vacuum system. A graphical presentation of the halfwidth of the beam was given at caesium temperatures from 330 K, corresponding to a vapour pressure of 3.3 mPa, and 420 K, when the vapour pressure was 1 Pa. For the metal tubes, the measured halfwidths were about 50% greater than expected from Equation 9.12 at the lowest caesium pressures but this percentage decreased towards the higher pressures. For the glass array a similar pressure dependence on the halfwidth is seen, but at the lower pressures where the halfwidth is almost independent of pressure, the measurements are unusually only about 80% of those expected theoretically.

We have already mentioned Bellman and Raj's (1997) method of using precursors to obtain a beam of lithium tantalate [$LiTaO_3$] in Section 5.5 and their fotoform array in Section 10.4.8. We therefore mention that they also made measurements of the angular distribution of their beam produced by this array using Rutherford backscattering spectroscopy to measure the thickness of deposited material. Experimental difficulties included the uncertainty in the vapour pressure of lithium tantalate and the possibility of collisions in the apparatus beyond the collimating array. These problems made measurements, including those for the throughput which have therefore not been included in Section 11.3.1, uncertain. The measured angular distributions are on average more than six times broader than predicted by Equation 9.12.

Guevremont et al.'s (2000) studies of the throughput of argon through 2-mm-thick glass plane capillary arrays were discussed in Section 11.3.1. Additional measurements were made of the beam profile and the intensity at different close distances from the array surface. Since the beam formed by an individual tube is divergent, it follows that the quantity that is proportional to the beam intensity decreases with distance from the array. The beam profile became more rounded with increasing distance from the array surface, as is to be expected since, depending on the halfwidth of the beam, at sufficient distance from the array more tubes contribute to the intensity at its centre than at the periphery.

We have already commented in Sections 11.2.2 and 11.2.4 upon Rugamas et al.'s (2000) measurements of the throughput and angular distribution of the beam formed by a single tube for eight different gases over a wide range of pressures. They also made measurements of the beam formed by a small diameter capillary array for seven of these gases. The 900-μm-diameter array was 5.08 mm thick and consisted of 91 tubes that were 50.8 μm in diameter, which leads to a Γ of 100 and a calculated open area of 29%. As in the case of the single tube, we found no systematic deviation of the ratio of measured to calculated beam halfwidth with gas species. There is a slight tendency for the ratio to increase with gas pressure, but this is not sufficient to be analyzable so we have collected 118 results together for all pressures such that $\lambda > d$ to yield a factor by which the measured halfwidth is greater than the theoretical of 4.9 ± 0.6. It is therefore particularly interesting to compare these extensive measurements of halfwidths from an array with those the same authors made using a single tube that are discussed in Section 11.2.4. The factor of 2.3 by which the mean halfwidths of the beam from a small array are greater than those from a tube, and that this factor is independent of gas pressure does suggest either scattering within the beams formed by individual channels or scattering from residual gas at the higher ambient pressures that the array produces. This uncertainty might be resolved by repeating some measurements with a baffled pump to reduce the pumping speed and hence increase the ambient pressure.

We note in conclusion that the halfwidths of beams produced by arrays and formed from vapours tend to be much closer to what is expected theoretically. Hence, we draw the tentative conclusion that the reason why gas beams have a larger halfwidth than expected is that the theory is based on an atom being emitted from a surface at the point where it impinged. However, it seems probable that it migrates along the surface in the direction of the tube exit since it moves under the influence of the incoming atoms. This is known as surface diffusion or creep. By contrast, atoms from vapours, which are usually metallic, do not show this effect and so are emitted from the point where they impinge. The absence of creep is apparent from the extensive use of atomic beams to produce evaporated thin films of the high positional precision needed for the production of integrated circuits. Only a few measurements of the angular distribution of vapour beams formed by single tubes have been made compared with those made by arrays. Since these eliminate the uncertainties in the finite size of arrays, further investigations are necessary. Ideally a gas and a vapour beam should be studied with the same capillary tube in the same apparatus in order to test this hypothesis.

11.4   BEAM FORMED BY FOCUSSING CAPILLARY ARRAYS

Since this is a short section, we discuss the three quantities of interest, namely, throughput, axial intensity, and halfwidth for each author in turn. We treat both cylindrical and spherical arrays in this separate section since the axial intensity of the beam formed by a spherical array and the beam halfwidth of both types of array can be compared with the relevant equations given in Chapter 9 without the problem of comparing single tube theory with measurements on an array of capillaries.

Aubert et al. (1971a) were the first to demonstrate the focusing action of a spherical array by measuring the beam profile at different distances from it. Its construction is described in Section 10.7. Maximum beam intensity was obtained 15 mm from the front surface of the array, but unlike the measurements of Lucas (1973b) the maximum intensity did not coincide with the position of minimum halfwidth. Unfortunately, insufficient information is given to compare their measurements with theory.

Hence, the first reported measurements on a focusing array that can be compared with the equations given in Chapter 9 are those made by Lucas (1973b) on the helium beam formed by a fused glass capillary array described in Section 10.7 that had an effective diameter of 23 mm and which was curved to a radius of 50 mm. Each capillary was 5 μm in diameter and 2.5 mm long, so $\Gamma = 500$.

The throughput measurements showed agreement with Equation 9.1 over a pressure range of three orders of magnitude, but with a possible tendency for theory to underestimate the throughput at the highest pressures. The axial intensity measurements were made from pressures where a linear dependence of axial intensity on pressure is expected according to Equation 9.4 up to higher pressures where a square root dependence is expected as given by Equation 9.5. There is again agreement within the experimental uncertainties, with evidence that the measured axial intensity is greater than predicted at the higher pressures.

These discrepancies are, however, slight by comparison with the behaviour of the halfwidth when compared with the equations given in Chapter 9. The experimental halfwidth H_E is found to be given by the empirical relationship $H_E = 0.54 + 6.1\ H$, where H is the theoretical halfwidth. This expression was found to be valid over the pressure range where the halfwidth is almost independent of pressure as given by Equation 9.9 into the range where the halfwidth is proportional to the square root of the pressure as given by Equation 9.10. The experimental results follow the shape of the theoretical curve if the constant 0.54 is attributed to the sum of any array aberration and the resolution of the probes used to sample the beam. The former term is used to describe any deviation from the axes of the capillaries intersecting at a point.

Ma et al. (1996) have made some measurements on a focussing glass capillary array that was 13 mm in diameter and 500 μm thick and curved to a radius of 95 mm. Individual capillaries had an internal diameter of 10 μm so that $\Gamma = 50$. The open area of the array was about 50%. Measurements of the relative axial intensities of nitrogen were made up to an input pressure of 13 Pa which only covers the range where the axial intensity is expected to be directly proportional to the pressure according to Equation 9.3. The shape of their curve shows this approximate linear dependence at low pressures but then increases much more slowly than theory predicts. Only one measurement is shown of the angular distribution of the molecular nitrogen beam at an unstated input pressure, from which a beam halfwidth of just under 17° can be estimated. If the pressure is consistent with that used for their axial intensity measurements, at which the halfwidth is predicted to be almost independent of pressure, the measurement is over eight times that predicted by Equation 9.12. Unfortunately the authors do not compare their measurements with those of Lucas (1973b) on a focussing array, neither have they replied to our query about the pressure of their angular distribution measurement.

In conclusion, further studies of the angular distribution of the beams formed by focussing arrays are clearly highly desirable in order to assist with the design of atomic beam systems. The most likely cause of the large values of the beam halfwidth for focussing arrays compared with what is predicted is atom–atom collisions within the converging beam.

11.5 INDIRECT COMPARISON BETWEEN THEORY AND EXPERIMENT

Up to now, we have been able to compare measurements of each of the three beam parameters, namely, the throughput, axial intensity and halfwidth independently with theory even though considerable reduction in data has sometimes been necessary. We have reserved for this separate section those comparisons where two measurements have been combined, or some other reason makes direct comparison with the equations presented in Section 9.2 impossible.

The most usual reason for results being included in this section is that they are presented to yield the so-called peaking factor κ, which is given by Equation 2.15 in Section 2.3. It is a measure of the axial intensity produced by an array to that produced by an orifice. We have not used this term previously, since we do not consider it to be a useful quantity. $I(0)$ largely governs the wanted signal that is obtained from the atomic beam whereas N is a measure of both the unwanted background and the rate of usage of source material. The peaking factor means that $I(0)$ and N cannot be established individually, so we are obliged to use it when describing some measurements. If the value of κ derived from measurements does not agree with theory, there is uncertainty as to whether $I(0)$, N or both are in error. Hence we do not consider that peaking factors are such a useful means of comparing experiment with theory as are separate comparisons of the axial intensity and the throughput.

Nevertheless, several measurements have been presented in the form of peaking factors, so we discuss them because they do add something to the general comparison of experiment with theory. As shown by the measurements of Giordmaine and Wang (1960) that were discussed in Sections 11.3.1, 11.3.2 and 11.3.3, misleading conclusions can sometimes be drawn since two measurements each differing from theory can combine to give agreement with it. Their results for the axial intensity and halfwidth for their source 'B' when plotted in terms of the square root of the throughput seem as plausible as those for sources 'A' and 'C', but as discussed in Section 11.3.1, the throughput is seriously in error.

Becker (1961a) processes his results for the peaking factor in terms of a dependence upon the throughput. However, for ammonia flowing through six different arrays the proportionality of κ to $N^{-1/3}$ over three decades neither follows the theoretical independence upon N at low pressures nor the $N^{-1/2}$ dependence to be expected theoretically at higher pressures nor the relationship noted below (Jones et al. 1969). Hence further analysis is pointless.

We have already referred to Jones et al.'s (1969) throughput measurements in Section 11.2.2. They present measurements of helium peaking factors as a function of pressure, both for a single tube and for the four arrays they describe. For their single tube, the measured peaking factors at pressures such that $\lambda > d$ are an average of 16% more than given by combining Equations 2.15, 9.2, and 9.8. Their peaking factors at all pressures, including those where $\lambda < d$ are a reasonable fit to a $p^{-1/4}$ law. From Equations 2.15, 9.1, 9.3 and 9.5 the peaking factor should be independent of p at low pressures and proportional to $p^{-1/2}$ at high pressures. Jones et al.'s peaking factor measurements for their crinkly foil source, described in Section 10.2 and for the fused silica membrane source discussed in Section 10.6.3, are not considered further because of the problems they had with measuring the throughputs. However, they also measured the peaking factor for two fused glass capillary arrays described in Section 10.4.7, but only one had $\Gamma > 10$, namely, $\Gamma = 22.73$. For this array, the measured peaking factors averaged to 94% of those calculated in the same way as for their single tube.

Beijerinck and Verster (1975) derive the peaking factor for the collimated holes structures described in Section 10.5. For the array with the smallest diameter, with 16-µm-diameter channels, the measured peaking factor is found to be 71% of the theoretical. For the array with 50-µm-diameter channels, the corresponding factor is 54%, whereas the array with 140-µm-diameter channels gives the best agreement with theory at 83%.

We now discuss one other experiment where comparison between measurements and the Section 9 equations is less certain because of the way the data is presented.

Woznyj et al. (1981) present data for the angular distribution of benzene molecules at six different throughputs. The authors have employed considerable reduction of the data to produce corresponding angular distributions of the beam. Unfortunately comparison with the equations of Chapter 9 requires the derivation of theoretical halfwidths from their measured throughputs and these have different dependencies on pressure and temperature. We, however, conclude tentatively that their beam is found to be nearly four times that predicted at the lowest throughput but reduces to a factor 2 at the highest. We note from the authors' own comparison of angular distributions with theoretical predictions that the measured distribution tended to be significantly greater than expected at larger angles.

11.6   CONCLUSIONS FROM ALL PREVIOUS COMPARISONS

Throughput measurements, whether through a single capillary tube or an array, are in general in good agreement with theory. There is a sufficient range of measurements available to give confidence in Equation 9.1, and no systematic deviations due to either the dimensions of the array nor the species of the atom beam are evident. Measurements are straightforward, and any deviation from theory is more likely to be of experimental than theoretical origin. Deviations are most common when the array is markedly irregular, such as that obtained by the use of crinkly foil, and this is presumably due to the difficulty in determining the dimensions of the array channels.

Measurements of the axial intensity also show quite reasonable agreement with Equation 9.6. Although the scatter of experimental measurements about the theory is, as is to be expected, larger than for the throughput, there is no indication that there is any mean deviation from the theory, and those deviations that do occur can be assumed to be of experimental origin. Since throughput measurements are dependable, reliable measurements of peaking factors also confirm the axial intensity measurements.

When measurements of the halfwidth are compared with Equation 9.12, the position is quite different. Measurements that are less than, or agree with the theoretical prediction are so rare, that it is a reasonable assumption that they are in error. Most measurements on beams formed from gases are typically scattered around a value of about twice the theoretical, and much larger values than this are not uncommon, even when the throughput and the axial intensity are in good agreement with theory. We have already commented that the fewer measurements on beams from a vapour source show a tendency of better agreement with theory. That all results do have a large scatter indicates that there are experimental difficulties, and many of these must lie with the problem of reducing the observations on an array to those of a single capillary tube. This is why we recommend that measurements be made on single tubes, in order that one can be quite certain that the gas beam formed by these is in disagreement with theory.

There is clearly further need of experimental evidence to test the suggestion that the unexpectedly wide halfwidths produced by focussing arrays is due to collisions between the array exit and the point of measurement. It is particularly important in all cases to measure the entire angular distribution. From Equation 2.10, the shape must be in disagreement with theory in order that its integral over the solid angle should give the correct throughput. The probability is, however, large that this will be the case.

11.7 COMPARISON WITH ATOMIC BEAMS PRODUCED BY FREE JET SOURCES

The free jet source is characterised by a much higher pressure in the gas source before the impedance than the beam-forming devices we have been considering so far. The impedance is in the form of a nozzle through which the gas flows at super-sonic velocities. Because of the different nature of both the theoretical treatment of beams produced by free jet sources, and also the experimental techniques, we do not discuss these in detail. However, we make a basic comparison of the gas beam produced by a focussing capillary array with that produced by such sources. Measurements are compared with those of Lucas (1973b) in helium since it represents a high quality beam that is readily achievable.

We basically address the question: if a high density gas beam is required in the presence of as little background gas as possible, which technique achieves this? This question is surprisingly difficult to answer, since the many experimental papers where free-jet sources are used tend not to measure all the quantities which are directly comparable with those we discuss above. Review articles have offered little assistance; however, Miller (1988) addresses this problem, but nevertheless sums up the position thus:

'Nothing has been more perplexing in the development of free jet beams than the prediction and measurement of beam intensities.'

Deckers and Fenn (1963) made a detailed study of beams of water vapour, methane and nitrogen. The throughput can be estimated to be about 10^{20} molecules s^{-1} from their quoted ambient pressures and pumping speeds in each of the two chambers, which is comparable to the highest throughputs produced by the array when beam densities of 10^{11} atoms mm^{-3} were obtained, which are comparable to those measured by the authors. The halfwidth of their beam was under $7°$, which is again comparable to that produced by an array.

Zankel and Vosicki (1975) produced a curtain of sodium vapour at a density of 10^9 atoms mm^{-3} of very low angular divergence at a distance of 1 m from the source. The rate of usage of sodium was quoted as 1 g s^{-1}, which is equivalent to a throughput of 3×10^{22} atoms s^{-1}. The helium beam produced by Lucas (1973b) was about two orders of magnitude greater than this with over a factor 1000 less throughput. However, it is a much more convergent beam, so this is an unfair comparison. So using the criterion adopted by Zankel and Vosicki, namely, a beam divergence of $0.04°$, we can use the limiting halfwidth given by Equation 9.9 to see if a beam can be designed to this specification. We find that a capillary array would need a Γ of about 2400, which is outside what is currently commercially available. However, a 12-mm-thick array with 5-μm-diameter capillaries, for example, has been shown to be technically feasible by Knight et al. (1996, 1997), as described in Section 10.6.7. We note, however, that Zankel and Vosicki do not quote any measurements of their beam divergence.

Beijerinck and Verster (1981), whose work on thermal atomic beams has been referred to extensively (Beijerinck and Verster 1975), have also studied supersonic beams of argon, and molecules of oxygen, nitrogen, carbon dioxide and methane. Their measured peaking factors decreased in the above sequence from slightly more than 2 for argon down to 1.2 for the polyatomic molecules. Hence the peaking factor is hardly an improvement over an orifice, though of course the axial intensity (and hence the throughput) is greater than that produced by an orifice that is running under input pressures that enable the beam properties to be predicted from kinetic theory considerations.

The object of Beijerinck et al.'s (1985) study was to compare measurements of the beam produced by a Campargue-type source (Campargue 1984) with their detailed theoretical predictions. Hence they only had modest pumping facilities, so with a pumping speed of only 110 l s^{-1} in the region of measurement they obtained a peak helium beam density that was about 5.4 times that of the background gas when the beam halfwidth was 3.6 mm. These data were obtained at an input pressure of 2.05 MPa to their 50-μm-diameter nozzle, leading to an axial intensity of 9.2×10^{19} s^{-1} sr^{-1}. Direct comparison with a thermal beam is not possible without having specific data for the angular divergence and throughput but Beijerinck et al.'s beam properties would be readily achievable with a focussing capillary array.

We have already mentioned Garvey and Kuppermann's (1986) arc source used to produce atomic hydrogen in Section 6.3.3. They also ran argon in the same system. Since they quoted the rate of usage of argon gas, N can be calculated. The value obtained is over 90 times greater than that obtained by Lucas (1973b) in helium when the axial beam density was 10^{11} atoms mm^{-3}, which is 40 times less than Garvey and Kuppermann. Hence these authors obtain a comparable ratio of $I(0)/N$ to that obtainable with a focussing capillary array. Their source is also suitable for the production of beams of species such as H_3 molecules and the authors suggest that the energy of the beams could be selected over a wide range using a velocity selector.

Singy et al. (1989) describe in detail the design and characteristics of a nozzle source that produces a beam of atomic hydrogen at a temperature of 30 K. At the exit of the beam formation part of their apparatus, they obtained a density of 2×10^9 atoms mm^{-3} at a beam halfwidth of about 5° with a gas throughput of 1.24×10^{19} atoms s^{-1}. It appears that comparable values could be obtained with a focussing capillary array as described by Lucas (1973b), even with the longer focus that would be needed to accommodate Singy et al.'s sextupole magnets and velocity selector.

The source briefly described by Field et al. (1991), which was used to obtain beams of carbon dioxide, nitrogen, nitrous oxide, oxygen, and sulphur hexafluoride, was of the Campargue type. It operated at an input pressure of about 100 kPa to yield a nitrogen beam density in the collision region estimated to be 10^{10} mm^{-3} when the beam diverged at 2.9°. With a total pumping speed of 8350 l s^{-1}, the background gas pressure was 10 Pa. Comparison with the data of Lucas (1973b) show that a focussing array would easily be able to obtain a similar beam halfwidth and density.

A detailed study of the properties of helium, neon, argon and nitrogen beams produced by a Campargue type supersonic source has been made by Gőtte et al. (2000), who discuss the design of their apparatus. A thorough investigation of how the beam properties varied with quantities such as the input pressure up to 1 MPa and various dimensional variables was made. Their best results are tabulated for the above gases except for argon, in which cluster formation made the results uncertain. For helium a beam density of over 8×10^9 mm^{-3} was obtained in the measurement region, which was 160 mm from the source. The densities were lower in neon and nitrogen but with a total pumping speed of 1700 l s^{-1} the ambient density for all gases was better than 1% of that in the beam. The beam shape was only determined for neon, for which a beam halfwidth of just over 1° was measured. Since the measurement region was 50 mm from the plane of the final collimating aperture where a focussing capillary array could be placed, we can make comparison with the measurements of Lucas (1973b), who readily obtained this density at an input pressure of about 7 Pa, a comparable beam halfwidth and probably a much lower gas throughput, which is not given by Gőtte et al.

It is particularly interesting to compare Barr et al.'s (2012) beam measurements with those of a focussing capillary array, since their object was to obtain a small scale Campargue type helium beam and they also measured all the quantities needed for comparison. Input pressures of 12 MPa behind a nozzle that was about 4 μm in diameter produced an axial intensity of 10^{19} atoms sr^{-1} s^{-1} when their beam halfwidth was about 2°. The focussing array described by Lucas (1973b) produced a comparable halfwidth, axial intensity and throughput. Barr et al. made their trial measurements some distance from the beam source, so it is not clear whether their intended application of a scanning helium microscope could be achieved much closer to a focussing capillary array.

We conclude that if the special features of a free jet source are not required, it would appear, from the comparisons that have been possible, that the best designed sources offer no real advantage over a focussing capillary array in terms of either a larger beam intensity or a high beam intensity for a given throughput.

The measurement of all beam parameters in the free jet expansion case has now been shown to be possible in the case of the most recent experiments so that comparison of the beam produced by a capillary array can be more conclusive. It is suggested that a particularly useful comparison could be made if a focussing array were substituted for the free jet source in an apparatus so that a more direct comparison of the beam properties in terms of the measured signal from the beam interaction could be obtained.

12 Other Indications

12.1 INTRODUCTION

We have collected together in this short chapter some additional material that is relevant to atomic beams. The question of the extent to which partial specular reflection occurs is important to the development of the theory of beam formation. Some care is needed with applying the correct mean velocity of atoms either within an oven, in the beam or when impinging on surfaces. Although the differences are not large, they would be relevant in precision work, such as measurements of the velocity distribution. We also need to return to the question of the distribution of atoms within a capillary tube. We need to consider the end effects, namely, that the extent that the density of atoms just within a tube is less than that of the atoms in the input container V_P and that the density in the exit plane of the tube is greater than V_0, the vacuum of the system into which the atomic beam emerges. There are differing expressions for the dissociation of hydrogen on hot surfaces, which are important in estimating the degree of dissociation.

12.2 SPECULAR REFLECTION

We mentioned in Section 1.2 Hurlbut's (1959) measurements of the beams reflected from surfaces. These indicated that the assumption that atoms striking a surface leave it without any memory of their incident direction is not always valid. As discussed in more detail in Section 8.1, this assumption has largely remained the basis of the calculations of beam properties.

Hurlbut (1959) made measurements with air, argon and nitrogen on surfaces of polished aluminium and steel as well as window-quality glass and PTFE. He showed that the majority of atoms are scattered at random but that a measurable departure from specular reflection was found for glass and PTFE. Logan and Stickney (1966) justify theoretically some experimental results which show that partial specular reflection is the more probable.

Libuda et al.'s (2000) measurements are mentioned in Section 11.3.2. They also studied the scattered beam of argon and oxygen from an alumina (Al_2O_3) film on NiAl (110) and on a multilayer of ice on an alumina film for an incident beam angle of 35°. With the exception of the scattered argon beam from ice, which shows a near isotropic distribution at a temperature of 90 K, the other three distributions at 298 K have halfwidths of roughly 60° centred on the angle of specular reflection. Complete trapping desorption is believed to be responsible for the argon results on ice, whereas other results are believed to be a mixture of this component with direct inelastic scattering.

12.3 BEAM VELOCITY

Equation 2.3 in Section 2.2 gives the well-known expression for the mean velocity of atoms in a gas. Whether this expression is also valid for atoms in a typical atomic beam requires further discussion. For all but the most accurate beam designs, the difference is not large, so we include the discussion here rather than in Chapter 2, which deals with the basic equations needed for beam calculations.

A physicist well known for studies of velocities drew Stern's (1920b) attention to the fact that atoms with higher velocities than the mean leave an oven preferentially. As 'Herr Einstein' pointed out to Stern, the effect is to modify the c^2 term in the mean velocity expression to c^3. This results in a mean beam velocity that is higher by $3\pi/8$ or 18%. In a little known paper that does not directly cite Stern, Comsa (1969) argued that, in practical cases, the higher velocity would not be observed in the beam. Comsa and David (1985) expand the discussion and indicate that the expression for impinging molecules should include the c^3 term.

Detailed measurements of the velocity distribution of beams, for example, by Miller and Kusch (1955), whose oven for potassium and thallium is discussed in Sections 4.3 and 5.11 respectively and Walraven and Silvera (1982), who dissociated hydrogen in a microwave discharge (Section 6.3.5) show however a good fit to the modified Maxwell distribution as proposed by Einstein. Unfortunately measurements of velocity distributions are not discussed by Comsa and David. Some more recent papers that include measurements of beam velocities unfortunately refer to the 'Maxwell distribution', without discussing whether the c^2 or c^3 expression is intended.

12.4 ATOM DISTRIBUTION WITHIN A TUBE

We have included measurements that are not directly concerned with the properties of the atomic beam in this section.

We have discussed in Section 8.5.4 the flow of atoms through a tube when collisions take place within it. It was mentioned that the density of atoms in the entrance plane of the tube must be less than that of the measured ambient gas pressure p in V_P because of the flow and for the same reasons the density in the exit plane must be greater than the zero of V_0, as assumed, as a simplifying assumption in the theoretical discussion. End effects are obviously greater for tubes with a small value of Γ, since then the flow is greater than for tubes preferred for producing atomic beams. Since the calculations make assumptions about these densities, it is clearly of interest to determine if these can be established by experiment.

De Leeuw and Gadamer (1967) measured the radial distribution of the density of air just downstream of the exit plane of a tube that was 10 mm in diameter and 10 mm long. They used an electron beam that was about 1 mm in diameter as a probe that was directed across the tube and observed the fluorescence induced by a small section of this with a photomultiplier with its angular resolution limited by apertures. The volume of air sampled was about 1 mm³. Their measurements were made at distances of 750 µm and 1.5 mm from the exit plane of the tube at an air pressure such that the mean free path was 2.3 times the tube diameter. These show a fairly flat distribution across the central section of the tube but decreasing toward the tube walls as is to be expected. They obtained good agreement with their calculations, which are based on the same principle as that used in the derivation of the Clausing equation.

Though not of course immediately useful in determining the density distribution within a tube, de Leeuw and Gadamer's (1967) probing method is of interest in suggesting a fluorescence method that could be developed, for example, by using laser fluorescence within a tube.

Kurepa and Lucas (1981) made measurements of the density along the walls of four tubes having Γ values of 5, 10 and 20. They used tubes with a figure-of-eight cross section, with the larger part either 1 mm or 2 mm in diameter and the smaller a sliding fit to a needle probe, which contained a small slot in the side to sample the gas within the tubes which had lengths of 10 mm and 20 mm. Measurements were made either in helium or in air enriched with helium, since the gas was sampled with a leak detector. It was estimated that the amount of gas from any tube passing into the sampling system was less than 0.2% of that flowing through a tube.

They obtained the expected linear variation of atom density along the tube length when $\lambda \geq d$. However, the reduction in density at the tube entrance Z_P (Section 8.5.4) was substantially less than that predicted by Ivanov and Troitskii (1963) and Jones et al. (1969). Hence it comes closest to the assumptions used by Giordmaine and Wang (1960), Becker (1961b) and Zugenmaier (1966), namely that it is the same as that corresponding to the measured pressure. The latter's exit density expression Z_V comes closest to the measurements, so the beam properties obtained by his theoretical expressions are the best still available. Measurements were also made by Kurepa and Lucas (1981) at higher pressures such that $\lambda < d$ from which they concluded that the upper limit for free molecular flow is when $\lambda = d$ as is normally, but not always, assumed.

Kurepa and Lucas' measurements of the density distribution within the tube are of interest to compare with Flory and Cutler's (1993) calculations discussed in Section 8.3.5. Whereas the measured density distribution at high pressures departs from linearity in a convex direction, the theoretical shape of the Clausing function is in the opposite direction. Hence, the conclusion of Davies and Lucas (1983) referred to in Section 8.5.3, namely, that the two should be complementary, may be invalid in the pressure region used to form atomic beams.

Steckelmacher and Lucas (1983) postulated that the sampling probe used by Kurepa and Lucas (1981) had a directional effect, but they did not calculate what correction was needed to the measurements. Neither are the experiments they cite in support of their assertion of direct relevance to these measurements. In addition their suggested modification cannot be easily implemented. The absence either of any calculations of the effect or of further measurements in 30 years suggests that any directional effect is not generally considered significant.

The direct simulation Monte-Carlo calculations of Gallis and Torczynski (2012) of the pressure variation of argon along a tube with $\Gamma = 50$ show the same pressure changes from linear at low pressures to the curvature similar to that measured by Kurepa and Lucas (1981) at high pressures. Gallis and Torczynski's curves also exhibit an exit density that is greater than zero, and this is more marked when partial specular reflection is assumed. It would clearly be of interest for further calculations to enable direct comparison with experiment.

12.5 DISSOCIATION OF HYDROGEN ON HOT SURFACES

In order that Section 6.2.3 remains consistent with the other sections of Chapter 6 in describing the techniques that have been used for dissociating gases, we discuss in this section what information is available on the temperature of the furnace required to obtain highly dissociated hydrogen atoms. Unfortunately there is considerable uncertainty about this. The uncertain hydrogen pressure within the furnace may also be a contributory factor.

Hendrie's (1954) measurements agree with his calculated curve and show no dissociation at a temperature of 2000 K at which temperature Koschmieder and Raible (1975) measured 30% dissociation. They also measured higher dissociations than Hendrie at a temperature of 2600 K. Boh et al. (1998) believe they obtained 100% dissociation in their furnace at temperatures of 2000 K and a pressure at the beam exit of 2.5 mPa. These values are to be compared with the measurements and calculations of Eibl et al. (1998), who measured complete dissociation at their lowest pressures at temperatures as low as 1800 K. At this temperature Hendrie found no dissociation, and Koschmieder and Raible measured 20%. Unfortunately Eibl et al. do not discuss Hendrie's and Koschmieder and Raible's work. Tschersich (2000) made a number of measurements on the atomic hydrogen beam formed by the furnace described by Tschersich and von Bonin (1998). The dissociation was not measured directly, but reduction of the data led it to be in the range of 83% to 98% at a temperature of 2600 K, with the latter obtained at the lowest pressures.

The differences in the degrees of dissociation appear to be due to the pressure dependence of the equilibrium constant in the reaction $H + H \Leftrightarrow H_2$ and to whether there is equilibrium between the atoms and molecules within the furnace. There must be a balance between the rate of production of atoms from molecules and their leaving to form the beam. It also seems intuitive that since the molecules only dissociate on a hot surface, the pressure in the furnace needs to be low enough for the mean free path of the hydrogen atoms to be greater than the smallest dimension of the furnace. For these reasons, the equilibrium between the atom and molecular concentrations, as is assumed in the theoretical treatment, is not achieved in practice. The estimated pressure in Hendrie's furnace tube was between about 100 Pa and 500 Pa whereas Eibl et al. based their measurements and calculations on much lower estimated pressures, namely, between 67 μPa and 57 mPa. As a guide, the mean free path of hydrogen molecules at a temperature of 2000 K is about 1 mm (namely, a typical furnace smallest dimension) at a pressure of 270 Pa. Since there is considerable uncertainty of the dependence of the degree of dissociation on pressure and particularly temperature, it is important to measure rather than estimate it.

Brennan and Fletcher (1959) present a detailed review of previous work on the understanding of the mechanism for dissociation of hydrogen on a hot tungsten surface. The two mechanisms suggested are that either hydrogen atoms bound to a tungsten surface site to form W – H dissociate according to the equation

$$W - H \rightarrow W + H$$

or that dissociation of hydrogen molecules occurs on such a surface

$$H_2 + W \rightarrow W - H + H,$$

within each case an equilibrium condition

$$H_2 + 2W \rightarrow 2(W - H).$$

The authors investigated the pressure dependence of the dissociation of hydrogen on a tungsten filament at temperatures between 1200 K and 1800 K. Their measurements included both clean and contaminated filaments and led them to conclude that the former mechanism is correct.

Apparently unaware of the work of Brennan and Fletcher (1959), Miyake et al. (1999) suggest that the mechanism for the dissociation of hydrogen on hot tungsten is that the molecules dissociate on the surface and then atoms evaporate into the bulk solid from where they are emitted thermionically. However, Winkler (2000) disagrees with their interpretation and shows that existing classical adsorption–desorption processes explain their results without invoking desorption from subsurface sites.

13 Concluding Observations

13.1 INTRODUCTION

In this chapter we initially draw together in one place some important information from previous chapters that is otherwise widely distributed. This includes in Section 13.2 some myths on atomic beam production and in Section 13.3 some recommendations.

We begin by stating that it may not be immediately evident from Chapters 4 and 5 that every element in the periodic table that is sufficiently abundant either has been or presumably could be formed into an atomic beam. Some are clearly more difficult to form into a beam than others, depending upon the temperature needed for evaporation. So for some elements obtaining a collimated beam has not proved possible. Sometimes the reactivity of the element may cause problems, particularly if a stable beam is needed for many hours of operation. All elements that are naturally diatomic gases can be formed into dissociated beams, as is described in Chapter 6, but obtaining high concentrations of highly dissociated atomic nitrogen has proved particularly challenging. Forming beams of other radicals and molecules does not normally appear to be problematic. However, it is reported by Periquet et al. (2000) that forming a molecular beam of the nucleobase guanine [$C_5H_5N_5O$] was not readily possible because of isomerisation or decomposition.

For all beams, any problems discussed by the authors are included in the appropriate section. The general design of ovens, which is likely to be applicable for the production of any beam, is discussed in Chapter 3. Chapter 10 describes the numerous techniques for constructing multichannel collimators, but their employment is not as widespread as it could be. As is discussed in Section 6.5, they have been successfully employed, for example, to form collimated and dissociated beams of hydrogen using tungsten foil arrays or to collimate already dissociated hydrogen in glass capillary arrays. The equations needed for the design of a collimator are brought together in an easily applicable form in Chapter 9 so that the optimum design of a collimator may be made from the variables at the experimenter's disposal. Measurements on collimated beams are critically discussed in Chapter 11 in order to give an experimental indication of the reliability of the calculations for various atoms and beam-forming impedance configurations.

13.2 SOME ATOMIC BEAM MYTHS

We have encountered numerous misconceptions in the papers we cite, but we have ignored them unless they are material to the particular information we discuss. However, we include in this section some more generally held erroneous beliefs. Some of the early mythology of atomic beams may still be believed, so we draw attention to these cases, starting with the 'cloud' in front of the beam-forming impedance mentioned explicitly by Kratzenstein (1935), who is referred to by Becker (1961a) and Pauly and Toennies (1965). Cloud formation is also mentioned by Hostettler and Bernstein (1960), Troitskii (1962) and Angel and Giles (1972). This cloud is caused by collisions between the beam and the background gas, and these will certainly occur if the vacuum is inadequate, but Kratzenstein's pumps only had a speed of under $2 \, l \, s^{-1}$, which is minute by today's standards.

The report of the improvement on an aperture to form an atomic beam by a metallic foil collimator by Zacharias and Haun (1954) unfortunately contained the statement that

'... an upper limit is placed on the length by the condition that the length be short compared with the mean free path for atoms in the oven exit canal. If this condition is not fulfilled, then atoms will collide with each other and the point at which such collisions are most probable will serve as a new source ...'

Unfortunately the condition that the mean free path must be greater than the length of the beam-forming impedance has been stated in several of the reviews included in Section 1.2. Experiments that were made with short collimators when much longer ones could have been constructed have presumably been influenced by the same fallacy. Some of the reviews and original papers were published a decade after both the theoretical treatment of collisions within the impedance and experiments showed that this was not the case.

As mentioned in Section 2.5, another myth when interatomic collisions do occur within a tube follows from Giordmaine and Wang (1960). When considering the axial intensity they state that increasing the length of a tube beyond a certain amount that depends upon a number of factors 'has a negligible effect upon the source characteristics'. Becker (1961a) introduced the term effective length, but he used it to define the region where $\lambda \geq d$, which has always been assumed in our discussion. It was also used by Troitskii (1962) and as mentioned in Section 1.2, Anderson et al.'s (1965) review also uses the term. Zugenmaier (1966) states that a tube length exceeding an effective length does not increase the axial intensity. Gray and Sawin (1992) define a critical length of a tube such that increasing its length further attenuates the throughput without changing the directivity of the source. Hence a longer tube is not expected to be an improvement over a shorter one, which is supported by neither numerical calculations nor experiments. There is no advantage in reducing Γ beyond the maximum that is achievable. We do not therefore include expressions for the effective length.

We also include as a myth that a beam produced by a free jet source is always superior to a thermal atomic beam. The conclusion from Section 11.7 is that this is not generally the case. It is unfortunate that reviewers such as Kudryavtsev et al. (1993), mentioned in Section 1.2, draw misleading comparisons. Designers of thermal atomic beams may well describe the use of free jet sources as a brute force and ignorance source. The term 'brute force' is used because gas admitted into a vacuum system at near atmospheric pressure or above could hardly be described as a gentle approach and 'ignorance' because measurement of all the properties of the beam have rarely been made in comparison with the large number in the case of thermal beams.

13.3   SOME RECOMMENDATIONS

In the previous section we expressed concern that some of the reviews and original papers that stated that the mean free path of atoms must be longer than the tube length were published a decade after both the theoretical treatment of collisions within the impedance and experiments showed that this was not the case. This suggests that theoretical work, for example, that pioneered by Giordmaine and Wang (1960), which took collisions within the impedance into account, should be supported by at least some numerical evaluation. Experimenters are usually reluctant to undertake non-algebraic computations.

Quoting actual beam properties only in the form of measured intensity is insufficient. We use the term 'beam-forming impedance' to indicate that more intensity can be obtained by increasing the flow from the gas reservoir by reducing this impedance. Complete beam specification must include not only its axial intensity but also the throughput that produces it as well as at least the beam halfwidth, and preferably the complete angular distribution.

As already discussed in Section 11.5, peaking factors are not now usually considered a useful way of presenting either calculations or measurements. We therefore recommend that, if any data are presented in this way, the beam axial intensity and throughput are also quoted. The design criteria discussed in Chapter 9 do not include peaking factors, neither are terms such as 'conductance', 'transmission probability' and 'Knudsen number' used, since they are not normally explicitly needed in atomic beam design. Whenever possible, reported measurements should give the beam parameters as a minimum with full details of the conditions under which they were obtained, such as the gas species and pressure and the temperature of evaporation or dissociation as well as the impedance dimensions. We use the term 'minimum' since it is clear that just quoting beam halfwidths is insufficient. The full angular distribution is required both experimentally and theoretically in order to investigate where the disagreement between them, particularly for gaseous beams, occurs. Further theoretical work needs to consider both the effects of partial specular reflection on the beam properties and surface diffusion, namely, the possibility of atoms migrating before being reemitted from the collimator surface. It is also disappointing that further measurements have not been made on the shape of the beam formed by focussing capillary arrays, since these beams are much wider than predicted. They are nevertheless still very useful. Authors may of course reduce their data however they wish, but just quoting mean free paths, for example, without stating the value of atomic diameter used to calculate them, leads to avoidable uncertainty.

Unfortunately, as we personally know only too well, most measurements on atomic beams did not involve dedicated experiments but were made on apparatus under development that was intended eventually to utilise the beam. It follows from our conclusions that fully understanding beam characteristics still remains an unfulfilled goal that justifies further thorough work on both gases and vapours and through single tubes as well as arrays. Multiplying the theoretical beam halfwidth by a factor two in order to obtain a good estimate of what is likely to be obtained in practice for a beam formed from a gaseous species and probably a bit less if formed from a vapour is hardly a satisfactory approach. Added to which, the factor to use for a focussing array is still even more uncertain.

Useful comparisons, for example, on the effects of the hydrogen dissociation fraction on microwave discharge tube construction, input power, and cleaning as undertaken by Donnelly et al. (1992) and described in Section 6.3.5, are unfortunately rare. Since the handling of materials that are unstable in air is inconvenient, it would be useful also to have a comparison of the direct evaporation of the substance to be formed into the beam with that produced by the use of precursors. For alkali atoms these have their own treatment in Section 4.2, but some use of precursors will also be found in Chapter 5.

In addition, many accounts of the experimental details of beam formation are very brief. We have included all that should assist anyone wanting to produce a beam of the same material, so that a paper merely stating that 'the atomic beam was formed using a resistively heated oven' is rejected when further details are available from other authors even though it might indicate that there were no experimental difficulties.

Theoretical developments have progressed to the stage that the Clausing function and transmission probability are both known accurately and simple approximations are available to both when needed. The challenge remains to predict the angular distribution of an atomic beam both with and without interatomic collisions. The properties of the beam formed by an orifice are accurately predictable. In the case of a tube, the importance of partial specular reflection, end effects, and probably most importantly surface diffusion needs to be investigated, preferably in combination.

13.4　FINALLY

Many of the reviews mentioned in Section 1.2 will also assist in the design and application of a beam once it has left the beam-forming impedance. Their primary purpose is however usually to discuss the scientific results that have been achieved with their use.

We conclude by stating that the design and implementation of atomic beam devices that simultaneously have

- no short- or long-term fluctuations; and
- reliability for many hours or days of operation; and
- freedom from substantially any contamination of the beam; and
- no excessive environmental demands on, for example, electric power, cooling water, and use of source material; and
- no significant disturbance of the rest of the system by, for example, stray electric or magnetic fields and radiation; and
- high beam quality.

require considerable technical skills. We have included over 800 relevant references and so hope that drawing upon the experience of these numerous others will considerably aid this process. It will certainly be realised that the atomic beam is never more than part of an apparatus.

References

To aid retrieval, references are primarily ordered alphabetically by the first or only author's surname, irrespective of initials, followed by secondary ordering by the date of publication.

Single page numbers are terminated with '#'.

Abernathy C R. 1994. The role of hydrogen in UHV growth of III–V semiconductors. *Materials Sci. Forum* **148–149**, 3–26.

Abernathy C R, Jordan A S, Pearton S J, Hobson W S, Bohling D A and Muhr G T. 1990. Growth of high quality AlGaAs by metalorganic molecular beam epitaxy using trimethylamine alane. *Appl. Phys. Lett.* **56**, 2654–6.

Adam M Y, Hellner L, Dujardin G, Svensson A, Martin P and Combet Farnoux F. 1989. Single and double photoionisation of atomic silver between 43.5 and 15.5 nm. *J. Phys. B: At. Mol. Opt. Phys.* **22**, 2141–50.

Adams J Q, Phipps T E and Wahlbeck P G. 1968. Effusion. III. Angular number distributions of gaseous CsCl from right-circular cylindrical orifices into vacuum. *J. Chem. Phys.* **49**, 1609–16.

Adamson S and McGilp J F. 1986. Measurement of gas flux distributions from single capillaries using a modified, uhv-compatible ion gauge, and comparison with theory. *Vacuum* **36**, 227–32.

Adamson S, O'Carroll C and McGilp J F. 1988a. The spatial distribution of flux produced by single capillary gas dosers. *Vacuum* **38**, 341–4.

Adamson S, O'Carroll C and McGilp J F. 1988b. The angular distribution of thermal molecular beams formed by single capillaries in the molecular flow regime. *Vacuum* **38**, 463–7.

Ad'yasevich B P and Antonenko V G. 1963. Preparation of glass collimators. *Instrum. Exper. Techn.* **2**, 308–10 [*Prib. Tekh. Éksp.* **2**, 126–8 1963].

Ad'yasevich B P, Antonenko V G, Polunin Y P and Fomenko D E. 1965. A polarized ion source. *Plasma Phys. (J. Nucl. Energy C)* **7**, 187–94 [*Atomnaya Energiya* **17**, 17–? 1964].

Afzal F A and Giutronich J E. 1974. A graphite tube electric furnace capable of fast heating and cooling, and its use in measuring electrical properties of magnesium oxide single crystal. *J. Phys. E: Sci. Instrum.* **7**, 579–82.

Aguilar A, de Andrés J, Romero T, Albertí M, Lucas J M, Bocanegra J M, Sogas J and Gadea F X. 2001. Crossed beams and theoretical study of the (NaRb)$^+$ collisional system. In *Atomic and Molecular Beams—The State of the Art 2000*, R Campargue (ed) (Berlin, Germany: Springer), pp. 599–611.

Ahmed N, Nadeem A, Nawaz M, Bhatti S A, Iqbal M and Baig M A. 2005. Resistively heated high temperature atomic beam source. *Rev. Sci. Instrum.* **76**, 063105-1–4.

Akin'shin V D, Borisov S F, Porodnov B T and Suetin P E. 1975. Flow of rarefied gases in a capillary screen at different temperatures. *J. Appl. Mech. Tech. Phys.* (2) 183–6 [*Zh. Prikla. Mekh. i Tekh. Fiz.* 45–9 1974].

Akulov Y A, Mamyrin B A and Shikhaliev P M. 1997. Production of atomic hydrogen in an rf gas discharge and mass spectrometer diagnostics of the process. *Tech. Phys.* **42**, 584–5 [*Zh. Tekh. Fiz.* **67**, 140–2 1997].

Alcock N W. 1990. *Bonding and Structure: Structural Principles in Inorganic and Physical Chemistry* (Chichester, UK: Ellis Horwood).

Aleksakhin I S and Zayats V A. 1974. Experiments on the electron impact excitation of beryllium atoms. *Opt. Spectrosc.* **36**, 717 [*Opt. Spektrosk.* **36**, 1229–30 1974].

Alleau T, Devin B, Durand J P and Lesueur R. 1967. Convertisseurs thermioniques à deux alcalins utilisant les composés d'insertion du graphite. *Int. Conf. Thermionic Electrical Power Generation* (London: IEEE) Session 3B.

Allen G C, Baerends E J, Vernooijs P, Dyke J M, Ellis A M, Fehér M and Morris A. 1988. High temperature photoelectron spectroscopy: A study of U, UO and UO_2. *J. Chem. Phys.* **89**, 5363–72.

Allen J D, Boggess G W, Goodman T D, Wachtel A S and Schweitzer G K. 1973. A high-temperature photoelectron spectrometer. *J. Electron. Spectrosc.* **2**, 289–94.

Allen J F and Misener A D. 1939. The properties of flow of liquid He II. *Proc. Roy. Soc. A* **172**, 467–91.

Allison A C and Smith F J. 1971. Transport properties of atomic hydrogen. *Atomic Data* **3**, 317–21.

Anderson J B. 1974. Molecular beams and nozzle sources. In *Molecular Beams and Low Density Gasdynamics*, P P Wegener (ed) (New York: Dekker), pp. 1–71.

Anderson J B, Andres R P and Fenn J B. 1965. High intensity and high energy molecular beams. In *Advances in Atomic and Molecular Physics*, vol. 1, D R Bates and I Estermann (eds) (New York: Academic), pp. 345–89.

Anderson J B, Andres R P and Fenn J B. 1966. Supersonic nozzle beams. In *Adv. Chem. Phys.*, vol. 10, J Ross (ed) (New York: Interscience), pp. 275–317.

Angel G C and Giles R A. 1972. The velocity distribution of atoms issuing from a multichannel glass capillary array and its implication on the measurement of atomic beam scattering cross sections. *J. Phys. B: Atom. Molec. Phys.* **5**, 80–8.

Aquilanti V, Liuti G, Luzzatti E, Vecchio-Cattivi F and Volpi G G. 1972. Production and detection of an energy selected beam of hydrogen atoms in the range 0.01–0.20 eV. *Z. phys. Chem.* **79**, 200–8.

Archer N J. 1977. The preparation and properties of pyrolytic boron nitride. In *High Temperature Chemistry of Inorganic and Ceramic Materials*, F P Glasser and P E Potter (eds) (London: The Chemical Society), pp. 167–80.

Aubert D, Baldy A and Chantrel H. 1971a. Production de jets moléculaires de haute densité et excitation du spectre d'émission. *Entropie* (42) 60–4.

Aubert D, Baldy A and Chantrel H. 1971b. Procédé de fabrication de collimateurs multicanaux focalisants. *Entropie* (42) 64–5.

Aushev V E, Zaika N I and Mokhnach A V. 1982. Properties of a lithium atomic beam. *Sov. Phys. Tech. Phys.* **27**, 878–80 [*Zh. Tekh. Fiz.* **52**, 1438–41 1982].

Aydin R, Ertmer W and Johann U. 1982. Laser-rf double-resonance hyperfine structure measurements of the metastable 3d 4s $^1D_2$ state of ^{43}Ca. *Z. Physik A* **306**, 1–5.

Bacal M, Doucet H J, Labaune G, Lamain H, Jacquot C and Verney S. 1982. Cesium supersonic jet for D—production by double electron capture. *Rev. Sci. Instrum.* **53**, 159–67.

Back C G, White M D, Pejcev V and Ross K J. 1981. The ejected electron spectrum of zinc vapour autoionising levels excited by low-energy electron impact. *J. Phys. B: Atom. Molec. Phys.* **14**, 1497–507.

Baldacchini G, Marchetti S and Montelatici V. 1982. Multitube collimators driven at high pressure for molecular spectroscopy. *J. Appl. Phys.* **53**, 3888–9.

Ballard A and Bonin K. 2001. Glass slits for collimating particle beams. *Rev. Sci. Instrum.* **72**, 1657–9.

Balooch M and Olander D R. 1975. Reactions of modulated molecular beams with pyrolytic graphite. III. *Hydrogen J. Chem. Phys.* **63**, 4772–86.

Bañares L and Ureña A G. 1989. Simple oven design for highly reactive metal beam applications. *J. Phys. E: Sci. Instrum.* **22**, 1046–7.

Barashkin S T, Neudachin I G and Porodnov B T. 1980. A beam of free molecules emerging from a packet of capillaries. *J. Appl. Mech. Tech. Phys.* **7** 547–51 [*Zh. Prikl. Mekh. Tekh. Fiz.* **5**, 31–7 1979].

Barashkin S T, Porodnov B T and Chemagin M F. 1978. Experimental investigation of total flow and directional diagrams in discharge of gas to vacuum through capillaries of different lengths. *J. Appl. Mech. Tech. Phys.* **4**, 491–5 [*Zh. Prikl. Mekh. Tekh. Fiz.* **4**, 74–80 1977].

Barr M, O'Donnell K M, Fahy A, Allison W and Dastoor P C. 2012. A desktop supersonic free-jet beam source for a scanning helium microscope (SHeM). *Meas. Sci. Technol.* **23**, 105901-1–7.

Bauer G and Springholz G. 1992. Molecular-beam epitaxy—aspects and applications. *Vacuum* **43**, 357–65.

Baum G, Granitza B, Hesse S, Leuer B, Raith W, Rott K, Tondera M and Witthuhn B. 1991. An optically pumped, highly polarized cesium beam for the study of spin-dependent electron scattering. *Z. Physik D* **22**, 431–6.

Beck D, Engelke F and Loesch H J. 1968. Reaktive Streuung in Molekularstrahlen: Cl + Br$_2$. *Ber. Bunsenges. Phys. Chem.* **72**, 1105–7.

Becker E W, Ehrfeld W, Hagmann P, Maner A and Münchmeyer D. 1986. Fabrication of microstructures with high aspect ratios and great structural heights by synchrotron radiation lithography, galvoforming, and plastic moulding (LIGA) process. *Microelectron. Eng.* **4**, 35–56.

Becker G. 1961a. Zur Erzeugung starker Molekularstrahlen im Hochvakuum mit Düsen. *Z. angew. Phys.* **13**, 59–64.

Becker G. 1961b. Zur Theorie der Molekularstrahlerzeugung mit langen Kanälen. *Z. Physik* **162**, 290–312.

Beetz C P, Boerstler R, Steinbeck J, Lemieux B and Winn D R. 2000. Silicon-micromachined microchannel plates. *Nucl. Instrum. Methods A* **442**, 443–51.

Beijerinck H C W and Verster N F. 1975. Velocity distribution and angular distribution of molecular beams from multichannel arrays. *J. Appl. Phys.* **46**, 2083–91.

Beijerinck H C W and Verster N F. 1981. Absolute intensities and perpendicular temperatures of supersonic beams of polyatomic gases. *Physica* **111C**, 327–52.

Beijerinck H C W, Stevens M P J M and Verster N F. 1976. Monte-Carlo calculation of molecular flow through a cylindrical channel. *Physica* **83C**, 209–19.

Beijerinck H C W, van Gerwen R J F, Kerstel E R T, Martens J F M, van Vliembergen E J W, Smits M R T and Kaashoek G H. 1985. Campargue-type supersonic beam sources: Absolute intensities, skimmer transmission and scaling laws for mono-atomic gases He, Ne and Ar. *Chem. Phys.* **96**, 153–73.

Bell S C, Junker M, Jasperse M, Turner L D, Lin Y J, Spielman I B and Scholten R E. 2010. A slow atom source using a collimated effusive oven and a single-layer variable pitch coil Zeeman slower. *Rev. Sci. Instrum.* **81**, 013105-1–7.

Bellamy E H and Smith K F. 1953. The nuclear spins and magnetic moments of ^{24}Na, ^{42}K, ^{86}Rb, ^{131}Cs and ^{134}Cs. *Phil. Mag.* **44**, 33–45.

Bellman R and Raj R. 1994. Growth of epitaxial lithium tantalate on sapphire by chemical beam epitaxy from lithium hexaethoxytantalate. *Ferroelectrics* **152**, 7–12.

Bellman R and Raj R. 1997. Design and performance of a new type of Knudsen cell for chemical beam epitaxy using metal-organic precursors. *Vacuum* **48**, 165–73.

Benvenuti G, Halary-Wagner E, Brioude A and Hoffmann P. 2003. High uniformity deposition with chemical beams in high vacuum. *Thin Solid Films* **427**, 411–16.

Berkowitz J. 1972. Photoelectron spectroscopy of high-temperature vapors. I. TlCl, TlBr and TlI. *J. Chem. Phys.* **56**, 2766–74.

Berkowitz J. 1975. PES of high temperature vapors. VII. S$_2$ and Te$_2$. *J. Chem. Phys.* **62**, 4074–9.

Berkowitz J and Chupka W A. 1958. Polymeric gaseous molecules in the vaporization of alkali halides. *J. Chem. Phys.* **29**, 653–7.

Berkowitz J, Batson C H and Goodman G L. 1979. PES of higher temperature vapors: Lithium halide monomers and dimers. *J. Chem. Phys.* **71**, 2624–36.

Berman A S. 1965. Free molecule transmission probabilities. *J. Appl. Phys.* **36**, 3356#.

Bermudez V M. 1996. Simple, efficient technique for exposing surfaces to hydrogen atoms. *J. Vac. Sci. Technol. A* **14**, 2671–3.

Bernhardt A F. 1976. Isotope separation by laser deflection of an atomic beam. *Appl. Phys.* **9**, 19–34.

Bertl W, Healey D, Zmeskal J, Hasinoff M D, Blecher M and Wright D H. 1995. A compact hydrogen recycling system using metal hydrides. *Nucl. Instrum. Methods A* **355**, 230–5.

Beskok A and Karniadakis G E. 1999. A model for flows in channels, pipes and ducts at micro and nano scales. *Microscale Thermophys. Eng.* **3**, 43–77.

Bessey W H and Simpson O C. 1942. Recent work in molecular beams. *Chem. Rev.* **30**, 239–79.

Bethe H A. 1937. Nuclear Physics B. Nuclear dynamics, theoretical. *Rev. Mod. Phys.* **9**, 69–244.

Bhaskar N D and Kahla C M. 1990. Cesium gettering by graphite—Improvement in the gettering efficiency. *IEEE Trans. Ultrason. Ferroelectr. Frequ. Contr.* **37**, 355–8.

Bhatia M S, Dongare A S, Mago V K and Lal B. 2000. Filamentless operation of a high-power electron bombardment furnace used for refractory metals atom beam generation. *Rev. Sci. Instrum.* **71**, 3031–6.

Bickes R W, Newton K R, Herrmann J M and Bernstein R B. 1976. Utilization of an arc-heated jet for production of supersonic seeded beams of atomic nitrogen. *J. Chem. Phys.* **64**, 3648–57.

Bischler U and Bertel E. 1993. Simple source of atomic hydrogen for ultrahigh vacuum applications. *J. Vac. Sci. Technol. A* **11**, 458–60.

Boh J, Eilmsteiner G, Rendulic K D and Winkler A. 1998. Adsorption and abstraction of atomic hydrogen (deuterium) on Al (100). *Surface Sci.* **395**, 98–110.

Boivin R F and Srivastava S K. 1998. Electron-impact ionization of Mg. *J. Phys. B: At. Mol. Opt. Phys.* **31**, 2381–94.

Bonhoeffer K F. 1924. Das Verhalten von aktivem Wasserstoff. *Z. phys. Chem.* **113**, 119–219.

Borisov S F, Porodnov B T and Suetin P E. 1972. Experimental investigation of a gas flow in capillaries. *Sov. Phys. Tech. Phys.* **17**, 1039–42 [*Zh. Tekh. Fiz.* **42**, 1310–4 1972].

Bornscheuer K H, Lucas S R, Choyke W J, Partlow W D and Yates J T. 1993. Reflector atomic hydrogen source: A method for producing pure atomic hydrogen in ultrahigh vacuum. *J. Vac. Sci. Technol. A* **11**, 2822–6.

Borovik A A, Rojas H L and King G C. 1995. A compact metal vapour source for use in electron spectroscopy studies. *Meas. Sci. Technol.* **6**, 334–6.

Boutry G A, Évrard R and Richard J C. 1964. Contribution à l'étude des propriétés photoélectriques du césium pur, préparé et conservé dans l'ultravide. *C. R. Acad. Sc. Paris* **258**, 143–6.

Boys C V. 1887. On the production, properties and some suggested uses of the finest threads. *Phil. Mag.* **23**, 489–99.

Brackmann R T and Fite W L. 1961. Condensation of atomic and molecular hydrogen at low temperatures. *J. Chem. Phys.* **34**, 1572–9.

Brandt E H. 1989. Levitation in physics. *Science* **243**, 349–55.

Brennan H and Fletcher P C. 1959. The atomization of hydrogen on tungsten. *Proc. Roy. Soc. A* **250**, 389–408.

Brewer P D, Chow D H and Miles R H. 1996. Atomic antimony for molecular beam epitaxy of high quality III-V semiconductor alloys. *J. Vac. Sci. Technol. B* **14**, 2335–8.

Brink G O, Fluegge R A and Hull R J. 1968. Microwave discharge source for atomic and molecular beam production. *Rev. Sci. Instrum.* **39**, 1171–2.

Brinkmann R T and Trajmar S. 1981. Effective path length corrections in beam-beam scattering experiments. *J. Phys. E: Sci. Instrum.* **14**, 245–55.

Brix P, Eisinger J T, Lew H and Wessel G. 1953. The Zeeman effect of the Cr ground state. *Phys. Rev.* **93**, 647–9.

Bröhl W H and Hartmann H. 1981. On constant molecular gas flow through cylindrical tubes. A note concerning the information content of Clausing's equations. *Vacuum* **31**, 117–8.

Bromberg E E A, Proctor A E and Bernstein R B. 1975. Pure-state molecular beams: Production of rotationally, vibrationally, and translationally selected CsF beams. *J. Chem. Phys.* **63**, 3287–94.

Brown D O, Cvejanović D and Crowe A. 2003. The scattering of 40 eV electrons from magnesium: A polarization correlation study for the 3^1P state and differential cross sections for elastic scattering and excitation of the 3^1P and 3^3P states. *J. Phys. B: At. Mol. Opt. Phys.* **36**, 3411–23.

Brown R L. 1967. Effects of impurities on the production of oxygen atoms by a microwave discharge. *J. Phys. Chem.* **71**, 2492–5.

Brunger M J and Buckman S J. 2002. Electron-molecule scattering cross sections I Experimental techniques and data for diatomic molecules. *Phys. Repts.* **357**, 215–458.

Brunton A N, Martin A P, Fraser G W and Feller W B. 1999. A study of 8.5 µm microchannel plate X-ray optics. *Nucl. Instrum. Methods A* **431**, 356–65.

BS 5884. 1999. Magnetic materials—Methods for the determination of the relative magnetic permeability of feebly magnetic materials.

BS EN 13556. 2003. Round and sawn timber. Nomenclature of timbers used in Europe.

BS ISO 15510. 2010. Stainless steels—Chemical composition.

BS ISO 80000-1. 2009. Quantities and units Part 1: General.

Buckley H. 1927. On the radiation from the inside of a circular cylinder. *Phil. Mag.* **4**, 753–62.

Buckley H. 1928. On the radiation from the inside of a circular cylinder. Part II. *Phil. Mag.* **6**, 447–57.

Buckman S J, Gulley R J, Moghbelalhossein M and Bennett S J. 1993. Spatial profiles of effusive molecular beams and their dependence on gas species. *Meas. Sci. Technol.* **4**, 1143–53.

Budrevich A, Tsipinyuk B and Kolodney E. 1996. Critical behaviour of super-heated (1900–2000 K) C_{60} vapours. *J. Phys. B: At. Mol. Opt. Phys.* **29**, 4965–74.

Bulgin D, Dyke J, Goodfellow F, Jonathan N, Lee E and Morris A. 1977. A high temperature furnace for use in photoelectron spectroscopy. *J. Electr. Spectrosc.* **12**, 67–76.

Burden M S and Walley P A. 1969. The evaporation of metals and elemental semiconductors using a work-accelerated electron beam source. *Vacuum* **19**, 397–402.

Büttgenbach S and Meisel G. 1971. Hyperfine structure measurements in the ground states $^4F_{3/2}$, $^4F_{5/2}$, $^4F_{7/2}$ of Ta^{181} with the atomic beam magnetic resonance method. *Z. Physik* **244**, 149–62.

Büttgenbach S, Meisel G, Penselin S and Schneider K H. 1970. A new method for the production of atomic beams of highly refractory elements and first atomic beam magnetic resonances in Ta^{181}. *Z. Physik* **230**, 329–36.

Cahn R W. 1964. Making fuel for inertially confined fusion reactors. *Nature* **311**, 408#.

Campargue R. 1984. Progress in overexpanded supersonic jets and skimmed molecular beams in free-jet zones of silence. *J. Phys. Chem.* **88**, 4466–74.

Campargue R (ed). 2001. *Atomic and Molecular Beams* (Berlin, Germany: Springer).

Carette J D, Pandolfo L and Dubé D. 1983. New developments in the calculation of the molecular flow conductance of a straight cylinder. *J. Vac. Sci. Technol. A* **1**, 143–6 *Erratum* 1574#.

Carlson K D, Gilles P W and Thorn R J. 1963. Molecular and hydrodynamical effusion of mercury vapor from Knudsen cells. *J. Chem. Phys.* **38**, 2725–35.

Carter G M and Pritchard D E. 1978. Recirculating atomic beam oven. *Rev. Sci. Instrum.* **49**, 120–1.

Casavecchia P. 2000. Chemical reaction dynamics with molecular beams. *Rep. Prog. Phys.* **63**, 355–414.

Casella A M, Loyalka S K and Hanson B D. 2009. Computation of free-molecular flow in nuclear materials. *J. Nucl. Materials* **394**, 123–30.

Cercignani C. 1988. *The Boltzmann Equation and its Applications*, Ch 5 (London: Springer).

Chai Y G. 1984. Tin phosphide as a phosphorus beam source for molecular beam epitaxy. *Appl. Phys. Lett.* **45**, 985–7.

Chalek C L and Gole J L. 1976. Chemiluminescence spectra of ScO and YO: Observation and analysis of the A$'^2\Delta$ - X $^2\Sigma^+$ band system. *J. Chem. Phys.* **65**, 2845–59.

Chapman S and Cowling T G. 1970. *The Mathematical Theory of Non-Uniform Gases* (Cambridge, UK: University Press).

Childs W J. 1974. Hyperfine and Zeeman studies of metastable atomic states by atomic-beam magnetic-resonance Case Studies. In *Atomic Physics III*, E W McDaniel and M R C McDowell (eds) (Amsterdam: North Holland), pp. 215–304.

Choong P T. 1971. Modified Monte Carlo technique for clausing factor calculations. *J. Comp. Phys.* **7**, 358–60.

Clampitt R and Hanley P E. 1988. Oxidation of cold copper films with oxygen radicals. *Supercond. Sci. Technol.* **1**, 5–6.

Clausing P. 1926. Over de stationnaire strooming van een zeer verdund gas door een ronde cylinderbois van willekeurige lengte Verslag van de gewone vergaderingen der Afdeeling. *Natuurkunde* **35**, 1023–35.

Clausing P. 1930. Über die Strahlformung bei der Molekularströmung. *Z. Physik* **66**, 471–6 (AEC-TR-2446).

Clausing P. 1932. Über die Strömung sehr verdünnter Gase durch Röhren von beliebiger Länge. *Ann. Phys.* **12**, 961–89 (Translation: The flow of highly rarefied gases through tubes of arbitrary length. *J. Vac. Sci. Technol.* **8**, 636–46 1971: AEC-TR-2447).

Clausnitzer G. 1963. A source of polarized protons. *Nucl. Instrum. Methods* **23**, 309–24.

Cline H E. 1981. Directionally solidified thin-film eutectic alloys. *J. Appl. Phys.* **52**, 256–60.

Cline H E. 1982. Submicron eutectic thin film structure. *J. Appl. Phys.* **53**, 4896–902.

Clough P N and Geddes J. 1981. Some aspects of chemical interaction of atomic beams. *J. Phys. E: Sci. Instrum.* **14**, 519–29.

Cochrane E C A, Benton D M, Forest D H and Griffith J A R. 1998. Hyperfine structure and isotope shifts in natural vanadium. *J. Phys. B: At. Mol. Opt. Phys.* **31**, 2203–13.

Cole R J. 1977a. Transmission probability of free molecular flow through a tube. *Proc. 10th Int. Symp. Rarefied Gas Dynamics, Aspen, Co. 1976 Prog. Astronautics and Aeronautics*, vol. 51, pp. 261–72.

Cole R J. 1977b. Complementary variational principles for Knudsen flow rates. *J. Inst. Maths. Applics.* **20**, 107–15.

Cole R J and Pack D C. 1975. Some complementary bivariational principles for linear integral equations of Fredholm type. *Proc. Roy. Soc. A* **347**, 239–52.

Collins E R, Glavish H F and Whineray S. 1963. An ion source for polarized protons and deuterons. *Nucl. Instrum. Methods* **25**, 67–76.

Comsa G. 1969. The rate of impingement of a molecular beam. *Vacuum* **19**, 277–9.

Comsa G and David R. 1985. Dynamical parameters of desorbing molecules. *Surface Science Repts.* **5**, 145–98.

Craddock M K. 1961. The polarized proton source of the Harwell proton linear accelerator. *Helv. Phys. Acta Suppl.* **VI**, 59–76.

Crane J K. 1980. Molecular beam levitation. *Conference on Inertial Confinement Fusion*, February 26–28, 1980, OSA/IEEE, San Diego, CA, pp. 38–9.

Crane J K, Smith R D, Johnson W L, Jordan C W, Letts S A, Korbel G R and Krenik R M. 1982. The use of molecular beams to support microspheres during plasma coating. *J. Vac. Sci. Technol.* **20**, 129–33.

Cratchley D. 1965. Experimental aspects of fibre-reinforced metals. *Metallurgical Rev.* **10**, 79–144.

Crawford C K. 1972. High-efficiency high-temperature radiation heat shields. *J. Vac. Sci. Technol.* **9**, 23–6.

Crumley W H, Hayden J S and Gole J L. 1986. Laser induced excitation spectroscopy of copper trimer in various stages of supersonic expansion: Observation of fluorescence from dissociative levels. *J. Chem. Phys.* **84**, 5250–61.

Cushing G W, Navin J K, Valadez L, Johánek V and Harrison I. 2011. An effusive molecular beam technique for studies of polyatomic gas-surface reactivity and energy transfer. *Rev. Sci. Instrum.* **82**, 044102-1–11.

Cvejanovic D and Murray A J. 2002. Design and characterization of an atomic beam oven for combined laser and electron impact experiments. *Meas. Sci. Technol.* **13**, 1482–7.

Cvejanović D, Adams A, Imhof R E and King G C. 1975. An efficient atomic beam oven for use in low energy electron scattering experiments. *J. Phys. E: Sci. Instrum.* **8**, 809–10.

Czarnetzki U, Döbele H F and Schulz-von der Gathen V. 1988. Laser-induced fluorescence spectroscopy of beryllium vapour at UV and VUV wavelengths. *J. Phys. D: Appl. Phys.* **21**, 246–50.

Dagdigian P J, Cruse H W, Schultz A and Zare R N. 1974. Product state analysis of BaO from the reactions Ba + CO_2 and Ba + O_2. *J. Chem. Phys.* **61**, 4450–65.

Davies C M and Lucas C B. 1983. The failure of theory to predict the density distribution of gas flowing through a tube under free molecular conditions. *J. Phys. D: Appl. Phys.* **16**, 1–16.

Davis D H. 1960. Monte Carlo calculation of molecular flow rates through a cylindrical elbow and pipes of other shapes. *J. Appl. Phys.* **31**, 1169–76.

Davis D H, Levenson L L and Milleron N. 1964. Effect of 'rougher than rough' surfaces on molecular flow through short ducts. *J. Appl. Phys.* **35**, 529–32.

Davis L, Feld B T, Zabel C W and Zacharias J R. 1949b. The hyperfine structure and nuclear moments of the stable chlorine isotopes. *Phys. Rev.* **76**, 1076–85.

Davis L, Nagle D E and Zacharias J R. 1949a. Atomic beam magnetic resonance experiments with radioactive elements Na^{23}, K^{40}, Cs^{135} and Cs^{137}. *Phys. Rev.* **76**, 1068–75.

Dayton B B. 1957. Gas flow patterns at entrance and exit of cylindrical tubes. *Trans. 3 CVT Nat. Vac. Symp. 1956* (New York: Pergamon), pp. 5–11.

Dayton B B. 1958. Strömungsbilder von Gasen am Ein- und Ausgang zylindrischer Rohre. *Vakuum-Technik.* **7**, 7–13 [Translation of Dayton (1957)].

de Boer J H, Broos J and Emmens H. 1930. Herstellung der Alkalimetalle durch Reduktion ihrer Verbindungen mit Zirkonium. *Z. anorg. Allge. Chem.* **191**, 113–21.

Deckers J and Fenn J B. 1963. High intensity molecular beam apparatus. *Rev. Sci. Instrum.* **34**, 96–100.

DeGraffenreid W, Ramirez-Serrano J, Liu Y M and Weiner J. 2000. Continuous, dense, highly collimated sodium beam. *Rev. Sci. Instrum.* **71**, 3668–76.

de Leeuw J H and Gadamer E O. 1967. Density distribution of a molecular flux from a short cylindrical tube. UTIAS Tech. Note No. 103.

De Leeuw D M, Mooyman R and De Lange C A. 1978. He(I) photoelectron spectroscopy of halogen atoms. *Chem. Phys. Lett.* **54**, 231–4.

Delves L M and Mohamed J L. 1985. *Computational Methods for Integral Equations* (Cambridge, UK: University Press).

DeMarco B, Rohner H and Jin D S. 1999. An enriched ^{40}K source for fermionic atom studies. *Rev. Sci. Instrum.* **70**, 1967–9.

DeMarcus W C. 1956. The problem of Knudsen flow. *Part I. General Theory*, pp. 1–18; *Part II Solution of Integral Equations with Probability Kernels*, pp. 1–27; Union Carbide Nuclear Company, Oak Ridge Gaseous Diffusion Plant Report K-1302, Oak Ridge, Tennessee.

DeMarcus W C. 1957. The problem of Knudsen flow. *Part III Solutions for One-Dimensional Systems*, pp. 1–50; *Part III Solutions for One-Dimensional Systems (Addendum)*, pp. 1–7; *Part IV Specular Reflection*, pp. 1–22; *Part V Application of the Theory of Radiative Transfer*, pp. 1–18; Union Carbide Nuclear Company, Oak Ridge Gaseous Diffusion Plant Report K-1302, Oak Ridge, Tennessee.

DeMarcus W C and Hopper E H. 1955. Knudsen flow through a circular capillary. *J. Chem. Phys.* **23**, 1344#.

DeMarcus W C and Jenkins H B. 1957. The problem of Knudsen flow. *Part VI Tortuosity* pp. 1–14 [Includes errata to parts I–V]; Union Carbide Nuclear Company, Oak Ridge Gaseous Diffusion Plant Report K-1302, Oak Ridge, Tennessee.

Dembczyński J, Ertmer W, Johann U, Penselin S and Stinner P. 1979. Laser-rf double-resonance studies of the hyperfine structure of metastable atomic states of ^{55}Mn. *Z. Physik A* **291**, 207–18.

Dembczyński J, Ertmer W, Johann U and Stinner P. 1980. High precision measurements of the hyperfine structure of seven metastable atomic states of ^{57}Fe by laser-RF double-resonance. *Z. Physik A* **294**, 313–7.

Derenbach H, Kossmann H, Malutzki R and Schmidt V. 1984. Photoionisation processes in the 5d, 6s and 6p shells of atomic lead and the 4d shell of atomic tin. *J. Phys. B: Atom. Molec. Phys.* **17**, 2781–94.

Dharmasena G, Copeland K, Young J H, Lasell R A, Phillips T R, Parker G A and Keil M. 1997. Angular dependence for v', j'-resolved states in $F + H_2 \rightarrow HF(v', j') + H$ reactive scattering using a new atomic fluorine beam source. *J. Phys. Chem. A* **101**, 6429–40.

Ding A, Karlau J and Weise J. 1977. Production of H-atom and O-atom beams by a cooled microwave discharge source. *Rev. Sci. Instrum.* **48**, 1002–4.

Dinklage A, Lokajczyk T, Kunze H J, Schweer B and Olivares I E. 1998. *In situ* density measurement for a thermal lithium beam employing diode lasers. *Rev. Sci. Instrum.* **69**, 321–2.

Dinneen T, Ghiorso A and Gould H. 1996. An orthotropic source of thermal atoms. *Rev. Sci. Instrum.* **67**, 752–5.

Donnelly A, Hughes M P, Geddes J and Gilbody H B. 1992. A microwave discharge atom beam source of high intensity. *Meas. Sci. Technol.* **3**, 528–32.

Doyle W M and Marrus R. 1963. Atomic beam studies of some radioactive isotopes of refractory group elements. *Nucl. Phys.* **49**, 449–55.

Drullinger R E. 1988. Heat pipe oven molecular beam source. US Patent 4 789 779.

Drullinger R E, Glaze D J and Sullivan D B. 1985. A recirculating oven for atomic beam frequency standards. *Proc. 39 Ann. Symp. Frequency Control Philadelphia*, pp. 13–17.

Dubois L H and Gole J L. 1977. Bimolecular, single collision reaction of ground and metastable excited states of titanium with O_2, NO_2, and N_2O: Confirmation of $D°_0$ (TiO). *J. Chem. Phys.* **66**, 779–90.

Duncan M A. 2012. Invited review article: Laser vaporization cluster sources. *Rev. Sci. Instrum.* **83**, 041101-1–19.

Dunoyer L. 1911a. Sur la théorie cinétique des gaz et la réalisation d'un rayonnement matériel d'origine thermique. *Comptes Rendus* **152**, 592–5.

Dunoyer L. 1911b. Sur la réalisation d'un rayonnement matériel d'origine purement thermique. Cinétique expérimentale. *Le Radium* **8**, 142–6.

Dushman S. 1962. *Scientific Foundations of Vacuum Technique*, 2nd edn. J M Lafferty (ed) (New York: Wiley).

Dyke J M, Elbel S, Morris A and Stevens J C H. 1986b. High-temperature photoelectron spectroscopy A study of atomic and molecular arsenic. *J. Chem. Soc., Faraday Trans. 2* **82**, 637–45.

Dyke J M, Fayad N K, Morris A and Trickle I R. 1979. Gas-phase He I photoelectron spectra of some transition metals: Cu, Ag, Au, Cr and Mn. *J. Phys. B: Atom. Molec. Phys.* **12**, 2985–91.

Dyke J M, Morris A and Stevens J C H. 1986a. High-temperature photoelectron spectroscopy: A study of atomic and molecular antimony. *Chem. Phys.* **102**, 29–36.

Edmonds T and Hobson J P. 1965. A study of thermal transpiration using ultrahigh-vacuum techniques. *J. Vac. Sci. Technol.* **2**, 182–97.

Egorov D, Lahaye T, Schöllkopf W, Friedrich B and Doyle J M. 2002. Buffer-gas cooling of atomic and molecular beams. *Phys. Rev. A* **66**, 043401-1–8.

Eibl C, Lackner G and Winkler A. 1998. Quantitative characterization of a highly effective atomic hydrogen doser. *J. Vac. Sci. Technol. A* **16**, 2979–89.

Eichenbaum A I and Moi M E. 1964. Cesium vapor dispenser. *Rev. Sci. Instrum.* **35**, 691–3.

Eichler B, Hübener S, Erdmann N, Eberhardt K, Funk H, Herrmann G, Köhler S, Trautmann N, Passler G and Urban F J. 1997. An atomic beam source for actinide elements: Concept and realization. *Radiochimica Acta* **79**, 221–33.

Eisenstadt M, Rothberg G M and Kusch P. 1958. Molecular composition of alkali fluoride vapors. *J. Chem. Phys.* **29**, 797–804.

Eisenstadt M M. 1965. Beam source for molecular and atomic hydrogen. *Rev. Sci. Instrum.* **36**, 1878–9.

Elam J W, Routkevitch D, Mardilovich P P and George S M. 2003. Conformal coating on ultra-high-aspect-ratio nanopores of anodic alumina by atomic layer deposition. *Chem. Mater.* **15**, 3507–17.

El-Gendi S E. 1969. Chebyshev solution of differential integral and integro differential equations. *Computer J.* **12**, 282–7.

Eminyan M, MacAdam K B, Slevin J and Kleinpoppen H. 1974. Electron-photon angular correlations in electron-helium collisions: Measurements of complex excitation amplitudes, atomic orientation and alignment. *J. Phys. B: Atom. Molec. Phys.* **7**, 1519–42.

Engel T and Rieder K H. 1982. Structural studies of surfaces with atomic and molecular beam diffraction. *Springer Tracts in Modern Physics*, vol. 91, *Structural Studies of Surfaces*, G Höhler (ed), pp. 55–180.

English T C and Zorn J C. 1974. Molecular beam spectroscopy. In *Methods of Experimental Physics*, vol. 3B, *Molecular Physics*, 2nd edn, D Williams (ed), pp. 669–846.

Ericson T, Copeland K, Keil M, Apelblat Y and Fan Y B. 1994. Fluoride salts as supersonic nozzle materials for hot fluorine. *Rev. Sci. Instrum.* **65**, 3587–8.

Eschard G and Polaert R. 1969. The production of electron-multiplier channel plates. *Philips Tech. Rev.* **30**, 252–5.

Essen L and Parry J V L. 1958. The caesium resonator as a standard of frequency and time. *Phil. Trans. Roy. Soc.* **250A**, 45–69.

Estermann I. 1946. Molecular beam technique. *Rev. Mod. Phys.* **18**, 300–23.

Estermann I (ed). 1959a. *Recent Research in Molecular Beams* (New York: Academic).

Estermann I. 1959b. Molecular beam research in Hamburg 1922–1933. In *Recent Research in Molecular Beams*, I Estermann (ed) (New York: Academic), pp. 1–7.

Estermann I, Simpson O C and Stern O. 1947. The free fall of atoms and the measurement of the velocity distribution in a molecular beam of cesium atoms. *Phys. Rev.* **71**, 238–49.

Exley R R, Meylan B A and Butterfield B G. 1977. A technique for obtaining clean cut surfaces on wood samples prepared for the scanning electron microscope. *J. Microsc.* **110**, 75–8.

Fain D E and Brown W K. 1974. Neon isotope separation by gaseous diffusion transport in the transition flow regime with regular geometries. Report K-1863 Oak Ridge, Tennessee, USA.

Fain D E and Brown W K. 1977. Neon isotope separation by gaseous diffusion transport in regular geometries. *10th Int. Symp. Rarefied Gas Dynamics*, Aspen, Colorado, USA, 1976, pp. 65–78.

Fantz U, Friedl R and Fröschle M. 2012. Controllable evaporation of cesium from a dispenser oven. *Rev. Sci. Instrum.* **83**, 12305-1-5.

Farrow R F C (ed). 1995. *Molecular Beam Epitaxy: Applications to Key Materials* (Park Ridge, NJ: Noyes).

Faubel M, Martinez-Haya B, Rusin L Y, Tappe U and Toennies J P. 1996. An intense fluorine atomic beam source. *J. Phys. D: Appl. Phys.* **29**, 1885–93.

Fehsenfeld F C, Evenson K M and Broida H P. 1965. Microwave discharge cavities operating at 2450 MHz. *Rev. Sci. Instrum.* **36**, 294–8.

Feuermann D, Gordon J M and Huleihil M. 2002. Solar fiber-optic mini-dish concentrators: First experimental results and field experience. *Solar Energy* **72**, 459–72.

Field D, Knight D W, Mrotzek G, Randell J, Lunt S L, Ozenne J B and Ziesel J P. 1991. A high-resolution synchrotron photoionization spectrometer for the study of low-energy electron-molecule scattering. *Meas. Sci. Technol.* **2**, 757–69.

Figger H and Wolber G. 1973. Precision measurement of the hyperfine structure of Lu[175] with the atomic beam magnetic resonance method. *Z. Physik* **264**, 95–108.

Fischer B E and Spohr R. 1988a. Teilchenspuren in der Mikrotechnik I. *Grundlagen Naturwissenschaften* **75**, 57–66.

Fischer B E and Spohr R. 1988b. Teilchenspuren in der Mikrotechnik II. *Anwendungsbeispiele Naturwissenschaften* **75**, 117–22.

Fisher D J. 1977. The resistance of ceramics to chemical corrosion. In *High Temperature Chemistry of Inorganic and Ceramic Materials*, F P Glasser and P E Potter (eds) (London: The Chemical Society), pp. 1–11.

Fite W L and Brackmann R T. 1958. Collisions of electrons with hydrogen atom. Ionization. *Phys. Rev.* **112**, 1141–51.

Fite W L and Brackmann R T. 1959. Ionization of atomic oxygen on electron impact. *Phys. Rev.* **113**, 815–6.

Fite W L and Datz S. 1963. Chemical research with molecular beams. *Ann. Rev. Phys. Chem.* **14**, 61–88.

Fite W L, Lo H H and Irving P. 1974. Associative ionization in U + O and U + O_2 collisions. *J. Chem. Phys.* **60**, 1236–50.

Fleischer R L, Price P B and Walker R M. 1963. Method of forming fine holes of near atomic dimensions. *Rev. Sci. Instrum.* **34**, 510–12.

Fleming J A. 1883. On a phenomenon of molecular radiation in incandescent lamps. *The Electrician* **11**, 65#.

Fleming J A. 1885a. On molecular shadows in incandescence lamps. *Proc. Phys. Soc.* **7**, 178–81.

Fleming J A. 1885b. On molecular shadows in incandescence lamps. *Phil. Mag.* **20**, 141–4 [Identical text to Fleming (1885a.)].

Fleming J A. 1890a. On electric discharge between electrodes at different temperatures in air and in high vacua. *Proc. Roy. Soc.* **47**, 118–26.

Fleming J A. 1890b. Problems in the physics of an electric lamp. *Notices of the Proceedings of the Meetings of the Royal Institution* **13**, 34–49.

Flory C A and Cutler L S. 1993. Integral equation solution of low-pressure transport of gases in capillary tubes. *J. Appl. Phys.* **73**, 1561–9.

Fluendy M A D and Lawley K P. 1973. *Chemical Applications of Molecular Beam Scattering* (London: Chapman and Hall).

Fluendy M A D, Horne D S, Lawley K P and Morris A W. 1970. Elastic scattering of alkali atoms from iodine atoms and molecules at thermal energies. *Mol. Phys.* **19**, 659–71.

Fluendy M A D, Martin R M, Muschlitz E E and Herschbach D R. 1967. Hydrogen atom scattering: Velocity dependence of total cross section for scattering from rare gases, hydrogen, and hydrocarbons. *J. Chem. Phys.* **46**, 2172–81.

Foord J S, Davies G J and Tsang W T. 1997. *Chemical Beam Epitaxy and Related Techniques* (Chichester, UK: Wiley).

Ford M J, Pejcev V, Smith D, Ross K J and Wilson M. 1990. The ejected-electron spectra of manganese and samarium vapour atoms arising from autoionizing and Auger transitions following electron impact excitation. *J. Phys. B: Atom. Molec. Phys.* **23**, 4247–62.

Fortagh J, Grossmann A, Hänsch T W and Zimmermann C. 1998. Fast loading of a magneto-optical trap from a pulsed thermal source. *J. Appl. Phys.* **84**, 6499–501.

Fraser R G J. 1931. *Molecular Rays* (Cambridge, UK: University Press).

Fraser R G J. 1937. *Molecular Beams* (London: Methuen).

Fraser R G J and Jewitt T N. 1937. The ionization potentials of the free radicals methyl and ethyl. *Proc. Roy. Soc. A* **160**, 563–74.

Freedman A and Stinespring C D. 1992. Halogenation of GaAs (100) and (111) surfaces using atomic beams. *J. Phys. Chem.* **96**, 2253–8.

Fricke G, Kopfermann H and Penselin S. 1959a. Messung der Hyperfeinstrukturaufspaltungen der beiden Yttrium-Grundzustände $^2D_{3/2}$ und $^2D_{5/2}$ mit der Atomstrahlresonanzmethode. *Z. Physik* **154**, 218–30.

Fricke G, Kopfermann H, Penselin S and Schlüpmann K. 1959b. Bestimmung der Hyperfeinstrukturaufspaltungen der Scandium-Grundzustände $^2D_{3/2}$ und $^2D_{5/2}$ und des Quadrupolmomentes des Sc^{45} - Kernes. *Z. Physik* **156**, 416–24.

Friedman L. 1955. Mass spectrum of lithium iodide. *J. Chem. Phys.* **23**, 477–82.

Frisch O R. 1959. Molecular beams. *Contemp. Phys.* **1**, 3–16.

Füstöss L. 1970. Calculation of transmission probability in molecular flow through straight tubes. *Vacuum* **20**, 279–83.

Gadamer E O, Muntz E P and Patterson G N. 1961. Application of an electron gun for the measurement of density and temperature in rarefied gas flows. UTIA Report No. 73, AFOSR 629.

Gallis M A and Torczynski J R. 2012. Direct simulation Monte Carlo-based expressions for the gas mass flow rate and pressure profile in a microscale tube. *Phys. Fluids* **24**, 012005-1–21.

Gardner W L. 1968. Method of making a vitreous off-axis light filter. US Patent 3 380 817.

Garelis E and Wainwright T E. 1973. Free molecule flow in a right circular cylinder. *Phys. Fluids* **16**, 476–81.

Garvey J F and Kuppermann A. 1984. An intense beam of metastable H_3 molecules. *Chem. Phys. Lett.* **107**, 491–5.

Garvey J F and Kuppermann A. 1986. Design and operation of a stable intense high-temperature arc-discharge source of hydrogen atoms and metastable trihydrogen molecules. *Rev. Sci. Instrum.* **57**, 1061–5.

Garvin H L, Green T M and Lipworth E. 1958a. Spins of some radioactive iodine isotopes. *Phys. Rev.* **111**, 534–9.

Garvin H L, Green T M, Lipworth E and Nierenberg W A. 1958b. Nuclear spin of astatine-211. *Phys. Rev. Lett.* **1**, 74–5.

Geddes J, Clough P N and Moore P L. 1974. Crossed molecular beam study of reactive scattering: $O + CS_2 \rightarrow SO + CS$. *J. Chem. Phys.* **61**, 2145–9.

Geddes J, McCullough R W, Higgins D P, Woolsey J M and Gilbody H B. 1994. Enhanced dissociation of molecular nitrogen in a microwave plasma with an applied magnetic field. *Plasma Sources Sci. Technol.* **3**, 58–60.

Geis M W, Efremow N N and Lincoln G A. 1986. Hot jet etching of GaAs and Si. *J. Vac. Sci. Technol. B* **4**, 315–7.

Geis M W, Efremow N N, Pang S W and Anderson A C. 1987. Hot-jet etching of Pb, GaAs, and Si. *J. Vac. Sci. Technol. B* **5**, 363–5.

Geis M W, Efremow N N and Pang S W. 1988. Dry etching patterning of electrical and optical materials. US Patent 4 734 152.

Gentry W R. 1988. Low-energy pulsed beam sources. In *Atomic and Molecular Beam Methods*, vol. 1, G Scoles (ed) (Oxford, UK: University Press), pp. 54–82.

Gerginov V and Tanner C E. 2003. Fluorescence of a highly collimated atomic cesium beam: Theory and experiment. *Opt. Comm.* **222**, 17–28.

Gericke W, Höricke M and von Kalben J. 1991. A detailed study of the molecular beam flux distribution of MBE effusion sources. *Vacuum* **42**, 1209–12.

Gerlach W. 1925. Über die Richtungsquantelung im Magnetfeld II Experimentelle Untersuchungen über das Verhalten normaler Atome unter magnetischer Kraftwirkung. *Ann. Physik* **76**, 163–97.

Gerlach W and Cilliers A C. 1924. Magnetische Atommomente. *Z. Physik* **26**, 106–9.

Gerlach W and Stern O. 1924. Über die Richtungsquantelung im Magnetfeld. *Ann. Physik* **74**, 673–99.

German O. 1963. The equations for the density of molecular flow. *Sov. Phys. Tech. Phys.* **7**, 834–7 [*Zh. Tekh. Fiz.* **32**, 1134–8 1962].

Ghadiri M R, Granja J R, Milligan R A, McRee D E and Khazanovich N. 1993. Self-assembling organic nanotubes based on a cyclic peptide architecture. *Nature* **366**, 324–7.

Giordmaine J A and Wang T C. 1960. Molecular beam formation by long parallel tubes. *J. Appl. Phys.* **31**, 463–71.

Glaze D J, Hellwig H, Allan D W, Jarvis S and Wainwright A E. 1974. Accuracy evaluation and stability of NBS primary frequency standards. *IEEE Trans.* **IM-23**, 489–501.

Glocker D A, Drumheller J P and Miller J R. 1982. Ion beam sputter deposition onto levitated and stalk mounted laser fusion targets. *J. Vac. Sci. Technol.* **20**, 1331–5.

Godfroid T, Dauchot J P and Hecq M. 2003. Atomic nitrogen source for reactive magnetron sputtering. *Surf. Coatings Tech.* **174–5**, 1276–81.

Gole J L and Zare R N. 1972. Determination of D_0^0(AlO) from crossed-beam chemiluminescence of Al + O_3. *J. Chem. Phys.* **57**, 5331–5.

Gole J L, Green G J, Pace S A and Preuss D R. 1982. The characterization of supersonic sodium vapor expansions including laser induced atomic fluorescence from trimeric sodium. *J. Chem. Phys.* **76**, 2247–66.

Gómez-Goñi J and Lobo P J. 2003. Comparison between Monte Carlo and analytical calculation of the conductance of cylindrical and conical tubes. *J. Vac. Sci. Technol. A* **21**, 1452–7.

González Ureña A, Verdasco Costales E and Sáez Rábanos V. 1990. Pulsed metastable atom source for low vapour-pressure metals. *Meas. Sci. Technol.* **1**, 250–4.

Gordon J P, Zeiger H J and Townes C H. 1955. The maser—New type of microwave amplifier, frequency standard and spectrometer. *Phys. Rev.* **99**, 1264–74.

Gorry P A and Grice R. 1979. Microwave discharge source for the production of supersonic atom and free radical beams. *J. Phys. E: Sci. Instrum.* **12**, 857–60.

Gossla M, Hahn T, Metzner H, Conrad J and Geyer U. 1995. Thin $CuInS_2$ films by three-source molecular beam deposition. *Thin Solid Films* **268**, 39–44.

Götte S, Gopalan A, Bömmels J, Ruf M W and Hotop H. 2000. A triply differentially pumped supersonic beam target for high-resolution collision studies. *Rev. Sci. Instrum.* **71**, 4070–7.

Götting R, Mayne H R and Toennies J P. 1986. Molecular beam scattering measurements of differential cross sections for $D+H_2(\nu = 0) \rightarrow HD+H$ at $E_{c.m.} = 1.5$ eV. *J. Chem. Phys.* **85**, 6396–419.

Gottwald B A. 1973. Über die Strahlformung bei der Molekularströmung I. Teil. *Vakuum-Technik* **22**, 106–15.

Govyadinov A, Emeliantchik I and Kurilin A. 1998. Anodic aluminum oxide microchannel plates. *Nucl. Instrum. Methods A* **419**, 667–75.

Grams M P, Crook A M, Turner J H and Doak R B. 2006. Microscopic fused silica capillary nozzles as supersonic molecular beam sources. *J. Phys. D: Appl. Phys.* **39**, 930–6.

Graper E B. 1973. Distribution and apparent source geometry of electron-beam-heated evaporation sources. *J. Vac. Sci. Technol.* **10**, 100–3.

Gray D C and Sawin H H. 1992. Design considerations for high-flux collisionally opaque molecular beams. *J. Vac. Sci. Technol. A* **10**, 3229–38.

Green G J and Gole J L. 1980. Single and multiple collision chemiluminescence studies of the Si-OCS and Ge-OCS reaction. A study of the SiS and GeS $a^3\Sigma^+$-$X^1\Sigma^+$ and SiS $b^3\Pi$-$X^1\Sigma^+$ intercombination systems and the nature of SiS* collisional quenching. *Chem. Phys.* **46**, 67–85.

Greenland P T, Lauder M A and Wort D J H. 1985. Atomic beam velocity distributions. *J. Phys. D: Appl. Phys.* **18**, 1223–32.

Griffin P F, Weatherill K J and Adams C S. 2005. Fast switching of alkali atom dispensers using laser-induced heating. *Rev. Sci. Instrum.* **76**, 093102-1–3.

Grimley R T, Muenow D W and Larue J L. 1972. On a mass spectrometric angular distribution study of the effusion of the potassium chloride vapor system from cylindrical orifices. *J. Chem. Phys.* **56**, 490–502.

Grosof G M and Hubbs J C. 1956. Low-frequency electrodeless discharge tube. *Rev. Sci. Instrum.* **27**, 171#.

Grover J R and Lilenfeld H V. 1972. Molecular beams of short lived radioactive nuclides. II. Preparation of beams by wall reactions. *Rev. Sci. Instrum.* **43**, 690–2.

Grover J R, Kiely F M, Lebowitz E and Baker E. 1971. Molecular beams of short lived radioactive nuclides. *Rev. Sci. Instrum.* **42**, 293–302.

Grunthaner F J, Bicknell-Tassius R, Deelman P, Grunthaner P J, Bryson C, Snyder E, Giuliani J L, Apruzese J P and Kepple P. 1998. Ultrahigh vacuum arcjet nitrogen source for selected energy epitaxy of group III nitrides by molecular beam epitaxy. *J. Vac. Sci. Technol. A* **16**, 1615–20.

Guevremont J M, Sheldon S and Zaera F. 2000. Design and characterization of collimated effusive gas beam sources: Effect of source dimensions and backing pressure on total flow and beam profile. *Rev. Sci. Instrum.* **71**, 3869–81.

Haberland H and Weber W. 1980. Optical interaction potentials from differential cross section measurements at thermal energies III: $He(2^1S)$ + Na. *J. Phys. B: Atom. Molec. Phys.* **13**, 4147–55.

Hackspill L. 1928. Sur quelques propriétés des métaux alcalins. *Helv. Chim. Acta* **11**, 1003–26.

Hahn T, Metzner H, Plikat B and Seibt M. 1998. Epitaxial growth of $CuInS_2$ on sulphur terminated Si(001). *Appl. Phys. Lett.* **72**, 2733–5.

Hanes G R. 1960. Multiple tube collimator for gas beams. *J. Appl. Phys.* **31**, 2171–5.

Hanner A W and Gole J L. 1980. Evidence for ultrafast V-E transfer in boron oxide (BO). *J. Chem. Phys.* **73**, 5026–39.

Harvey K C. 1982. Slow metastable atomic hydrogen beam by optical pumping. *J. Appl. Phys.* **53**, 3383–6.

Harvey K C and Fehrenbach C. 1983. Semiconductor detector for the selective detection of atomic hydrogen. *Rev. Sci. Instrum.* **54**, 1117–20.

Hau L V, Golovchenko J A and Burns M M. 1994. A new atomic beam source: The 'candlestick.' *Rev. Sci. Instrum.* **65**, 3746–50.

Hau L V. 2001. Taming light with cold atoms. *Physics World* **14**, 35–40.

Heddle D W O and Keesing R G W. 1968. Measurements of electron excitation functions. In *Advances in Atomic and Molecular Physics*, vol. 4, D R Bates and I Estermann (eds) (New York: Academic), pp. 267–98.

Hedley W H, Olt R G, DuFour H R and Wurstner A L. 1977. Fabrication and calibration of ultrafine capillaries having diameters between 0.5 and 60 μ. *Rev. Sci. Instrum.* **48**, 64–7.

Hedley W H, Olt R G, Holboke L E and Wurstner A L. 1978. Flow of gases and liquids through ultrafine capillaries having diameters between 1.5 and 60 micrometers. *J. Rheol.* **22**, 91–112.

Helmer J C. 1967a. Applications of an approximation to molecular flow in cylindrical tubes. *J. Vac. Sci. Technol.* **4**, 179–85.

Helmer J C. 1967b. Solutions of clausing's integral equation for molecular flow. *J. Vac. Sci. Technol.* **4**, 360–3.

Helmer J C, Jacobus F B and Sturrock P A. 1960. Focusing molecular beams of NH_3. *J. Appl. Phys.* **31**, 458–63.

Hendrie J M. 1954. Dissociation energy of N_2. *J. Chem. Phys.* **22**, 1503–7.

Henning H. 1978. The approximate calculation of transmission probabilities for the conductance of tubulations in the molecular flow regime. *Vacuum* **28**, 151–2.

Herman M A. 1982. Physical problems concerning effusion processes of semiconductors in molecular beam epitaxy. *Vacuum* **32**, 555–65.

Herman M A and Sitter H. 1989. *Molecular Beam Epitaxy: Fundamentals and Current Status* (London: Springer).

Herschbach D R. 1987. Molecular dynamics of elementary chemical reactions. *Chemica Scripta* **27**, 327–47.

Hershcovitch A, Kponou A and Niinikoski T O. 1987. Cold high-intensity atomic hydrogen beam source. *Rev. Sci. Instrum.* **58**, 547–56.

Hertel I V and Ross K J. 1968. An atomic beam oven with very low associated magnetic field. *J. Sci. Instrum. (J. Phys. E)* **1**, 1245–6.

Higgins D P, McCullough R W, Geddes J, Woolsey J M, Schlapp M and Gilbody H B. 1995a. New thermal energy sources of H, O, Cl and N atoms for material processing. *Key Eng. Mat.* **99–100**, 177–84.

Higgins D P, McCullough R W, Geddes J, Schlapp M, Woolsey J, Salzborn E and Gilbody H B. 1995b. A high efficiency nitrogen atom source for accelerator target studies. *Nucl. Instrum. Methods B* **103**, 508–10.

Hildebrandt B, Vanni H and Heydtmann H. 1984. Infrared chemiluminescence in the reactions of 0.45 eV hydrogen atoms with Cl_2, SCl_2 and PCl_3. *Chem. Phys.* **84**, 125–37.

Hils D, Kleinpoppen H and Koschmieder H. 1966. Remeasurement of the total cross section for excitation of the hydrogen $2\,^2S_{1/2}$ state by electron impact. *Proc. Phys. Soc.* **89**, 35–40.

Hirschfelder J O, Curtiss C F and Bird R B. 1964. *Molecular Theory of Gases and Liquids* (London: Chapman & Hall).

Hisadome K and Kihara M. 1992. An atomic beam collimator for Cs beam frequency standards. *Jpn. J. Appl. Phys.* **31**, 1232–5.

Hobson J P. 1969. Surface smoothness in thermal transpiration at very low pressures. *J. Vac. Sci. Technol.* **6**, 257–9.

Hobson J P, Hubbs J C, Nierenberg W A, Silsbee H B and Sunderland R J. 1956. Spins of rubidium isotopes of masses 81, 82, 83 and 84. *Phys. Rev.* **104**, 101–6.

Hodgson J A B and Haasz A A. 1991. Compact radio-frequency glow-discharge atomic hydrogen beam source. *Rev. Sci. Instrum.* **62**, 96–9.

Hoke W E, Lemonias P J and Weir D G. 1991. Evaluation of a new plasma source for molecular beam epitaxial growth of InN and GaN films. *J. Crystal Growth* **111**, 1024–8.

Holland D M P, Codling K and Chamberlain R N. 1981. Double ionisation in the region of the 5p thresholds of Ba and Yb. *J. Phys. B: Atom. Molec. Phys.* **14**, 839–50.

Holland L. 1956. *Vacuum Deposition of Thin Films* (London: Chapman & Hall).

Holland L. 1965. Thin film microelectronics. In *Evaporation Sources and Techniques* pp. 143–65 of Section 4 of Chapter 4, *Vacuum Deposition Apparatus and Techniques*, L Holland (ed) (London: Chapman & Hall).

Hollywood M T, Crowe A and Williams J F. 1979. Coherent excitation of the 2^1P state of He for large momentum transfer electron scattering. *J. Phys. B: Atom. Molec. Phys.* **12**, 819–34.

Hood S T, Dixon A J and Weigold E. 1978. A gas-discharge atomic hydrogen source for electron-scattering experiments. *J. Phys. E: Sci. Instrum.* **11**, 948–54.

Hopkins J B, Langridge-Smith P R R, Morse M D and Smalley R E. 1983. Supersonic metal cluster beams of refractory metals: Spectral investigations of ultracold Mo_2. *J. Chem. Phys.* **78**, 1627–37.

Horn A, Kammler T and Kappel M. 1999. Vorrichtung zur Erzeugung von Radikalen und/oder Reaktionsprodukten. German Patent DE 197 57 851 C 1.

Horton J R and Tasker G W. 1992. Microchannel electron multipliers. US Patent 5 086 248.

Horton L H and Young A W. 1970. The fabrication of fine slits for use in vacuum tubes. *J. Phys. E: Sci. Instrum.* **3**, 736–7.

Hostettler H U and Bernstein R B. 1960. Improved slotted disk type velocity selector for molecular beams. *Rev. Sci. Instrum.* **31**, 872–7.

Hottel H C and Keller J D. 1933. Effect of reradiation on heat transmission in furnaces and through openings. *Trans. A.S.M.E Iron and Steel* **IS-55-6**, 39–49.

Hubbs J C and Grosof G M. 1956. Spin of neon-21. *Phys. Rev.* **104**, 715–7.

Hubbs J C and Marrus R. 1958. Hyperfine structure measurements on neptunium-239. *Phys. Rev.* **109**, 287–9.

Hubbs J C, Marrus R, Nierenberg W A and Worcester J L. 1958. Hyperfine structure measurements on plutonium-239. *Phys. Rev.* **109**, 390–8.

Hubbs J C, Marrus R and Winocur J O. 1959. Zeeman investigations of curium-242. *Phys. Rev.* **114**, 586–9.

Huber B A, Bumbel A and Wiesemann K. 1983. A high-density effusive target of atomic hydrogen. *J. Phys. E: Sci. Instrum.* **16**, 145–50.

Hucks P, Flaskamp K and Vietzke E. 1980. The trapping of thermal atomic hydrogen on pyrolytic graphite. *J. Nucl. Materials* **93–94**, 558–63.

Huczko A. 2000. Template-based synthesis of nanomaterials. *Appl. Phys. A* **70**, 365–76.

Hudson J J, Sauer B E, Tarbutt M R and Hinds E A. 2002. Measurement of the electron electric dipole moment using YbF molecules. *Phys. Rev. Lett.* **89**, 023003-1–4.

Huels M A, Hahndorf I, Illenberger E and Sanche L. 1998. Resonant dissociation of DNA bases by subionization electrons. *J. Chem. Phys.* **108**, 1309–12.

Huet D, Lambert M, Bonnevie D and Dufresne D. 1985. Molecular beam epitaxy of $In_{0.53} Ga_{0.47}$ As and InP on InP by using cracker cells and gas cells. *J. Vac. Sci. Technol. B* **3**, 823–9.

Huggill J A W. 1952. The flow of gases through capillaries. *Proc. Roy. Soc. A* **212**, 123–36.

Hulteen J C and Martin C R. 1997. A general template-based method for the preparation of nanomaterials. *J. Mater. Chem.* **7**, 1075–87.

Hurlbut F C. 1959. Molecular scattering at the solid surface. In *Recent Research in Molecular Beams*, I Estermann (ed) (New York: Academic), pp. 145–56.

Iczkowski R P, Margrave J L and Robinson S M. 1963. Effusion of gases through conical orifices. *J. Phys. Chem.* **67**, 229–33.

Ikegami T. 1994. Angular distribution measurement of cesium atomic beam from long tube collimators. *Jpn. J. Appl. Phys.* **33**, 4795–6.

Imai F, Kunimori K and Nozoye H. 1995. Performance characteristics of an oxygen radical beam radio-frequency source. *J. Vac. Sci. Technol. A* **13**, 2508–12.

Ishikawa T, Paradis P F, Itami T and Yoda S. 2005. Non-contact thermophysical property measurements of refractory metals using an electrostatic levitator. *Meas. Sci. Technol.* **16**, 443–51.

Itaya K, Sugawara S, Arai K and Saito S. 1984. Properties of porous anodic aluminium oxide films as membranes. *J. Chem. Eng. Jpn.* **17**, 514–20.

Ivanov B S and Troitskii V S. 1963. Formation of directivity patterns of molecular beams. *Sov. Phys. Tech. Phys.* **8** 365–8 [*Zh. Tekh. Fiz.* **33**, 494–9 1963].

Jackson L C and Broadway L F. 1930. An application of the Stern-Gerlach experiment to the study of active nitrogen. *Proc. Roy. Soc. A* **127**, 678–89.

Jackson S C, Baron B N, Rocheleau R E and Russell T W F. 1985. Molecular beam distributions from high rate sources. *J. Vac. Sci. Technol. A* **3**, 1916–20.

Jacob W, Hopf C, von Keudell A, Meier M and Schwarz-Selinger T. 2003. Particle-beam experiment to study heterogeneous surface reactions relevant to plasma-assisted thin film growth and etching. *Rev. Sci. Instrum.* **74**, 5123–36.

Jaensch R and Kamke W. 2000. Vapor pressure of C_{60} revisited. *Mol. Materials* **13**, 163–72.

James G K, Forrest L F, Ross K J and Wilson M. 1985. The ejected-electron spectrum of Al I autoionising transitions resulting from 20-500 eV electron impact excitation. *J. Phys. B: Atom. Molec. Phys.* **18**, 775–90.

Jane F W. 1970. *The Structure of Wood*, 2nd edn (London: Black).

Jeffrey A. 2000. *Handbook of Mathematical Formulas and Integrals* (London: Academic) p. 241.

Johnson J C, Stair A T, and Pritchard J L. 1966. Molecular beams formed by arrays of 3–40 μ diameter tubes. *J. Appl. Phys.* **37**, 1551–8

Johnson T H. 1928. The reflection of hydrogen atoms from crystals. *J. Franklin Inst.* **206**, 301–15.

Jonah C D, Zare R N and Ottinger C. 1972. Crossed-beam chemiluminescence studies of some Group IIa metal oxides. *J. Chem. Phys.* **56**, 263–74.

Jones R H, Olander D R and Kruger V R. 1969. Molecular-beam sources fabricated from multichannel arrays. I. Angular distributions and peaking factors. *J. Appl. Phys.* **40**, 4641–9.

Jones R H, Olander D R, Siekhaus W J and Schwarz J A. 1972. Investigation of gas-solid reactions by modulated beam mass spectrometry. *J. Vac. Sci. Technol.* **9**, 1429–41.

Joyce B A. 1985. Molecular beam epitaxy. *Rep. Prog. Phys.* **48**, 1637–97.

Kalos F and Grosser A E. 1969. Free radical beam source. *Rev. Sci. Instrum.* **40**, 804–6.

Kämmerling B, Läuger J and Schmidt V. 1994. Angular distribution of photon-induced $M_{2,3}$-N_1N_1 Auger electrons in atomic calcium. *J. Electr. Spectrosc.* **67**, 363–71.

Kanik I, Johnson P V, Das M B, Khakoo M A and Tayal S S. 2001. Electron-impact studies of atomic oxygen: I. Differential and integral cross sections; experiment and theory. *J. Phys. B: At. Mol. Opt. Phys.* **34**, 2647–65.

Kapitanskii V R, Kostakov A I, Livshits A I, Notkin M E and Metter I M. 1971. Collimated-beam hydrogen maser. *Sov. Phys. Tech. Phys.* **16**, 272–5 [*Zh. Tekh. Fiz.* **41**, 362–7 1971].

Kappes M and Leutwyler S. 1988. Molecular beams of clusters. In *Atomic and Molecular Beam Methods*, vol. 1, G Scoles (ed) (Oxford, UK: University Press), pp. 380–415.

Kasabov S G. 1986. A sodium-atom beam source for UHV surface investigations and industrial applications. *J. Phys. E: Sci. Instrum.* **19**, 369–72.

Kasper E and Bean J C (eds). 1988. *Silicon Molecular Beam Epitaxy*, vol. 1 (Boca Raton: CRC).

Katayama H, Norimatsu T, Nakai S and Yamanaka C. 1990. Hydrocarbon coating by laser-induced chemical vapor deposition onto microsphere target levitated by a viscous gas jet. *J. Vac. Sci. Technol. A* **8**, 855–60.

Kaufman F and Kelso J R. 1960. Catalytic effects in the dissociation of oxygen in microwave discharges. *J. Chem. Phys.* **32**, 301–2.

Keeling L A, Chen L and Greenlief C M. 1998. Surface reactions of monoethylgermane on Si (100)-(2 x 1). *Surface Science* **400**, 1–10.

Keil M, Young J H and Copeland K. 1997. Method and apparatus for etching surfaces with atomic fluorine. US Patent 5 597 495.

Kellogg J M B, Rabi I I and Zacharias J R. 1936. The gyromagnetic properties of the hydrogens. *Phys. Rev.* **50**, 472–81.

Kellogg J B M and Millman S. 1946. The molecular beam magnetic resonance method. The radiofrequency spectra of atoms and molecules. *Rev. Mod. Phys.* **18**, 323–52.

Kelly A and Davies G J. 1965. The principles of the fibre reinforcement of metals. *Metallurgical Rev.* **10**, 1–77.

Kennard E H. 1938. *Kinetic Theory of Gases*, Ch 8 (New York: McGraw Hill).

Kern W and Puotinen D A. 1970. Cleaning solutions based on hydrogen peroxide for use in silicon semi-conductor technology. *RCA Rev.* **31**, 187–206.

Kersevan R and Pons J L. 2009. Introduction to MOLFLOW+: New graphical processing unit-based Monte Carlo code for simulating molecular flows and for calculating angular coefficients in the compute unified device architecture environment. *J. Vac. Sci. Technol. A* **27**, 1017–23.

Khryashchev L Y, Domelunksen V G, Kotlikov E N and Nikolaev A Y. 1988. Application of the optical pumping effect for measuring the intensity of an atomic beam. *Sov. Phys. Tech. Phys.* **33**, 813–7 [*Zh. Tekh. Fiz.* **58**, 1368–74 1988].

Kikuchi J, Fujimura S, Suzuki M and Yano H. 1993. Effects of H_2O on atomic hydrogen generation in hydrogen plasma. *Jpn. J. Appl. Phys.* **32**, 3120–4.

Kim E, Xia Y and Whitesides M. 1995. Polymer microstructures formed by moulding in capillaries. *Nature* **376**, 581–4.

King J G and Zacharias J R. 1956. Some new applications and techniques of molecular beams. *Advan. Electron. Electron. Phys.* **8**, 1–88.

Klein W. 1971. A molecular beam cesium source for photoemission experiments. *Rev. Sci. Instrum.* **42**, 1082–3.

Kleinpoppen H. 1961. Vermessung der Lamb-Verschiebung des $3^2S_{1/2}$-Zustandes gegenüber dem $3^2P_{1/2}$-Zustand am Wasserstoffatom. *Z. Physik* **164**, 174–89.

Kleinpoppen H, Krüger H and Ulmer R. 1962. Excitation and polarization of Balmer—α radiation in electron-hydrogen atom collisions. *Phys. Lett.* **2**, 78–9.

Knauer F and Stern O. 1926. Zur Methode der Molekularstrahlen II. *Z. Physik* **39**, 764–79.

Knight J C, Birks T A, Russell P St. J and Atkin D M. 1996, 1997. All-silica single-mode optical fiber with photonic crystal cladding. *Opt. Lett.* **21**, 1547–9; *errata* **22**, 484–5.

Knudsen M. 1909a. Die Gesetze der Molekularströmung und der inneren Reibungsströmung der Gase durch Röhren. *Ann. Physik* **28**, 75–130.

Knudsen M. 1909b. Die Molekularströmung der Gase durch Öffnungen und die Effusion. *Ann. Physik* **28**, 999–1016.

Knuth E L. 1964. Supersonic molecular beams. *Appl. Mechanics Rev.* **17**, 751–62.

Kobrin P H, Becker U, Southworth S, Truesdale C M, Lindle D W and Shirley D A. 1982. Autoionizing resonance profiles in the photoelectron spectra of atomic cadmium. *Phys. Rev. A* **26**, 842–56.

Koch N and Steffens E. 1999. High intensity source for cold atomic hydrogen and deuterium beams. *Rev. Sci. Instrum.* **70**, 1631–9.

Kolb C E and Kaufman M. 1972. Molecular beam analysis investigation of the reaction between atomic fluorine and carbon tetrachloride. *J. Phys. Chem.* **76**, 947–53.

Kolodney E, Tsipinyuk B and Budrevich A. 1995. The thermal energy dependence (10–20 eV) of electron impact induced fragmentation of C_{60} in molecular beams: Experiment and model calculations. *J. Chem. Phys.* **102**, 9263–75.

Konijn J et al [The RD 46 collaboration] 1998. Capillary detectors. *Nucl. Instrum. Methods A* **418**, 186–95.

Koschmieder H and Raible V. 1975. Intense atomic-hydrogen beam source. *Rev. Sci. Instrum.* **46**, 536–7.

Koschmieder H, Raible V and Kleinpoppen H. 1973. Resonance structure in the excitation cross section by electron impact of the 2s state in atomic hydrogen. *Phys. Rev. A* **8**, 1365–8.

Kotlikov E N and Khryashchev L Y. 1986. Measurement of atomic-beam absolute intensity in terms of resonance fluorescence observation. *Opt. Spectrosc.* **60**, 114–5 [*Opt. Spektrosk.* **60**, 184–6].

Kowalewska D, Bekk K, Göring S, Hanser A, Kälber W, Meisel G and Rebel H. 1991. Isotope shifts and hyperfine structure in polonium isotopes by atomic-beam laser spectroscopy. *Phys. Rev. A* **44**, R1442–5.

Krakowski R A and Olander D R. 1968. Dissociation of hydrogen on tantalum using a modulated molecular beam technique. *J. Chem. Phys.* **49**, 5027–41.

Kratzenstein M. 1935. Untersuchungen über die 'Wolke' bei Molekularstrahlversuchen. *Z. Physik* **93**, 279–91.

Krause M O. 1980. Photoionization of atomic silver between 17 and 41 eV. *J. Chem. Phys.* **72**, 6474–6.

Krause M O and Caldwell C D. 1987. Strong correlation and alignment near the Be 1s photoionization threshold. *Phys. Rev. Lett.* **59**, 2736–9.

Krause M O, Svensson W A, Carlson T A, Leroi G, Ederer D E, Holland D M P and Parr A C. 1985. Photoeffect in the 4d subshell of atomic silver between 14 and 140 eV. *J. Phys. B: Atom. Molec. Phys.* **18**, 4069–75.

Krikorian O H. 1976. Method for producing uranium atomic beam source. US patent 3 963 921.

Kubo A, Kitajima M, Yata M and Fukutani H. 2000. An intense pulsed atomic hydrogen beam source. *Jpn. J. Appl. Phys.* **39**, 6101–4.

Kudryavtsev N N, Mazyar O A and Sukhov A M. 1993. Apparatus and techniques for the investigation of methods of generating molecular beams. *Physics Uspekhi* **36**, 513–28 [*Usp. Fiz. Nauk* **163**, 75–93 1993].

Kuhl D E and Tobin R G. 1995. On the design of capillary and effusive gas dosers for surface science. *Rev. Sci. Instrum.* **66**, 3016–20.

Kuhn H. 1946. Atomic and molecular beams, some recent developments. *J. Sci. Instrum.* **23**, 249–56.

Kuijt G and Teunissen B. 1988. Fibre production for high-reliability systems. *Telecommunications* June 77–84, 97#.

Kurepa M V and Lucas C B. 1981. The density gradient of molecules flowing along a tube. *J. Appl. Phys.* **52**, 664–9.

Kurt O E and Phipps T E. 1929. The magnetic moment of the oxygen atom. *Phys. Rev.* **34**, 1357–66.

Kusch P and Hughes V W. 1959. Atomic and molecular beam spectroscopy. In *Handbuch der Physik*, vol. 37/1, *Atome III—Moleküle I*, S Flügge (ed) (Berlin, Germany: Springer), pp. 1–172.

Kuyatt C E. 1968. Measurement of electron scattering from a static gas target. In *Methods of Experimental Physics*, vol. 7A, *Atomic and Electron Physics—Atomic Interactions*, B Bederson and W L Fite (eds) (New York: Academic), pp. 1–43.

Lahmam-Bennani A and Duguet A. 1979. Jet Moléculaire formé par une buse à multicanaux: II. Etude complémentaire. *Rev. Phys. Appl.* **14**, 525–31.

Lamb W E and Retherford R C. 1950. Fine structure of the hydrogen atom. Part I. *Phys. Rev.* **79**, 549–72.

Lamb W E and Retherford R C. 1951. Fine structure of the hydrogen atom. Part II. *Phys. Rev.* **81**, 222–32.

Lambropoulos M and Moody S E. 1977. Design of a three-stage alkali beam source. *Rev. Sci. Instrum.* **48**, 131–4.

Lampton M. 1981. The microchannel image intensifier. *Scientific American* **245**, 46–55.

Langmuir I. 1912. The dissociation of hydrogen into atoms. *J. Am. Chem. Soc.* **34**, 860–77.

Laprade B N and Reinhart S T. 1989. Recent advances in small pore microchannel plate technology. *Soc. Photo Instrum. Eng. (SPIE)* **1072**, 119–29.

Larson H P and Stanley R W. 1967. Analysis of the He II 4686-Å (n = 4 to n = 3) line complex excited in an atomic-beam light source. *J. Opt. Soc. Amer.* **57**, 1439–49.

Lawson A W and Fano R. 1947. Note on the efficiency of radiation shields. *Rev. Sci. Instrum.* **18**, 727–9.

Leavitt P K and Thiel P A. 1990. A warning concerning the use of glass capillary arrays in gas dosing: Potential chemical reactions. *J. Vac. Sci. Technol. A* **8**, 148–9.

Lee Y T, McDonald J D, LeBreton P R and Herschbach D R. 1969. Molecular beam reactive scattering apparatus with electron bombardment detector. *Rev. Sci. Instrum.* **40**, 1402–8.

Lemonick A, Pipkin F M and Hamilton D R. 1955. Focussing atomic beam apparatus. *Rev. Sci. Instrum.* **26**, 1112–9.

Leonas V B. 1964. The present state and some new results of the molecular-beam method. *Sov. Phys. Usp.* **7**, 121–44 [*Usp. Fiz. Nauk* **82** 287–323 1964].

Leonas V B. 1966. The study of elementary chemical reactions by the molecular beam method. *Russian Chem. Phys.* **35**, 879–91.

Leonas V B. 1979. New methods in atomic beam studies. *Sov. Phys. Usp.* **22**, 109–15 [*Usp. Fiz. Nauk* **127**, 318–30 1979].

Leu A. 1927. Versuche über die Ablenkung von Molekularstrahlen im Magnetfeld. *Z. Physik* **41**, 551–62.

Leu A. 1928. Untersuchungen an Wismut nach der magnetischen Molekularstrahlmethode. *Z. Physik* **49**, 498–506.

Levdansky V V, Smolik J and Moravec P. 2008. Effect of surface diffusion on transfer processes in heterogeneous systems. *Int. J. Heat Mass Transfer* **51**, 2471–81.

Levinson J A, Shaqfeh E S G, Balooch M and Hamza A V. 2000. Ion-assisted etching and profile development of silicon in molecular and atomic chlorine. *J. Vac. Sci. Technol. B* **18**, 172–90.

Lew H. 1949. The hyperfine structure of the $^2P_{3/2}$ state of Al^{27}. The nuclear electric quadrupole moment. *Phys. Rev.* **76**, 1086–92.

Lew H. 1953. The hyperfine structure and nuclear moments of Pr^{141}. *Phys. Rev.* **91**, 619–30.

Lew H, Morris D, Geiger F E and Eisinger J T. 1958. Rotational spectra of RbF by the electric resonance method. *Can. J. Phys.* **36**, 171–83.

Lew H. 1967. Sources of atomic particles—Atoms. In *Methods of Experimental Physics*, vol. 4A, *Atomic Sources and Detectors*, 1st edn, V W Hughes and H L Schultz (eds) (New York: Academic), pp. 155–98.

Lew H and Title R S. 1960. Note on the hyperfine structure of the $2s^22p^2P_{1/2}$ state of boron 10 and 11. *Can. J. Phys.* **38**, 868–71.

Lewis B. 1964. The deposition of alumina, silica and magnesia films by electron bombardment evaporation. *Microelectron. Reliab.* **3**, 109–20.

Lewis L C. 1931. Die Bestimmung des Gleichgewichts zwischen den Atomen und den Molekülen eines Alkalidampfes mit einer Molekularstrahlmethode. *Z. Physik* **69**, 786–809.

Li Y, Chen X, Wang L, Guo L and Li Y. 2013. Molecular flow transmission probabilities of any regular polygon tubes. *Vacuum* **92**, 81–4.

Libuda J, Meusel I, Hartmann J and Freund H J. 2000. A molecular beam/surface spectroscopy apparatus for the study of reactions on complex model catalysts. *Rev. Sci. Instrum.* **71**, 4395–408.

Lieber A, Benjamin R, Lyons P and Webb C. 1975a. Micro-channel plate as a parallel-bore collimator for soft X-ray imaging. *Nucl. Instrum. Methods* **125**, 553–6.

Lieber A J, Benjamin R F, Sutphin H D and Webb C B. 1975b. Investigation of micro-channel plates as parallel-bore electron collimators for use in a proximity-focussed ultra-fast streak tube. *Nucl. Instrum. Methods* **127**, 87–92.

Lindgren I. 1958. Atomic beam resonance apparatus with six-pole magnets for radioactive isotopes. *Nucl. Instrum.* **3**, 1–16.

Lindgren I and Johansson C M. 1959. Nuclear magnetic dipole and electric quadrupole moments of radioactive bismuth isotopes. *Arkiv för Fysik* **15**, 445–62.

Lindsay D M and Gole J L. 1977. $Al+O_3$ chemiluminescence: Perturbations and vibrational population anomalies in the B $^2\Sigma^+$ state of AlO. *J. Chem. Phys.* **66**, 3886–98.

Lipworth E. 1967. Sources of atomic particles—A review of source techniques used in radioactive beam experiments. In *Methods of Experimental Physics*, vol. 4A, *Atomic Sources and Detectors*, 1st edn, V W Hughes and H L Schultz (eds) (New York: Academic), pp. 198–256.

Lisitano G, Ellis R A, Hooke W M and Stix T H. 1968. Production of quiescent discharge with high electron temperatures. *Rev. Sci. Instrum.* **39**, 295–7.

Liu R H, Vasile M J and Beebe D J. 1999. The fabrication of nonplanar spin-on glass microstructures. *J. Microelectromechanical Systems* **8**, 146–51.

Livshits A I, Metter I M and Rikenglaz L É. 1971. Angular distribution of a molecular beam from a cylindrical channel with first-order reactions in the channel walls. *Sov. Phys. Tech. Phys.* **16**, 276–82 [*Zh. Tekh. Fiz.* **41**, 368–75 1971].

Lobo P J, Becheri F and Gómez-Goñi J. 2004. Comparison between Monte Carlo and analytical calculation of Clausing functions of cylindrical and conical tubes. *Vacuum* **76**, 83–8.

Lobo R F M and Silva N T. 2001. Neutral C_{60} effusive source for atomic collisions with fullerene. *Rev. Sci. Instrum.* **72**, 3505–6.

Logan R M and Stickney R E. 1966. Simple classical model for the scattering of gas atoms from a solid surface. *J. Chem. Phys.* **44**, 195–201.

Long R L, Cox D M and Smith S J. 1968. Electron impact excitation of hydrogen Lyman-α radiation. *J. Res. N B S* **72A**, 521–35.

Lowe A T and Hosford C D. 1979. Magnetron sputter coating of microspherical substrates. *J. Vac. Sci. Technol.* **16**, 197–9.

Lucas C B. 1973a. The production of intense atomic beams. *Vacuum* **23**, 395–402.

Lucas C B. 1973b. Measurements of the atomic beam formed by a focussing capillary array. *J. Phys. E: Sci. Instrum.* **6**, 991–4.

Lucas C B, Wells C A and Wilson K. 1980. Unpublished.

Ludwig R and Micklitz H. 1983. Construction and calibration of a four-beam source for the preparation of rare-gas diluted metal-hydrogen films. *Rev. Sci. Instrum.* **54**, 1009–11.

Lund L M and Berman A S. 1966a. Flow and self-diffusion of gases in capillaries. Part I. *J. Appl. Phys.* **37**, 2489–95.

Lund L M and Berman A S. 1966b. Flow and self-diffusion of gases in capillaries. Part II. *J. Appl. Phys.* **37**, 2496–508.

Ma Y, Liu B Y H, Lee H S, Mauersberger K and Morton J. 1996. Focussing glass capillary array molecular beam inlet for a high sensitivity mass spectrometer system. *J. Vac. Sci. Technol. A* **14**, 2414–7.

Maire J C. 1961. La méthode des jets moléculaires. *Bull. Soc. Chim. France* **7**, 2476–90.

Mais W H. 1934. The scattering of a beam of potassium atoms in various gases. *Phys. Rev.* **45**, 773–80.

Majumder A, Jana B, Kathar P T, Das A K and Mago V K. 2009. Generation of a long wedge-shaped barium atomic beam and its density characterization. *Vacuum* **83**, 989–95.

Makarov G N. 2003. Studies on high-intensity pulsed molecular beams and flows interacting with a solid surface. *Physics Uspekhi* **46** 889–914 [*Usp. Fiz. Nauk* **173**, 913–40 2003].

Malhotra C P, Mahajan R L and Sampath W S. 2007. High Knudsen number physical vapor deposition: Predicting deposition rates and uniformity. *Trans. ASME J. Heat Transfer* **129**, 1546–53.

Malutzki R, Wachter A, Schmidt V and Hansen J E. 1987. Photoinduced 2p Auger spectra of atomic aluminium. *J. Phys. B: Atom. Molec. Phys.* **20**, 5411–22.

Marcus P M and McFee J H. 1959. Velocity distributions in potassium molecular beams. In *Recent Research in Molecular Beams*, I Estermann (ed) (New York: Academic), pp. 43–63.

Marinković B, Pejčev V, Filopović D and Vušković L. 1991. Elastic and inelastic electron scattering by cadmium. *J. Phys. B: At. Mol. Opt. Phys.* **24**, 1817–37.

Marino L. 2009. Experiments on rarefied gas flows through tubes. *Microfluid Nanofluid* **6**, 109–19.

Markus J. 1957. Etching klystron grids. *Electronics* **30**, April 248–52.

Martin A G, Dutta S B, Rogers W F and Clark D L. 1987. Laser spectroscopy of radioactive atoms using the photon-burst technique. *J. Opt. Soc. Amer.* **B4**, 405–12.

Martin L R and Kinsey J L. 1967. Crossed molecular beam reactions of tritium bromide. *J. Chem. Phys.* **46**, 4834–8.

Massey H S W. 1971. Electronic and ionic impact phenomena. In *Slow Collisions of Heavy Particles*, vol. III, H S W Massey, E H S Burhop and H B Gilbody (eds) (Oxford, UK: Clarendon), pp. 1342–4.

Massey H S W and Burhop E H S. 1952. *Electronic and Ionic Impact Phenomena* (Oxford, UK: Clarendon), pp. 388–9.

Masuda H, Hasegwa F and Ono S. 1997. Self-ordering of cell arrangement of anodic porous alumina formed in sulfuric acid solution. *J. Electrochem. Soc.* **144**, L127–30.

Mateječík Š, Foltin V, Stano M and Skalný J D. 2003. Temperature dependencies in dissociative electron attachment to CCl_4, CCl_2F_2, $CHCl_3$ and $CHBr_3$. *Int. J. Mass Spectrometry* **223–224**, 9–19.

Matijasevic V, Garwin E L and Hammond R H. 1990. Atomic oxygen detection by a silver-coated quartz deposition monitor. *Rev. Sci. Instrum.* **61**, 1747–9.

Mayer H. 1929. Über die Gültigkeitgrenzen des Kosinusgesetzes der Molekularstrahlung. *Z. Physik* **58**, 373–85.

McCallion P, Shah M B and Gilbody H B. 1992. Multiple ionization of magnesium by electron impact. *J. Phys. B: At. Mol. Opt. Phys.* **25**, 1051–60.

McCarroll B. 1970. An improved microwave discharge cavity for 2450 MHz. *Rev. Sci. Instrum.* **41**, 279–80.

McCartney P C E, Shah M B, Geddes J and Gilbody H B. 1998. Multiple ionization of lead by electron impact. *J. Phys. B: At. Mol. Opt. Phys.* **31**, 4821–31.

McClelland J J, Kelley M H and Celotta R J. 1989. Superelastic scattering of spin-polarized electrons from sodium. *Phys. Rev. A* **40**, 2321–9.

McCullough R W. 1997. *Characterisation and Applications of a New Reactive Atom Beam Source CP392 Applications of Accelerators in Research and Industry*, J L Duggett and I I Morgan (eds) (New York: AIP).

McCullough R W, Geddes J, Donnelly A, Liehr M, Hughes M P and Gilbody H B. 1993a. A new microwave discharge source for reactive atom beams. *Meas. Sci. Technol.* **4**, 79–82.

McCullough R W, Geddes J, Donnelly A, Liehr M and Gilbody H B. 1993b. A new reactive atom beam source for accelerator target studies. *Nucl. Instrum. Methods B* **79**, 708–10.

McCullough R W, Geddes J, Croucher J A, Woolsey J M, Higgins D P, Schlapp M and Gilbody H B. 1996. Atomic nitrogen production in a high efficiency microwave plasma source. *J. Vac. Sci. Technol. A* **14**, 152–5.

McDaniel E W. 1989. *Atomic Collisions: Electron and Photon Projectiles* (New York: Wiley).

McDowall P D, Grünzweig T, Hilliard A and Anderson M F. 2012. An atomic beam source for fast loading of a magneto-optical trap under high vacuum. *Rev. Sci. Instrum.* **83**, 055102-1–4.

McFadden D L, McCullough E A, Kalos F, Gentry W R and Ross J. 1972. Molecular beam study of polyatomic free radical reactions. $CH_3 + Cl_2$ and Br_2. *J. Chem. Phys.* **57**, 1351–2.

McFee J H, Miller B I and Bachmann K J. 1977. Molecular beam epitaxial growth of InP. *J. Electrochem. Soc.* **124**, 259–72.

McGowan R W, Giltner D M and Lee S A. 1995. Light force cooling, focusing, and nanometer-scale deposition of aluminium atoms. *Opt. Lett.* **20**, 2535–7.

McRaven C P, Sivakumar P and Shafer-Ray N E. 2007. Multiphoton ionization of lead monofluoride resonantly enhanced by the $X_1\ ^2\Pi_{1/2} \to B\ ^2\Sigma_{1/2}$ transition. *Phys. Rev. A* **75**, 024502-1–3.

Meikle S and Hatanaka Y. 1989. Measurements of the atomic nitrogen population produced by a microwave electron cyclotron resonance plasma. *Appl. Phys. Lett.* **54**, 1648–9.

Meshcheryakov N A, Perfil'ev B V, Gabe D R and Trubetskoi A I. 1967. Multichannel molecular-beam shaper. *Instrum. Exper. Technol.* (4) 911–2 [*Prib. Tekh. Éksp.* (4) 212–4 1967].

Michalak L. 1993. Calculation of the beam flux distribution from an effusion hole in the ion source of the mass spectrometer. *Int. J. Mass Spectrometry* **123**, 107–15.

Miles A. 1978. *Photomicrographs of World Woods* (London: HMSO).

Milisavljević S, Šević D, Pejčev V, Filopović D M and Marinković B P. 2004. Differential and integrated cross sections for the electron excitation of the 4 $^1P^0$ state of calcium atom. *J. Phys. B: At. Mol. Opt. Phys.* **37**, 3571–81.

Millar I C P, Washington D and Lamport D L. 1972. Channel electron multiplier plates in X-ray image intensification. In *Photo-Electronic Image Devices—5th Symposium*, J D McGee, D McMullan and E Kahan (eds) (London: Academic), pp. 153–63.

Miller D R. 1988. Free jet sources. In *Atomic and Molecular Beam Methods*, vol. 1, G Scoles (ed) (Oxford, UK: University Press), pp. 14–53.

Miller R C and Kusch P. 1955. Velocity distributions in potassium and thallium atomic beams. *Phys. Rev.* **99**, 1314–21.

Miller R C and Kusch P. 1956. Molecular composition of alkali halide vapors. *J. Chem. Phys.* **25**, 860–76.

Miller T M. 1974. Atomic beam velocity distributions with a cooled discharge source. *J. Appl. Phys.* **45**, 1713–20.

Minten A and Osberghaus O. 1958. Die Erzeugung von Intensiven Molekularstrahlen nicht kondensierender Gase. Wirkungsquerschnitte für den Stoß H_2–H_2 und H_2–N_2. *Z. Physik* **150**, 74–9.

Missert N, Hammond R, Mooij J E, Matijasevic V, Rosenthal P, Geballe T H, Kapitulnik A, Beasley M R, Laderman S S, Lu C, Garwin E and Barton R. 1989. *In situ* growth of superconducting YBaCuO using reactive electron-beam coevaporation. *IEEE Trans. Magnetics* **25**, 2418–21.

Mitchell A M J, Roots K G and Phillips G. 1960. Ammonia maser oscillator. *Electron. Technol.* **37**, 136–43.

Miyake Y, Shimomura K, Mills A P, Marangos J P and Nagamine K. 1999. Thermionic emission of hydrogen isotopes (Mu, H, D and T) from W and interpretation of a role of the hot W as an atomic hydrogen source. *Surface Science* **433–435**, 785–9.

Mohan A, Tompson R V and Loyalka S K. 2007. Efficient numerical solution of the Clausing problem. *J. Vac. Sci. Technol. A* **25**, 758–62.

Moisan M and Zakrzewski Z. 1991. Plasma sources based on the propagation of electromagnetic surface waves. *J. Phys. D: Appl. Phys.* **24**, 1025–48.

Moisan M, Margot J and Zakrzewski Z. 1995. Surface wave plasma sources. In *High Density Plasma Sources—Design, Physics and Performance*, O A Popov (ed) (Park Ridge, NJ: Noyes).

Moore B C. 1972. Gas flux patterns in cylindrical vacuum systems. *J. Vac. Sci. Technol.* **9**, 1090–9.

Mueller C R. 1970. Molecular-beam spectroscopy. In *Physical Chemistry An Advanced Treatise*, Vol. IV, D Henderson (ed) (New York: Academic), pp. 709–40.

Mungall A G, Daams H and Boulanger J S. 1981. Design, construction and performance of the NRC CsVI primary cesium clocks. *Metrologia* **17**, 123–45.

Murphy E J and Brophy J H. 1979. Atomic hydrogen beam source: A convenient, extended cavity, microwave discharge design. *Rev. Sci. Instrum.* **50**, 635–6.

Murphy D M. 1989. Wall collisions, angular flux, and pumping requirements in molecular flow through tubes and microchannel arrays. *J. Vac. Sci. Technol. A* **7**, 3075–91.

Murray A J. 2005. (e, 2e) ionization studies of alkaline-earth-metal and alkali-earth-metal targets: Na, Mg, K, and Ca, from near threshold to beyond intermediate energies. *Phys. Rev. A* **72**, 062711-1–14.

Murray A J and Cvejanovic D. 2003. Coplanar symmetric (e, 2e) measurements from calcium at low energy. *J. Phys. B: At. Mol. Opt. Phys.* **36**, 4875–88.

Murray A J, Hussey M J and Needham M. 2006. Design and characterization of an atomic beam source with narrow angular divergence for alkali-earth targets. *Meas. Sci. Technol.* **17**, 3094–101.

Myers T H and Schetzina J F. 1982. Molecular beam source for high vapor pressure materials. *J. Vac. Sci. Technol.* **20**, 134–6.

Nagata T, West J B, Hayaishi T, Itikawa Y, Itoh Y, Koizumi T, Murakami J, Sato Y, Shibata H, Yagishita A and Yoshino M. 1986. Single and double photoionisation of Sr atoms between 38 and 50 nm. *J. Phys. B: Atom. Molec. Phys.* **19**, 1281–90.

Nanbu K. 1985. Angular-distributions of molecular flux from orifices of various thicknesses. *Vacuum* **35**, 573–6.

Napier S A, Cvejanović D, Williams J F and Pravica L. 2008. Temporary negative-ion effects on photon emission from free zinc atoms excited by electron impact. *Phys. Rev. A* **78**, 022702-1–9.

Naumov A I. 1963. Experimental investigation of the directivity of an ammonia molecular beam. *Sov. Phys. Tech. Phys.* **8**, 88–91 [*Zh. Tekh. Fiz.* **33**, 127–31 1963].

Nelson E B and Nafe J E. 1949. The hyperfine structure of tritium. *Phys. Rev.* **75**, 1194–8.

Nemirovskii L N. 1968. Electron gun for floating-zone evaporation of germanium. *Instrum. Exper. Technol.* **6**, 1482–3 [*Prib. Tekh. Éksp.* **6**, 192–3 1968].

Nemirovskii L N and Seidman L A. 1971. Crucible-free cathode-ray evaporator. *Instrum. Exper. Technol.* **16**, 1736–7 [*Prib. Tekh. Éksp.* 133–4 1971].

Neubert A and Zmbov K F. 1983. The dissociation energies of gaseous SmLi, EuLi, TmLi and YbLi. *Chem. Phys.* **76**, 469–78.

Neudachin I G, Porodnov B T and Suetin P E. 1972. Formation of narrow molecular beams by cylindrical channels. *Sov. Phys. Tech. Phys.* **17**, 848–51 [*Zh. Tech. Fiz.* **42**, 1069–72 1972].

Neudachin I G, Porodnov B T and Suetin P E. 1974. Use of the distribution function to describe free molecular beams. *Sov. Phys. Tech. Phys.* **19**, 511–4 [*Zh. Tech. Fiz.* **44**, 812–7 1974].

Neumark D M, Wodtke A M, Robinson G N, Hayden C C and Lee Y T. 1985. Molecular beam studies of the $F + H_2$ reaction. *J. Chem. Phys.* **82**, 3045–66.

Neynaber R H, Marino L L, Rothe E W and Trujillo S M. 1961. Low-energy electron scattering from atomic oxygen. *Phys. Rev.* **123**, 148–52.

Nguyen B, Lahmam Bennani A and Rouault M. 1975. Jet moléculaire à haute densité formé par une buse à multicanaux. *J. Phys. E: Sci. Instrum.* **8**, 909–12.

Nierenberg W A. 1957. The measurement of the nuclear spins and static moments of radioactive isotopes. *Ann. Rev. Nucl. Sci.* **7**, 349–406.

Nishimura A, Ohba H and Shibata T. 1992. Velocity distributions in high density gadolinium atomic beam produced with axial electron beam gun. *J. Nucl. Sci. Technol.* **29**, 1054–60.

Noren C, Kanik I, Johnson P V, McCartney P, James G K and Ajello J M. 2001. Electron-impact studies of atomic oxygen: II. Emission cross section measurements of the O I $^3S^0 \to ^3P$ transition (130.4 nm). *J. Phys. B: At. Mol. Opt. Phys.* **34**, 2667–77.

Oblath S B and Gole J L. 1980. On the continuum emissions observed upon oxidation of aluminium and its compounds. *Combustion and Flame* **37**, 293–312.

Ochs S A, Coté R E and Kusch P. 1954. On the radiofrequency spectrum of the components of a sodium chloride beam. The dimerization of the alkali halides. *J. Chem. Phys.* **21**, 459–66.

Ogryzlo E A. 1961. Halogen atom reactions I. The electrical discharge as a source of halogen atoms. *Can. J. Chem.* **39**, 2556–62.

Ohba H, Araki M and Shibata T. 1994a. Time variation of surface temperature during electron beam evaporation. *Jpn. J. Appl. Phys.* **33L**, 693–5.

Ohba H, Nishimura A, Ogura K and Shibata T. 1994b. Removal of the plasma contained in an atomic beam produced by electron beam heating. *Rev. Sci. Instrum.* **65**, 657–60.

Ohba H, Ogura K, Nishimura A, Tamura K and Shibata T. 2000. Effect of electron beam on velocities of uranium atomic beams produced by electron beam heating. *Jpn. J. Appl. Phys.* **39**, 5347–51.

Ohba H and Shibata T. 1998. Generation of vapor stream using a porous rod in an electron beam evaporation process. *J. Vac. Sci. Technol. A* **16**, 1247–50.

Olander D R and Kruger V. 1970. Molecular beam sources fabricated from multichannel arrays. III. The exit density problem. *J. Appl. Phys.* **41**, 2769–76.

Oldenborg R C, Dickson C R and Zare R N. 1975. A new electronic band system of PbO. *J. Molec. Spectrosc.* **58**, 283–300.

Olsen L O, Smith C S and Crittenden E C. 1945. Techniques for evaporation of metals. *J. Appl. Phys.* **16**, 425–34.

Olt R G, Du Four H R and Gray M I. 1961. Quartz micro-tubing manufacture. US Patent 2 987 372.

Ota Y. 1977. Si molecular beam epitaxy (n on n^+) with wide range doping control. *J. Electrochem. Soc.* **124**, 1795–802.

Ota Y. 1983. Silicon molecular beam epitaxy. *Thin Solid Films* **106**, 3–136.

Ottinger C and Zare R N. 1970. Crossed beam chemiluminescence. *Chem. Phys. Lett.* **5**, 243–8.

Packard R E, Pekola J P, Price P B, Spohr R N R, Westmacott K H and Yu-Qun Z. 1986. Manufacture, observation, and test of membranes with locatable single pores. *Rev. Sci. Instrum.* **57**, 1654–60.

Pailloux A, Alpettaz T and Lizon E. 2007. Candlestick oven with a silica wick provides an intense collimated cesium atomic beam. *Rev. Sci. Instrum.* **78**, 023102-1–6.

Pamplin B R (ed) 1980. *Molecular Beam Epitaxy* (Oxford, UK: Pergamon).

Panish M B. 1980. Molecular beam epitaxy of GaAs and InP with gas sources for As and P. *J. Electrochem. Soc.* **127**, 2720–33.

Panish M B and Temkin H. 1993. *Gas Source Molecular Beam Epitaxy: Growth and Properties of Phosphorus Containing III–V Heterostructures* (London: Springer).

Panish M B, Temkin H and Sumski S. 1985. Gas source MBE of InP and $Ga_xIn_{1-x}P_yAs_{1-y}$: Materials properties and heterostructure lasers. *J. Vac. Sci. Technol. B* **3**, 657–65.

Pao Y and Tchao J. 1970. Knudsen flow through a long circular tube. *Phys. Fluids* **13**, 527–8.

Paolini B P and Khakoo M A. 1998. An intense atomic hydrogen source with a movable nozzle output. *Rev. Sci. Instrum.* **69**, 3132–5.

Park M, Hauge R H and Margrave J L. 1988. Reactions and photochemistry of atomic and diatomic nickel with water at 15 K. *High Temp. Science* **25**, 1–15.

Parker E H C (ed). 1980. *The Technology and Physics of Molecular Beam Epitaxy* (London: Plenum).

Parr A C. 1971. Photoionization of europium and thulium: Threshold to 1350 Å. *J. Chem. Phys.* **54**, 3161–7.

Parson J M and Lee Y T. 1972. Crossed molecular beam study of $F + C_2H_4$, C_2D_4. *J. Chem. Phys.* **56**, 4658–66.

Pasternack L and Dagdigian P J. 1976. Laser-fluorescence study of the reactions of alkaline earth atoms with BrCN: Spectroscopic observation of the alkaline earth monocyanides. *J. Chem. Phys.* **65**, 1320–34.

Patrick E L. 2006. Silicon carbide nozzle for producing molecular beams. *Rev. Sci. Instrum.* **77**, 043301-1–8.

Patterson D and Doyle J M. 2007. Bright, guided molecular beam with hydrodynamic enhancement. *J. Chem. Phys.* **126**, 154307-1–5.

Patton C J, Lozhkin K O, Shah M B, Geddes J and Gilbody H B. 1996. Multiple ionisation of gallium by electron impact. *J. Phys. B: At. Mol. Opt. Phys.* **29**, 1409–17.

Pauly H. 1961. Streuversuche an Molekularstrahlen und zwischenmolekulare Kräfte. *Fortschr. Phys.* **9**, 613–87.

Pauly H. 1988a. Other low-energy beam sources. In *Atomic and Molecular Beam Methods*, vol. 1, G Scoles (ed) (Oxford, UK: University Press), pp. 83–123.

Pauly H. 1988b. High-energy beam sources. In *Atomic and Molecular Beam Methods*, vol. 1, G Scoles (ed) (Oxford, UK: University Press), pp. 124–52.

Pauly H. 2000a. *Atom, Molecule and Cluster Beams*, vol. I, *Basic Theory, Production and Detection of Thermal Energy Beams* (Berlin, Germany: Springer).

Pauly H. 2000b. *Atom, Molecule and Cluster Beams*, vol. II, *Cluster Beams, Fast and Slow Beams, Accessory Equipment and Applications* (Berlin, Germany: Springer).

Pauly H and Toennies J P. 1965. The study of intermolecular potentials with molecular beams at thermal energies. In *Advances in Atomic and Molecular Physics*, vol. 1, D R Bates and I Estermann (eds) (New York: Academic), pp. 195–344.

Pauly H and Toennies J P. 1968. Beam experiments at thermal energies. In *Methods of Experimental Physics—Atomic and Electron Physics—Atomic Interactions*, vol. 7A, B Bederson and W L Fite (eds) (New York: Academic), pp. 227–341.

Pendlebury J M and Ring D B. 1972. Ground state hyperfine structures and nuclear quadrupole moments of ^{95}Mo and ^{97}Mo. *J. Phys. B: Atom. Molec. Phys.* **5**, 386–96.

Pendlebury J M and Smith K F. 1987. Molecular beams. *Contemp. Phys.* **28**, 3–32.

Peng X D, Viswanathan R, Smudde G H and Stair P C. 1992. A methyl free radical source for use in surface studies. *Rev. Sci. Instrum.* **63**, 3930–5.

Penselin S. 1978. Recent developments and results of the atomic-beam magnetic-resonance method. In *Progress in Atomic Spectroscopy Part A*, W Hanle and H Kleinpoppen (eds) (New York: Plenum), pp. 463–90.

Periquet V, Moreau A, Carles S, Schermann J P and Desfrançois C. 2000. Cluster size effects upon anion solvation of N-heterocyclic molecules and nucleic acid bases. *J. Electr. Spectrosc.* **106**, 141–51.

Perry J S A, Gingell J M, Newson K A, To J, Watanabe N and Price S D. 2002. An apparatus to determine the rovibrational distribution of molecular hydrogen formed by the heterogeneous recombination of H atoms on cosmic dust analogues. *Meas. Sci. Technol.* **13**, 1414–24.

Phaneuf R A, Meyer F W and McKnight R H. 1978. Single-electron capture by multiply charged ions of carbon, nitrogen, and oxygen in atomic and molecular hydrogen. *Phys. Rev. A* **17**, 534–45.

Phipps T E and Taylor J B. 1927. The magnetic moment of the hydrogen atom. *Phys. Rev.* **29**, 309–20.

Pine A S and Nill K W. 1979. Molecular-beam tunable-diode-laser sub-Doppler spectroscopy of Λ-doubling in nitric oxide. *J. Molec. Spectrosc.* **74**, 43–51.

Plekhotkina G L. 1981. High-temperature vacuum atomic and molecular source. *Instrum. Exper. Tech.* (6) 1552–3 [*Prib. Tekh. Éksp.* (6) 201–2 1981].

Pollard W G and Present R D. 1948. On gaseous self-diffusion in long capillary tubes. *Phys. Rev.* **73**, 762–74.

Poole H G. 1937. Atomic hydrogen II–Surface effects in the discharge tube. *Proc. Roy. Soc. A* **163**, 415–23.

Popov O A (ed) 1995. *High Density Plasma Sources—Design, Physics and Performance* (Park Ridge, NJ: Noyes).

Porodnov B T, Akin'shin V D, Kichaev V I, Borisov S F and Suetin P E. 1974b. Flow of low-density gases in a capillary grid. *Sov. Phys. Tech. Phys.* **19**, 515–8 [*Zh. Tekh. Fiz.* **44**, 818–23 1974].

Porodnov B T, Suetin P E, Borisov S F and Akinshin V D. 1974a. Experimental investigation of rarefied gas flow in different channels. *J. Fluid Mech.* **64**, 417–37.

Porter R A R and Grosser A E. 1980. CH_2 molecular beam source. *Rev. Sci. Instrum.* **51**, 140–1.

Post C B and Eberly W S. 1947. Stability of austenite in stainless steels. *Trans. Amer. Soc. Metals* **39**, 868–90.

Prada-Silva G, Kester K, Löffler D, Haller G L and Fenn J B. 1977. Recycling molecular beam reactor. *Rev. Sci. Instrum.* **48**, 897–902.

Preece W H. 1885. On a peculiar behaviour of glow-lamps when raised to high incandescence. *Proc. Roy. Soc.* **38**, 219–30.

Prescher T, Richter M, Sonntag B and Wetzel H E. 1987. Electron heated high temperature atomic beam source for VUV photoelectron spectroscopy. *Nucl. Instrum. Methods* **A254**, 627–9.

Preuss D R, Pace S A and Gole J L. 1979. The supersonic expansion of pure copper vapor. *J. Chem. Phys.* **71**, 3553–60.

Pribytkov V A, Matveev O I and Dibrova A K. 1987. A high-temperature furnace with field-emission electron heating. *Instr. Exp. Tech.* **30**, 746–7 [*Prib. Tekh. Éksp.* **3**, 218–9 1987].

Prodell A G and Kusch P. 1957. Hyperfine structure of tritium in the ground state. *Phys. Rev.* **106**, 87–9.

Rabi I I and Cohen V W. 1934. Measurement of nuclear spin by the method of molecular beams—The nuclear spin of sodium. *Phys. Rev.* **46**, 707–12.

Rabi I I, Kellogg J M B and Zacharias J R. 1934. The magnetic moment of the proton. *Phys. Rev.* **46**, 157–63.

Radlein D S A G, Whitehead J C and Grice R. 1975. Reactive scattering of oxygen atoms: O + I_2, ICl, Br_2. *Molec. Phys.* **29**, 1813–28.

Ramsey N F. 1956. *Molecular Beams* (Oxford, UK: Clarendon).

Ramsey N F. 1990a. *Molecular Beams* (Oxford, UK: Clarendon) [Reprint of Ramsey (1956)].

Ramsey N F. 1990b. Experiments with separated oscillatory fields and hydrogen masers. *Rev. Mod. Phys.* **62**, 541–52.

Ramsey N F. 1996. Thermal beam sources. In *Experimental Methods in the Physical Sciences Atomic, Molecular And Optical Physics: Atoms and Molecules*, vol. 29B, F B Dunning and R G Hulet (eds) (San Diego, CA: Academic), pp. 1–20.

Rapol U D, Wasan A and Natarajan V. 2001. Loading of a Rb magneto-optic trap from a getter source. *Phys. Rev. A* **64**, 023402-1–5.

Rassi D, Pejčev V and Ross K J. 1977. The ejected-electron spectrum of Li I autoionizing levels excited by low-energy electron impact on lithium vapour. *J. Phys. B: Atom. Molec. Phys.* **10**, 3535–42.

Ready J F. 1998. *Industrial Applications of Lasers*, Ch 16, *Applications for Material Removal: Drilling, Cutting, Marking* (San Diego, CA: Academic), pp. 387–95, 417–8.

Reichert E. 1963. Die Winkelverteilung im Bereich 30 bis 155° von elastisch an Golddampf gestreuten Elektronen mit Energien zwischen 150 und 1900 eV. *Z. Physik* **173**, 392–401.

Reid R C, Prausnitz J M and Poling B E. 1987. *The Properties of Gases and Liquids*, 4th edn (New York: McGraw Hill).

Remick R R and Geankoplis C J. 1973. Binary diffusion of gases in capillaries in the transition region between Knudsen and molecular diffusion. *Ind. Eng. Chem. Fundam.* **12**, 214–20.

Rhim W K, Chung S K, Barber D, Man K F, Gutt G, Rulison A and Spjut R E. 1993. An electrostatic levitator for high-temperature containerless materials processing in 1-g. *Rev. Sci. Instrum.* **64**, 2961–70.

Richley E A and Reynolds T W. 1964. Numerical solutions of free-molecule flow in converging and diverging tubes and slots. NASA TN D-2330.

Riley S J, Parks E K, Mao C R, Pobo L G and Wexler S. 1982. Generation of continuous beams of refractory metal clusters. *J. Phys. Chem.* **86**, 3911–3.

Roach T M and Henclewood D. 2004. Novel rubidium atomic beam with an alkali dispenser source. *J. Vac. Sci. Technol. A* **22**, 2384–7.

Roberts G C and Via G G. 1967. Monitored evaporant source. US Patent 3 313 914.

Roberts J A and King A H. 1971. Collimated hole structure with mask for producing high resolution images. US Patent 3 556 636.

Roberts J A and Surprenant N F. 1972. Collimated hole flow control device. US Patent 3 645 298.

Roberts J A and Roberts P R. 1970. Electric device having passage structure electrode. US Patent 3 506 885.

Roberts J A and Surprenant N F. 1969. Particulate and biological filters. US Patent 3 482 703.

Roberts J A and King A H. 1972. Method of producing high resolution images and structure for use therein. US Patent 3 678 564.

Rocke M J. 1982. Copper coated laser fusion targets using molecular beam levitation. *J. Vac. Sci. Technol.* **20**, 1325–7.

Rodebush W H. 1931. Molecular rays. *Rev. Mod. Phys.* **3**, 392–411.

Rodebush W H and Klingelhoefer W C. 1933. Atomic chlorine and its reaction with hydrogen. *J. Am. Chem. Soc.* **55**, 130–42.

Rosebury F. 1965. *Handbook of Electron Tube and Vacuum Techniques* (Reading, MA: Addison–Wesley).

Rosenkranz B and Bettmer J. 2000. Microwave-induced plasma-optical emission spectrometry—Fundamental aspects and applications in metal speciation analysis. *Trends in Analytical Chemistry* **19**, 138–56.

Rosenwaks S, Steele R E and Broida H P. 1975. Chemiluminescence of AlO. *J. Chem. Phys.* **63**, 1963–5.

Ross K J. 1993. A simple resistively heated vapour beam oven with an interchangeable crucible. *Vacuum* **44**, 863–4.

Ross K J and Sonntag B. 1995. High temperature metal atom beam sources. *Rev. Sci. Instrum.* **66**, 4409–33.

Ross K J and West J B. 1998. A resistively heated vapour beam oven for synchrotron radiation studies. *Meas. Sci. Technol.* **9**, 1236–8.

Rossi A M, Amato G, Boarino L and Novero C. 2000. Realisation of membranes for atomic beam collimator by macropore micromachining technique (MMT). *Materials Science and Engineering* **B69–70**, 66–9.

Roulet H and Alexandre E. 1981. A UHV compatible and miniaturized Knudsen cell used as a controlled evaporation source. *J. Vac. Sci. Technol.* **19**, 253–4.

Rousseau A, Granier A, Gousset G and Leprince P. 1994a. Microwave discharge in H_2: Influence of H-atom density on the power balance. *J. Phys. D: Appl. Phys.* **27**, 1412–22.

Rousseau A, Tomasini L, Gousset G, Boisse-Laport C and Leprince P. 1994b. Pulsed microwave discharge: A very efficient H atom source. *J. Phys. D: Appl. Phys.* **27**, 2439–41.

Roy S, Raju R, Chuang H F, Cruden B A and Meyyappan M. 2003. Modeling gas flow through microchannels and nanopores. *J. Appl. Phys.* **93**, 4870–9.

Rubinsztein H and Gustafsson M. 1975. Nuclear spin measurements on neutron-deficient isotopes of the refractory elements. *Phys. Lett.* **58B**, 283–5.

Rubinsztein H, Lindgren I, Lindström L, Riedl H and Rosén A. 1974. Atomic-beam measurements on refractory elements—Nuclear spin measurements of ^{99}Mo and ^{99m}Tc. *Nucl. Instrum. Methods* **119**, 269–74.

Rudin H, Striebel H R, Baumgartner E, Brown L and Huber P. 1961. Eine Quelle polarisierter Deuteronen und Nachweis der Polarisation durch die (d, T)-Reaktion. *Helvetica Physica Acta* **34**, 58–84.

Rugamas F, Roundy D, Mikaelian G, Vitug G, Rudner M, Shih J, Smith D, Segura J and Khakoo M A. 2000. Angular profiles of molecular beams from effusive tube sources: I. Experiment. *Meas. Sci. Technol.* **11**, 1750–65.

Rusin L Y and Toennies J P. 2006. An improved source of intense beams of fluorine atoms. *J. Phys. D: Appl. Phys.* **39**, 4186–93.

Rusk J R and Gordy W. 1962. Millimeter wave molecular beam spectroscopy: Alkali bromides and iodides. *Phys. Rev.* **127**, 817–30.

Ruster W, Ames F, Kluge H J, Otten E W, Rehklau D, Scheerer F, Herrmann G, Mühleck C, Riegel J, Rimke H, Sattelberger P and Trautmann N. 1989. A resonance ionization mass spectrometer as an analytical instrument for trace analysis. *Nucl. Instrum. Methods A* **281**, 547–58.

Saleem M, Amin N, Hussain S, Rafiq M, Mahmood S and Baig M A. 2006. Alternate technique for simultaneous measurement of photoionization cross-section of isotopes by TOF mass spectrometer. *Eur. Phys. J. D* **38**, 277–83.

Samano E C, Carr W E, Seidl M and Lee B S. 1993. An arc discharge hydrogen atom source. *Rev. Sci. Instrum.* **64**, 2746–52.

Samson J A R and Pareek P N. 1985. Absolute photoionization cross sections of atomic oxygen. *Phys. Rev. A* **31**, 1470–6.

Santeler D J. 1986. New concepts in molecular gas flow. *J. Vac. Sci. Technol. A* **4** 338–43.

Santesson S and Nilsson S. 2004. Airborne chemistry: Acoustic levitation in chemical analysis *Anal. Bioanal. Chem.* **378**, 1704–9.

Schaeffer D L. 1970. A cesium ion source and an oxygen source for photoemission studies. *Rev. Sci. Instrum.* **41**, 274–5.

Schioppo M, Poli N, Prevedelli M, Falke S, Lisdat C, Sterr U and Tino G M. 2012. A compact and efficient strontium oven for laser-cooling experiments. *Rev. Sci. Instrum.* **83**, 103101-1–6.

Schlier C. 1957. Neuere Messungen mit Hilfe der Molekularstrahltechnik. *Fortschr. Phys.* **5** 378–420.

Schlier C (ed). 1970. *Proceedings of the International School of Physics 'Enrico Fermi' Course 44 Molecular Beams and Reaction Kinetics* (New York: Academic).

Schlindwein R, Seki S and Korschinek G. 1990. A compact injector for a small triton accelerator. *Rev. Sci. Instrum.* **61**, 631–2.

Schmidt E, Schröder H, Sonntag B, Voss H and Wetzel H E. 1984. $M_{2,3}$-shell Auger and autoionisation spectra of free Cr, Mn, Fe, Co, Ni and Cu atoms. *J. Phys. B: Atom. Molec. Phys.* **17**, 707–18.

Schmidt E, Schröder H, Sonntag B, Voss H and Wetzel H E. 1985. Resonant satellite photoemission of atomic Mn. *J. Phys. B: Atom. Molec. Phys.* **18**, 79–93.

Schmidt V. 1985. Photoionization of free metal atoms using synchrotron radiation. *Comments. At. Mol. Phys.* **17**, 1–14.

Schmidt V. 1997. *Electron Spectrometry of Atoms using Synchrotron Radiation* (Cambridge, UK: University Press).

Schönhense G. 1983. Magnetic-field-free vapor furnace for photoelectron spectroscopy. I. Temperatures up to 900 K. *Rev. Sci. Instrum.* **54**, 419–21.

Schumacher W, Barz F, Dreesen E, Hammon W, Hansen H H, Penselin S and Scholzen A. 1975. The polarized proton and deuteron source for the Bonn isochronous cyclotron. *Nucl. Instrum. Methods* **127**, 157–62.

Schwab G M and Friess H. 1933. Darstellung und einige Eigenschaften atomaren Chlors. *Naturwissenschaften* **21**, 222#.

Schwalm U. 1983. High-purity F atom beam source. *Appl. Phys. B* **30**, 149–52.

Schwarz-Selinger T, Dose V, Jacob W and von Keudell A. 2001. Quantification of a radical beam source for methyl radicals. *J. Vac. Sci. Technol. A* **19**, 101–7.

Schwarz-Selinger T, von Keudell A and Jacob W. 2000. Novel method for absolute quantification of the flux and angular distribution of a radical source for atomic hydrogen. *J. Vac. Sci. Technol. A* **18**, 995–1001.

Scoles G. 1988. *Atomic and Molecular Beam Methods*, vol. 1 (Oxford, UK: University Press).

Scoles G. 1992. *Atomic and Molecular Beam Methods*, vol. 2 (Oxford, UK: University Press).

Seccombe D P, Collins S A and Reddish T J. 2001. The design and performance of an effusive gas source of conical geometry for photoionization studies. *Rev. Sci. Instrum.* **72**, 2550–7.

Senitzky B and Rabi I I. 1956. Hyperfine structure of Rb[85,87] in the 5 P state. *Phys. Rev.* **103**, 315–21.

Sepehrad A, Marshall R M and Purnell H. 1979. A simple preparation of stable, nonreactive surfaces for gas kinetic studies of reactions of hydrogen atoms. *Int. J. Chem. Kinetics* **11**, 411–3.

Shah M B and Gilbody H B. 1981. Experimental study of the ionisation of atomic hydrogen by fast H^+ and He^{2+} ions. *J. Phys. B: Atom. Molec. Phys.* **14**, 2361–77.

Shah M B, Bolorizadeh M A, Patton C J and Gilbody H B. 1996. Simple metallic atom source for crossed beam collision studies. *Meas. Sci. Technol.* **7**, 709–11.

Shah M B, Elliott D S and Gilbody H B. 1985. Ionisation and charge transfer in collisions of H^+ and He^{2+} with lithium. *J. Phys. B: Atom. Molec. Phys.* **18**, 4245–58.

Shah M B, Elliott D S and Gilbody H B. 1987. Pulsed crossed-beam study of the ionisation of atomic hydrogen by electron impact. *J. Phys. B: Atom. Molec. Phys.* **20**, 3501–14.

Shah M B, McCallion P, Okuno K and Gilbody H B. 1993. Multiple ionization of iron by electron impact. *J. Phys. B: At. Mol. Opt. Phys.* **26**, 2393–401.

Shank S M, Soave R J, Then A M and Tasker G W. 1995. Fabrication of high aspect ratio structures for microchannel plates. *J. Vac. Sci. Technol. B* **13**, 2736–40.

Shannon S P and Codling K. 1978. Partial photoionisation cross section measurements for atomic cadmium and mercury. *J. Phys. B: Atom. Molec. Phys.* **11**, 1193–202.

Sharipov F and Seleznev V. 1998. Data on internal rarefied gas flows. *J. Phys. Chem. Ref. Data* **27**, 657–706.

Shen L Y L. 1978. Angular distribution of molecular beams from modified Knudsen cells for molecular-beam epitaxy. *J. Vac. Sci. Technol.* **15**, 10–12.

Sherwood J E and Ovenshine S J. 1959. Nuclear spins of I^{128} and I^{130}. *Phys. Rev.* **114**, 858–61.

Shi Y, Lee Y T and Kim A S. 2012. Knudsen diffusion through cylindrical tubes of varying radii: Theory and Monte Carlo simulations. *Transp. Porous Med.* **93**, 517–54.

Shimada T, Koide J, Cho K A and Koma A. 1999. Velocity distribution of organic molecules emitted from effusion cells measured by time-of-flight technique. *J. Vac. Sci. Technol. A* **17**, 615–8.

Shuttleworth T, Burgess D E, Hender M A and Smith A C H. 1979. Inelastic scattering of electrons by lithium atoms. *J. Phys. B: Atom. Molec. Phys.* **12**, 3967–78.

Shuttleworth T, Newell W R and Smith A C H. 1977. Electron impact excitation of the sodium 3^2S–3^2P transition. *J. Phys. B: Atom. Molec. Phys.* **10**, 1641–51.

Sibener S J, Buss R J, Ng C Y and Lee Y T. 1980. Development of a supersonic $O(^3P_J)$, $O(^1D_2)$ atomic oxygen nozzle beam source. *Rev. Sci. Instrum.* **51**, 167–82.

Singy D, Schmelzbach P A, Grüebler W and Zhang W Z. 1989. Production of intense polarized hydrogen atomic beams by cooling the atoms to low temperature. *Nucl. Instrum. Methods A* **278**, 349–67.

Slabospitskii R P, Kiselev I E, Karnaukhov I M, Lopatko I D and Taranov A Y. 1970. Polarized negative deuteron injector for a tandem accelerator. *Sov. Phys. Tech. Phys.* **14**, 1129–33 [*Zh. Tekh. Fiz.* **39**, 1506–12 1969].

Slayter G. 1960. Method of making a metal element. US Patent 2 961 758.

Slevin J and Stirling W. 1981. Radio frequency atomic hydrogen beam source. *Rev. Sci. Instrum.* **52**, 1780–2.

Slowe C, Vernac L and Hau L V. 2005. High flux source of cold rubidium atoms. *Rev. Sci. Instrum.* **76**, 103101-1–10.

Smirnov Y M. 1994. Excitation cross-sections of the praseodymium atom. *J. Phys. II France* **4**, 23–35.

Smirnov Y M. 2000. Excitation cross sections for odd triplet levels of La II in e-La collisions. *Opt. Spectrosc.* **89**, 336–43 [*Opt. Spektrosk.* **89**, 368–77 2000].

Smirnov Y M. 2002. Excitation of even triplet levels of Y II in the e–Y collisions. *Opt. Spectrosc.* **93**, 351–6 [*Opt. Spektrosk.* **93**, 383–8 2002].

Smith A C H, Caplinger E, Neynaber R H, Rothe E W and Trujillo S M. 1962. Electron impact ionization of atomic nitrogen. *Phys. Rev.* **121**, 1647–9.

Smith K F. 1955. *Molecular Beams* (London: Methuen).

Snider G L, Then A M, Soave R J and Tasker G W. 1994. High aspect ratio dry etching for microchannel plates. *J. Vac. Sci. Technol. B* **12**, 3327–31.

Sonntag B and Wuilleumier F. 1983. Photoemission from atoms and molecules. *Nucl. Instrum. Methods* **208**, 735–52.

Sparrow E M and Haji-Sheikh A. 1964. Velocity profile and other local quantities in free-molecule tube flow. *Phys. Fluids* **7** 1256–61.

Sparrow E M, Jonsson V K and Lundgren T S. 1963. Free-molecule tube flow and adiabatic wall temperatures. *J. Heat Transfer (Trans ASME)* **85**, 111–8.

Spence D and Steingraber O J. 1988. Factors determining dissociation fractions in atomic beams generated by 'straight-through' microwave discharge sources. *Rev. Sci. Instrum.* **59**, 2464–7.

Stafford G H, Dickson J M, Salter D C and Craddock M K. 1962. A source for the production of polarized protons. *Nucl. Instrum. Methods* **15**, 146–54.

Stan C A and Ketterle W. 2005. Multiple species atom source for laser-cooling experiments. *Rev. Sci. Instrum.* **76**, 063113-1–5.

Stanley R W. 1966. Gaseous atomic-beam light source. *J. Opt. Soc. Amer.* **56**, 350–6.

Steckelmacher W. 1966. A review of the molecular flow conductance for systems of tubes and components and the measurement of pumping speed. *Vacuum* **16**, 561–84.

Steckelmacher W. 1974. Flow of rarefied gases in vacuum systems and problems of standard-ization of measuring techniques. *Jpn. J. Appl. Phys. Suppl.* **2**, 117–25.

Steckelmacher W. 1978a. The effect of cross-sectional shape on the molecular flow in long tubes. *Vacuum* **28**, 269–75.

Steckelmacher W. 1978b. Molecular flow conductance of long tubes with uniform elliptical cross-section and the effect of different cross-sectional shapes. *J. Phys. D: Appl. Phys.* **11**, 473–8.

Steckelmacher W. 1986. Knudsen flow 75 years on: The current state of the art for flow of rarefied-gases in tubes and systems. *Rep. Prog. Phys.* **49**, 1083–1107.

Steckelmacher W and Lucas M W. 1983. Gas flow through a cylindrical tube under free molec-ular conditions. *J. Phys. D: Appl. Phys.* **16**, 1453–60.

Steckelmacher W, Strong R and Lucas M W. 1978. A simple atomic or molecular beam as target for ion-atom collision studies *J. Phys. D: Appl. Phys.* **11**, 1553–66.

Stefanov B, Petrov P and Pirgov P. 1988. Electron beam evaporation and welding: Plasma formation and liquid pool instabilities. *Vacuum* **38**, 1029–33.

Stern O. 1920a. Eine direkte Messung der thermischen Molekulargeschwindigkeit. *Z. Physik* **2**, 49–56.

Stern O. 1920b. Nachtrag zu meiner Arbeit: Eine direkte Messung der thermischen Molekulargeschwindigkeit. *Z. Physik* **2**, 417–21.

Stickney R E, Keating R F, Yamamoto S and Hastings W J. 1967. Angular distribution of flow from orifices and tubes at high Knudsen numbers. *J. Vac. Sci. Technol.* **4**, 10–18.

Stinespring C D, Freedman A and Kolb C E. 1986. An ultrahigh vacuum compatible fluorine atom source for gas-surface reaction studies. *J. Vac. Sci. Technol. A* **4**, 1946–7.

Stockdale J A, Schumann L, Brown H H and Bederson B. 1977. High-temperature atomic beam source. *Rev. Sci. Instrum.* **48**, 938–9.

Streets D G and Berkowitz J. 1976. Photoelectron spectroscopy of Se_2 and Te_2. *J. Electr. Spectrosc.* **9**, 269–87.

Stroke H H, Jaccarino V, Edmonds D S and Weiss R. 1957. Magnetic moments and hyperfine-structure anomalies of Cs^{133}, Cs^{134}, Cs^{135} and Cs^{137}. *Phys. Rev.* **105**, 590–603.

Sudraud P, Ballongue P, Varoquaux E and Avenel O. 1987. Focused ion-beam milling of a submicrometer aperture for a hydrodynamic Josephson-effect experiment. *J. Appl. Phys.* **62**, 2163–8.

Sugaya T and Kawabe M. 1991. Low-temperature cleaning of GaAs substrate by atomic hydrogen irradiation. *Jpn. J. Appl. Phys.* **30**, L402–4.

Swenumson R D and Even U. 1981. Continuous flow reflux oven as the source of an effusive molecular Cs beam. *Rev. Sci. Instrum.* **52**, 559–61.

Szwemin P and Niewiński M. 2002. Comparison of transmission probabilities calculated by Monte Carlo simulation and analytical methods. *Vacuum* **67**, 359–62.

Tabet J, Eden S, Feil S, Abdoul-Carime H, Farizon B, Farizon M, Ouaskit S and Märk T D. 2010. Absolute molecular flux and angular distribution measurements to characterize DNA/RNA vapor jets. *Nucl. Instrum. Methods B* **268**, 2458–66.

Talley W K and Whittaker S. 1969. Monte Carlo analysis of Knudsen flow. *J. Comp. Phys.* **4**, 389–410.

Tamura K, Adachi H and Shibata T. 1999a. Charge transfer cross sections for dysprosium and cerium. *Jpn. J. Appl. Phys.* **38**, 2973–7.

Tamura K, Adachi H, Ogura K, Ohba H and Shibata T. 1999b. Charge transfer cross sections for uranium. *Jpn. J. Appl. Phys.* **38**, 6512–6.

Tang S P, Chien K R and Sabety-Dzvonik M J. 1981. High intensity and high energy carbon beam source. *Carbon* **19**, 403–4.

Taylor E H and Datz S. 1955. Study of chemical reaction mechanisms with molecular beams. The reaction of K with HBr. *J. Chem. Phys.* **23**, 1711–8.

Taylor G F. 1924. A method of drawing metallic filaments and a discussion of their properties and uses. *Phys. Rev.* **23**, 655–60.

Taylor J B. 1929. Das magnetische Moment des Lithiumatoms. *Z. Physik* **52**, 846–52.

Taylor J B. 1931. Molecular beams. *Ind. Eng. Chem.* **23**, 1228–31.

Thakur K B, Sahu G K, Tamhankar R V and Patel K. 2001. High power, high uniformity strip electron gun design, simulation and performance. *Rev. Sci. Instrum.* **72**, 207–15.

Thakur K B, Sahu G K, Gaur S J, Das R C, Tak A K, Patankar R A, Bhowmick G K, Manohar K G, Jagatap B N and Venkatramani N. 2005. Integrated atom density measurement in a zirconium atomic beam generated using a high-power strip electron beam source by diode laser absorption technique. *Vacuum* **77**, 443–9.

Timp G, Behringer R E, Tennant D M, Cunningham J E, Prentiss M and Berggren K K. 1992. Using light as a lens for submicron, neutral-atom lithography. *Phys. Rev. Lett.* **69**, 1636–9.

Ting Y. 1957. Hyperfine structure and quadrupole moment of lanthanum-139. *Phys. Rev.* **108**, 295–304.

Tison S A. 1993. Experimental data and theoretical modeling of gas flows through metal capillary leaks. *Vacuum* **44**, 1171–5.

Toennies J P. 1968. Molecular beam studies of chemical reactions. In *Chemische Elementarprozesse*, H Hartmann (ed) (Berlin, Germany: Springer), pp. 157–218.

Toennies J P. 1974. Molecular beam scattering experiments on elastic, inelastic, and reactive collisions In *Physical Chemistry an Advanced Treatise*, vol. 6A, *Kinetics of Gas Reactions*, W Jost (ed) (New York: Academic), pp. 227–381.

Toennies J P, Welz W and Wolf G. 1979. Molecular beam scattering studies of orbiting resonances and the determination of van der Waals potentials for H-Ne, Ar, Kr, and Xe and for H_2-Ar, Kr, and Xe. *J. Chem. Phys.* **71**, 614–42.

Tompa G S, Lopes J L and Wohlrab G. 1987. Compact efficient modular cesium atomic beam oven. *Rev. Sci. Instrum.* **58**, 1536–7.

Tonucci R J, Justus B L, Campillo A J and Ford C E. 1992. Nanochannel array glass. *Science* **258**, 783–5.

Touchard F, Biderman J, de Saint Simon M, Thibault C, Huber G, Epherre M and Klapisch R. 1981. Production of ionic and atomic beams of alkaline elements. *Nucl. Instrum. Methods* **186**, 329–34.

Townsend S J. 1965. Free-molecule flow through axi-symmetric tubes. UTIAS Report No. 106 AFOSR-65-0749.

Trischka J W. 1962. Molecular beams. In *Methods of Experimental Physics*, vol. 3, *Molecular Physics*, D Williams (ed) (New York: Academic), pp. 589–636.

Troitskii V S. 1962. Directivity of a molecular beam formed by gas flow in a channel. *Sov. Phys. Tech. Phys.* **7**, 353–62 [*Zh. Tekh. Fiz.* **32**, 488–502 1962].

Tschersich K G. 2000. Intensity of a source of atomic hydrogen based on a hot capillary. *J. Appl. Phys.* **87**, 2565–73.

Tschersich K G and von Bonin V. 1998. Formation of an atomic hydrogen beam by a hot capillary. *J. Appl. Phys.* **84**, 4065–70.

Tu M F, Ho J J, Hsieh C C and Chen Y C. 2009. Intense SrF radical beam for molecular cooling experiments. *Rev. Sci. Instrum.* **80**, 113111-1-5.

Tunna L, Barclay P, Cernik R J, Khor K H, O'Neill W and Seller P. 2006. The manufacture of a very high precision x-ray collimator array for rapid tomographic energy dispersive diffraction imaging (TEDDI). *Meas. Sci. Technol.* **17**, 1767–75.

Usiskin C M and Siegel R. 1960. Thermal radiation from a cylindrical enclosure with specified wall heat flux. *J. Heat Trans. (Trans ASME)* **82**, 369–74.

Valentini J J, Coggiola M J and Lee Y T. 1977. Supersonic atomic and molecular halogen nozzle beam source. *Rev. Sci. Instrum.* **48**, 58–63.

Vályi L. 1968. A source of polarized proton and deuteron beams. *Nucl. Instrum. Methods* **58**, 21–8.

Vályi L. 1977. *Atom and Ion Sources* (London: Wiley).

van Audenhove J. 1965. Vacuum evaporation of metals by high frequency levitation heating. *Rev. Sci. Instrum.* **36**, 383–5.

van Essen D and Heerens W C. 1976. On the transmission probability for molecular gas flow through a tube. *J. Vac. Sci. Technol.* **13**, 1183–7.

Van Zyl B and Gealy M W. 1986. New molecular-dissociation furnace for H and O atom sources. *Rev. Sci. Instrum.* **57**, 359–64.

Varian S F and Varian R H. 1952. Method of making a grid structure. US Patent 2 619 438.

Varon J and Goldstein I S. 1981. Molecular beam levitator for sputter coating of microspheres. *Rev. Sci. Instrum.* **52**, 975–8.

Venema A. 1973. The flow of highly rarefied gases. *Philips Tech. Rev.* **33**, 43–9.

von Ehrenstein D. 1961. Messung der Hyperfeinstrukturaufspaltung des $^4F_{9/2}$ Grundzustandes in Co^{59}-I-Spektrum und Bestimmung des Quadrupolmomentes des Co^{59}-Kernes. *Ann. Physik* **7**, 342–52.

von Smoluchowski M. 1910. Zur kinetischen Theorie der Transpiration und Diffusion verdünnter Gase. *Ann. Physik* **33**, 1559–70.

Voronov G S and Martakova N K. 1968. Pulsed source of high-density atomic hydrogen. *Instrum. Exper. Technol.* (2) 266–8 [*Prib. Tekh. Éksp.* (2) 23–5 1968].

Vostrikov A A, Dubov D Y and Agarkov A A. 1996. Electron-induced radiation from C_{60} fullerene in the gas phase. *JETP Lett.* **63**, 963–7 [*Pis'ma Zh. Éksp. Teor. Fiz.* **63**, 915–9 1996].

Voulot D, McCullough R W, Thompson W R, Burns D, Geddes J, Cosimini G J, Nelson E, Chow P P and Klaassen J. 1998. Determination of the atomic nitrogen flux from a radio frequency plasma nitride source for molecular beam epitaxy systems. *J. Vac. Sci. Technol.* A **16**, 3434–7.

Voulot D, McCullough R W, Thompson W R, Burns D, Geddes J, Cosimini G J, Nelson E, Chow P P and Klaassen J. 1999. Characterisation of an RF atomic nitrogen plasma source. *J. Crystal Growth* **201/202**, 399–401.

Wagner L C and Grimley R T. 1972. A study of ionization processes by the angular distribution technique. The AgCl system. *J. Phys. Chem.* **76**, 2819–28.

Wagner L C and Grimley R T. 1974. A mass spectrometric study of the bismuth vapor system by the angular distribution technique. *Chem. Phys. Lett.* **29**, 594–9.

Wagner K G. 1983. Widerstandsbeheizter Hochtemperaturofen für PES-Untersuchungen an freien Atomen im UHV mittels Synchrotronstrahlung. *Vakuum-Technik* **32**, 67–9.

Wahlbeck P G and Phipps T E. 1968. Effusion. II. Angular number distributions of gaseous cadmium from a right-circular cylindrical orifice into vacuum. *J. Chem. Phys.* **49**, 1603–8.

Wainer E, Rose S H and Harkulich T M. 1971. Channel multiplier of aluminium oxide produced anodically. US Patent 3 626 233.

Walker J D and St John R M. 1974. Design of a high density atomic hydrogen source and determination of Balmer cross sections. *J. Chem. Phys.* **61**, 2394–407.

Walkiewicz M R, Fox P J and Scholten R E. 2000. Candlestick rubidium beam source. *Rev. Sci. Instrum.* **71**, 3342–4.

Walraven J T M and Silvera I F. 1982. Helium-temperature beam source of atomic hydrogen. *Rev. Sci. Instrum.* **53**, 1167–81.

Walsh J W T. 1920a. Radiation from a perfectly diffusing circular disk (Part I). *Proc. Phys. Soc.* **32**, 59–71.

Walsh J W T. 1920b. The radiation from a perfectly diffusing circular disk (Part II). *Proc. Phys. Soc.* **32**, 315–25.

Wang D, Li Y, Li S and Zhao H. 1994. HeI photoelectron spectroscopy (UPS) of iodine atoms. *Chem. Phys. Lett.* **222**, 167–70.

Wang D, Li C, Qian X and Gamblin S D. 1998. An HeI photoelectron spectrum of bromine atoms—The use of $SiBr_4$ as a bromine atom source. *J. Electr. Spectrosc.* **97**, 59–61.

Wang K C and Wahlbeck P G. 1967. Effusion. I. Angular number distributions of gaseous CsCl from a near-ideal orifice into vacuum. *J. Chem. Phys.* **47**, 4799–809.

Ward J W, Bivins R L and Fraser M V. 1970. Monte Carlo simulation of specular and surface diffusional perturbations to flow from Knudsen cells. *J. Vac. Sci. Technol.* **7**, 206–10.

Ward J W, Mulford R N R and Kahn M. 1967. Study of some of the parameters affecting Knudsen effusion I. Experimental tests of the validity of the cosine law as a function of cell and sample geometries and materials. *J. Chem. Phys.* **47**, 1710–17.

Wartenberg H v and Schultze G. 1930. Über aktiven Wasserstoff. II. Die Wandkatalyse. *Z. physik. Chem. (Leipzig) B* **6**, 261–6.

Washington D, Duchenois V, Polaert R and Beasley R M. 1971. Technology of channel plate manufacture. *Acta Electronica* **14**, 201–24.

Way K R, Yang S C and Stwalley C. 1976. Arc-heated high-intensity source of hydrogen atoms. *Rev. Sci. Instrum.* **47**, 1049–55.

Wehlitz R, Lukić D, Koncz C and Sellin I A. 2002. Setup for measurements of partial ion yields at the Synchrotron Radiation Center. *Rev. Sci. Instrum.* **73**, 1671–3.

Weinreich G and Hughes V W. 1954. Hyperfine structure of Helium-3 in the metastable triplet state. *Phys. Rev.* **95**, 1451–60.

Weinreich G, Grosof G M and Hughes V W. 1953. Hyperfine structure of the metastable triplet state Helium-3. *Phys. Rev.* **91**, 195–6.

Werner F, Korzec D and Engemann J. 1994. Slot antenna 2.45 GHz microwave plasma source. *Plasma Sources Sci. Technol.* **3**, 473–81.

Wessel G and Lew H. 1953. Hyperfine structures of silver and gold by the atomic beam magnetic resonance method. *Phys. Rev.* **92**, 641–6.

Williams W D and Giordano N. 1984. Fabrication of 80Å metal wires. *Rev. Sci. Instrum.* **55**, 410–2.

Wilmoth R G. 1972. Speed distribution measurements of N_2 and Ar molecular beams produced by a multichannel source. *J. Vac. Sci. Technol.* **9**, 1121–3.

Wilsch H. 1972. Remark on cooling of atomic hydrogen beams. *J. Chem. Phys.* **56**, 1412–3.

Winkler A. 2000. Comment on the paper by Y. Miyake, K. Shimomura, A. P. Mills Jr., J. P. Marangos and K. Nagamine, 'Thermionic emission of hydrogen isotopes (Mu, H, D and T) from W and interpretation of a role of the hot W as an atomic hydrogen source.' *Surface Science* **470**, 186–8.

Wise H and Wood B J. 1967. Reactive collisions between gas and surface atoms. *Adv. Atom. Mol. Phys.* **3**, 291–354.

Wise T, Roberts A D and Haeberli W. 1993. A high-brightness source for polarized atomic hydrogen and deuterium. *Nucl. Instrum. Methods A* **336**, 410–22.

Withers D. 1999. *Radio Spectrum Management* (London: Institution of Electrical Engineers).

Witteveen G J. 1977. Low-consumption atomic beam source. *Rev. Sci. Instrum.* **48**, 1131–2.

Wittke J P and Dicke R H. 1956. Redetermination of the hyperfine splitting in the ground state of atomic hydrogen. *Phys. Rev.* **103**, 620–31.

Wiza J L. 1979. Microchannel plate detectors. *Nucl. Instrum. Methods* **162**, 587–601.

Wnuk J D, Gorham J M, Smith B A, Shin M and Fairbrother D H. 2007. Quantifying the flux and spatial distribution of atomic hydrogen generated by a thermal source using atomic force microscopy to measure the chemical erosion of highly ordered pyrolytic graphite. *J. Vac. Sci. Technol. A* **25**, 621–5.

Wolber G, Figger H, Haberstroh R A and Penselin S. 1969. Hyperfine structure separations in the ground state multiplet of the stable carbon isotope ^{13}C. *Phys. Lett.* **29A**, 461–2.

Wood B J and Wise H. 1962. The kinetics of hydrogen atom recombination on Pyrex glass and fused quartz. *J. Phys. Chem.* **66**, 1049–53.

Wood R W. 1920. An extension of the Balmer series of hydrogen and spectroscopic phenomena of very long vacuum tubes. *Proc. Roy. Soc. A* **97**, 455–70.

Wood R W. 1921. Hydrogen spectra from long vacuum tubes. *Phil. Mag.* **42**, 729–45.

Wood R W. 1922a. Atomic hydrogen and the Balmer series spectrum. *Phil. Mag.* **44**, 538–46.

Wood R W. 1922b. Spontaneous incandescence of substances in atomic hydrogen gas. *Proc. Roy. Soc. A* **102**, 1–9.

Woodgate G K and Hellwarth R W. 1956. Hyperfine structure of radioactive Ag^{113}_{47}. *Proc. Phys. Soc. A* **69**, 581–7.

Woodgate G K and Martin J S. 1957. Hyperfine structure in Mn^{55}_{25}. *Proc. Phys. Soc.* **70**, 485–8.

Woznyj M, Feldmeier F and Hofmann A. 1981. Determination of molecular beam characteristics for condensable molecules by optical interferences. *J. Appl. Phys.* **52**, 3116–20.

Wrede E. 1927. Über die Ablenkung von Molekularstrahlen elektrischer Dipolmoleküle im inhomogenen elektrischen Feld. *Z. Physik* **44**, 261–8.

Xu N, Du Y, Ying Z, Ren Z and Li F. 1997. An arc discharge nitrogen atom source. *Rev. Sci. Instrum.* **68**, 2994–3000.

Yagi S and Nagata T. 2000. Absolute total and partial cross-sections for ionization of Ba and Eu atoms by electron impact. *J. Phys. Soc. Jpn.* **69**, 1374–83.

Yagi S and Nagata T. 2001. Absolute total and partial cross sections for ionization of free lanthanide atoms by electron impact. *J. Phys. Soc. Jpn.* **70**, 2559–67.

Yamamoto K, Hara J and Hirose K. 1980. Free molecular flow through a long circular tube. *J. Phys. Soc. Jpn.* **49**, 1157–61.

Yasunaga H. 1976. Calibrated source of atomic cesium beams. *Rev. Sci. Instrum.* **47**, 726–9.

Yaws C L. 1994. *Handbook of Vapor Pressure*, vol. 1, C_1 *to* C_4 *Compounds*, vol. 2, C_5 *to* C_7 *Compounds* (Houston, TX: Gulf Publishing Co) [Now Oxford, UK: Butterworth-Heinemann].

Yaws C L. 1995a. *Handbook of Vapor Pressure*, vol. 4, *Inorganic Compounds and Elements* (Houston, TX: Gulf Publishing Co) [Now Oxford, UK: Butterworth-Heinemann].

Yaws C L. 1995b. *Handbook of Viscosity*, vol. 1, *Organic Compounds* C_1 *to* C_4, vol. 2, *Organic Compounds* C_5 *to* C_7, vol. 3, *Organic Compounds* C_8 *to* C_{28} (Houston, TX: Gulf Publishing Co) [Now Oxford, UK: Butterworth-Heinemann].

Yaws C L. 1997. *Handbook of Viscosity*, vol. 4, *Inorganic Compounds and Elements* (Houston, TX: Gulf Publishing Co) [Now Oxford, UK: Butterworth-Heinemann].

Yi W, Jeong T, Jin S, Yu S, Lee J and Kim J M. 2000. Novel fabrication method of microchannel plates. *Rev. Sci. Instrum.* **71**, 4165–9.

Young W S, Rodgers W E and Knuth E L. 1969. An arc heater for supersonic molecular beams. *Rev. Sci. Instrum.* **40**, 1346–7.

Yu-Jahnes L S, Brogan W T, Anderson A C and Cima M J. 1992. A high-flux atomic oxygen source for the deposition of high T_c superconducting films. *Rev. Sci. Instrum.* **63**, 4149–53.

Zacharias J R and Haun R D. 1954. Well-collimated atomic beam ovens. *Quarterly Progress Report, Research Laboratory of Electronics, Massachusetts Institute of Technology* 34–7.

Zankel K and Vosicki B. 1975. Generator of a dense atomic gas curtain. *J. Phys. E: Sci. Instrum.* **8**, 360–4.

Zavitsanos P D and Carlson G A. 1973. Experimental study of the sublimation of graphite at high temperatures. *J. Chem. Phys.* **59**, 2966–73.

Zingaro R A and Cooper W C. 1974. *Selenium* (New York: van Nostrand Reinhold).

Zorn J C. 1964. Resource Letter MB-1 on experiments with molecular beams. *Am. J. Phys.* **32**, 721–32.

Zugenmaier P. 1966. Zur Theorie der Molekularstrahlerzeugung mit Hilfe zylindrischer Rohre. *Z. angew. Physik* **20**, 184–8.

Subject Index

Name Index